Field Day and the Translation of Irish Identities

Field Day and the Translation of Irish Identities

Performing Contradictions

Aidan O'Malley

palgrave
macmillan

First published 2011 by
PALGRAVE MACMILLAN

Palgrave Macmillan in the UK is an imprint of Macmillan Publishers Limited, registered in England, company number 785998, of Houndmills, Basingstoke, Hampshire RG21 6XS.

Palgrave Macmillan in the US is a division of St Martin's Press LLC, 175 Fifth Avenue, New York, NY 10010.

Palgrave Macmillan is the global academic imprint of the above companies and has companies and representatives throughout the world.

Palgrave® and Macmillan® are registered trademarks in the United States, the United Kingdom, Europe and other countries.

ISBN 978–0–230–22969–3 hardback

This book is printed on paper suitable for recycling and made from fully managed and sustained forest sources. Logging, pulping and manufacturing processes are expected to conform to the environmental regulations of the country of origin.

A catalogue record for this book is available from the British Library.

Library of Congress Cataloging-in-Publication Data

O'Malley, Aidan, 1968–
Field Day and the translation of Irish identities: performing contradictions / Aidan O'Malley.
p. cm.
Based on the author's thesis (doctoral)—European University Institute, 2004.
Includes bibliographical references and index.
ISBN 978–0–230–22969–3 (hardback)
1. English drama—Irish authors—History and criticism. 2. English drama—20th century—History and criticism. 3. Field Day Theatre Company. 4. Translating and interpreting—Ireland—History—20th century. 5. Drama—Translations into English—History and criticism. 6. Theater—Political aspects—Northern Ireland—History—20th century. 7. National characteristics, Irish, in literature. 8. Group identity in literature. I. Title.

PR8789.O43 2011
822.009—dc22 2011006608

10 9 8 7 6 5 4 3 2 1
20 19 18 17 16 15 14 13 12 11

Printed and bound in Great Britain by
CPI Antony Rowe, Chippenham and Eastbourne

Contents

Acknowledgements

In a previous incarnation this book was a PhD thesis and I owe a debt of gratitude to my joint supervisors, Luisa Passerini and Richard Kearney, and to the other two members of my committee, Bo Stråth and, in particular, Declan Kiberd, who has continued to be a source of great support. Many important friendships were formed at the European University Institute, Florence and, in a variety of formal and informal ways, the international, interdisciplinary environment I enjoyed there has moulded my perspective on Field Day. I am also very grateful to have been awarded a postdoctoral fellowship at the Humanities Institute of Ireland, University College Dublin, where parts of this book were developed.

Thanks to my friends and colleagues who read and commented on sections of the book manuscript: James Carney, Charlotte Geniez, Ina Gjurgjan, Michael Griffin, Roisín Higgins and Barra Ó Seaghdha. I am indebted to my editors Christabel Scaife, Ben Doyle and Paula Kennedy for the faith they have shown in this project and for the judicious guidance they have provided. Colette O'Daly and the staff at the National Library of Ireland were always helpful, and it was a pleasure to do research there. I would also like to express my gratitude to Frank McGuinness for providing me with a copy of *Uncle Vanya*.

Over the course of writing this book my life has been fairly peripatetic with the result that I've had to call on my mother and sister to help me maintain a Dublin base; I'd like to thank them for this and for their support more generally. This book is dedicated to the memory of my father.

Permission to reproduce extracts from the Field Day Papers and the Friel Papers was kindly granted by the National Library of Ireland. Permission to reproduce the cover image was kindly granted by Rod Tuach and the National Library of Ireland. By kind permission of Brian Friel, Derek Mahon, Thomas Kilroy and The Gallery Press, Loughcrew, Oldcastle, County Meath, Ireland, I have cited from: Brian Friel, *Three Sisters* (1981), Brian Friel, *The Communication Cord* (1989), Brian Friel, *The Freedom of the City* (1992), Derek Mahon, *High Time* (1985), Thomas Kilroy, *Double Cross* (1994). Excerpts from *Translations* from Brian Friel, *Plays 1*. Copyright © 1996 by Brian Friel. Reprinted by permission of Faber and Faber Ltd, a division of Farrar, Straus and Giroux, LLC. Excerpts from *Making History* from Brian Friel, *Plays 2*. Copyright © 1999 by Brian Friel. Reprinted by permission of Faber and Faber Ltd, a division of Farrar, Straus and Giroux, LLC. Excerpts from Seamus Heaney, *The Cure at Troy*. Copyright © 1990 by Seamus Heaney. Reprinted by permission of Faber and Faber Ltd, a division of Farrar, Straus and Giroux, LLC. Permission to cite from Tom Paulin, *The Riot Act* (1985) was kindly granted by Faber and Faber. Permission to cite from Terry Eagleton, *Saint Oscar*, in *Saint Oscar and Other Plays* (1997), pp. 15–63 was kindly granted by Wiley-Blackwell. Citations from Athol Fugard, *Boesman and Lena*, in *Selected Plays* (1987), pp. 191–255 were kindly granted by permission of Oxford University Press. Permission to cite from © David Rudkin, *Cries from Casement as His Bones are Brought to Dublin* (1974), was kindly granted by David Rudkin and Casarotto Ramsay & Associates Ltd, 7–12 Noel Street, London. Permission to cite from © Frank McGuinness, *Uncle Vanya* (2010), was kindly granted by Frank McGuinness and Casarotto Ramsay & Associates Ltd, 7–12 Noel Street, London. Citations from Stewart Parker, *Northern Star* and *Pentecost*, © Stewart Parker, *Plays 2*, Methuen Drama, an imprint of A&C Black Publishers Ltd. Citations from Thomas Kilroy, *The Madame MacAdam Travelling Theatre*, © Thomas Kilroy, *The Madame MacAdam Travelling Theatre*, Methuen Drama, an imprint of A&C Black Publishers Ltd.

List of Abbreviations

Texts of plays produced by Field Day

BL Athol Fugard, *Boesman and Lena*, in Athol Fugard, *Selected Plays* (Oxford and Cape Town: Oxford University Press, 1987), pp. 191–255.
CC Brian Friel, *The Communication Cord* (Oldcastle: Gallery Books, 1989).
CFC David Rudkin, *Cries from Casement as His Bones are Brought to Dublin* (London: British Broadcasting Corporation, 1974).
CT Seamus Heaney, *The Cure at Troy: A Version of Sophocles' Philoctetes* (London: Faber and Faber in association with Field Day, 1990).
DC Thomas Kilroy, *Double Cross* (Oldcastle: Gallery Books, 1994).
HT Derek Mahon, *High Time: A Comedy in One Act Based on Molière's 'The School for Husbands'* (Dublin: The Gallery Press, 1985).
MH Brian Friel, *Making History*, in Brian Friel, *Plays 2* (London: Faber and Faber, 1999), pp. 241–339.
MMTT Thomas Kilroy, *The Madame MacAdam Travelling Theatre* (London: Methuen, 1991).
NS Stewart Parker, *Northern Star*, in Stewart Parker, *Plays: 2* (London: Methuen, 2000), pp. 1–82.
P Stewart Parker, *Pentecost*, in Stewart Parker, *Plays: 2* (London: Methuen, 2000), pp. 169–245.
RA Tom Paulin, *The Riot Act: A Version of Sophocles' Antigone* (London and Boston: Faber and Faber, 1985).
SO Terry Eagleton, *Saint Oscar*, in Terry Eagleton, *Saint Oscar and Other Plays* (Oxford: Blackwell, 1997), pp. 15–63.
T Brian Friel, *Translations*, in Brian Friel, *Plays 1* (London: Faber and Faber, 1996), pp. 377–451.
TS Brian Friel, *Anton Chekhov's Three Sisters: A Translation* (Oldcastle: Gallery Books, 1981).
UV Frank McGuinness, *Uncle Vanya* (Unpublished).

Archives

FDA *Field Day Archive*, National Library of Ireland.
FP *Friel Papers*, National Library of Ireland.

Introduction. In Other Words: Locating a Touring Theatre Company

> [W]hat Field Day is doing, in a sense, is interrogating the relationship between culture and authority [...] by looking at language [...]. And by doing that, we're actually saying, this is an exemplary way by which, and through which, we can achieve a greater more flexible generosity of spirit and hospitality towards all contributions to the country[,] this is an exemplary mode of operation for the politicians, for the political system.
>
> — Seamus Deane[1]

In the course of an interview for the 1988 BBC documentary *History Boys on the Rampage*, Thomas Kilroy suggested that Field Day should produce 'a good road movie'.[2] While this portion of the interview was not broadcast, the programme employed, as a linking device, footage of a red van taking the Field Day cast and crew from one venue to another, splashing through a bleak and deserted rural Ireland. Road movies are frequently structured around a hazy and elusive final objective; what matters is the journey. Field Day's loose organisation and the twists and turns of the route it has taken suggest that it has not been driving in a predetermined direction towards a definable destination. Its odyssey has encompassed fourteen plays by different dramatists, fifteen pamphlets on a diverse selection of topics, a book of poetry, a five-volume anthology of 1,500 years of Irish writing, a set of magazine articles, a series of academic monographs and collections of essays, and an annual review, to name only the main ventures of the enterprise. In short, the company has been, and continues to be, a banner under which a wide array of voices have found expression in a variety of forms to generate debates about the histories of Irish cultural identities and their significance on the cusp of the twenty-first century. Underpinning these interventions is an emphasis on the need to shake up the stagnant formulas into which arguments about Irish experience had degenerated. Rather than demarcating a fixed point in Irish Studies, Field Day has essentially been about movement, and this book takes Kilroy's description, from the same interview, of 'Field Day as process' as a starting-point.[3]

It is argued here that this 'process' can be conceptualised and analysed through what Mulhern saw as 'Field Day's founding theme of cultural translation'.[4] The 'necessary and impossible task of translation', as Derrida described it, renders translation nothing other than a process: as it can never be definitively completed but must invariably be redone, it is an ever-renewing starting-point.[5] Working from such a perspective, contemporary research in cultural translation studies has focused on the ways in which the liminal, unpredictable and unstable act of translation de-centres notions of originality and of the self-sufficient subject.[6] As Apter put it:

> [c]ast as an act of love, and as an act of disruption, translation becomes a means of re-positioning the subject in the world and in history; a means of rendering self-knowledge foreign to itself; a way of denaturalizing citizens, taking them out of the comfort zone of national space, daily ritual, and pre-given domestic arrangements.[7]

Apter's book, which according to its subtitle heralds *A New Comparative Literature*, is part of a recent Princeton University Press series 'Translation/ Transnation' that uses translation as a way of questioning notions of national and cultural identities. The approaches adopted by some of the volumes in this series seem to have little connection with the act of translating between languages. Nonetheless, they are explorations of issues that are brought into play through translation; as Bermann notes in *Nation, Language and the Ethics of Translation* (another book in this series), translation 'requires attention to cultural values, to economic and political inequalities, to individual choices and, perhaps most obviously, to otherness in its linguistic and cultural forms. In the process it foregrounds some explicitly ethical questions.'[8] This speaks directly to the approach employed in this study of Field Day's work, which is informed by a commensurate, interdisciplinary, apprehension of what happens in the act of translation. Fundamentally, translation is interaction with the other, and so at the heart of this perspective is an examination of ethical issues. Current scholarship on cultural translation, indeed, has turned its attention to what Spivak has called 'the irreducible work of translation, not from language to language but from body to ethical semiosis'.[9] Inscribed in this vision of translation is a clear strain of poststructuralist thinking; as Weller has pointed out, Derrida, Blanchot, de Man and Benjamin are all 'concerned principally with the ethics of translation', and 'their theories of translation are grounded in an ethics of alterity'.[10] Never a straightforward or predictable matter, this translational ethics is generated by the impossibility of ever fully apprehending the other; it is a response to the moment of ethical decision called for in the encounter with the other, who is both untranslatable and yet calls for translation, and who is neither univocal nor necessarily speaking in a language that is already known. J. Hillis Miller perfectly captures this unpredictable, potentially enabling, translational moment: '[t]his other calling on us to respond, this future that comes into being by way of response, can only get here, arrive on this shore, speaking in tongues, in a multitude of overlapping and contradictory voices.'[11] Looked at in this manner, an ethical translation strives to perform the contradictions that exceed simple binary constructions of self and other through forms of rethinking in (the) other('s) words. The ethical decisions that are called for in this motion constitute, as Keenan has argued, the political dimensions of poststructuralism: a politics of radical indeterminacy and instability.[12] Not inconsequently, such a politics reverberates with the disjunctures fostered by colonialism, as can be traced in Derrida's work. Derrida is, as Spivak claimed, 'the rare philosopher who thinks that philosophical "concepts [cannot] transcend idiomatic differences"',[13] and as he most explicitly details in *Monolingualism of the Other*, this sensitivity to linguistic and other forms of difference is an ethical response to, amongst other things, his status as an Algerian-born, French-speaking Jew.[14]

The politics of Irish identities encompass a comparable set of disjunctions; Deane, indeed, has claimed that Algeria provides the closest analogy to Irish colonial history.[15] The negotiation of what Kinsella termed Ireland's 'dual tradition' – 'the outcome of colonial upset and adjustment' – has found expression in a variety of different forms in Irish literature in English since the eighteenth century that have registered and/or sought to elide the essentially translated nature of Irish identity.[16] For Casanova, these dislocations render Ireland a paradigmatic case in *The World Republic of Letters*, another key text in recent comparative literary studies.[17] The Irish Literary Revival, for instance, was fuelled by a desire to formulate an English that might be suitable for the emerging new Irish state. On the other hand, while they took issue with this rather conservative project and with the methods of Yeats and the Abbey, Joyce's and Beckett's anarchistic explorations of the limits of language – the relationships between words and things – were nonetheless also informed, to some degree, by the contradictions inherent in being Irish in English. This troubling ambivalence lies at the core of Field Day's perspective: '[w]e write English,' Deane noted, 'but we write it haunted by the ghost of a lost language.'[18] Cast thus as a spectre, Irish offers no comfortable, clear-cut, recourse; rather, as Deane put it in 1985, it embodies the disjunctions of translation: 'Irish is in a sense a foreign language to most of us and yet it's a foreign language which is also native.'[19] What is apparent from these statements is that Field Day's interest in translation was not just about providing translations into Irish or English, nor did it revolve around an exploration of the ways in which Hiberno-English might express late twentieth-century Irish experience. To be sure, elements of the company's output speak to these issues, but, more generally, Field Day's project involved foregrounding the inherently translated nature of Irish experience and so it revisited the processes of language change in order to reinsert their inherent disjunctions into contemporary constructions of Irish identities.[20]

'The history of a language is often a story of possession and dispossession, territorial struggle and the establishment or imposition of a culture.'[21] This is the first sentence of the first Field Day pamphlet, Tom Paulin's *A New Look at the Language Question*. Language change and conflict are woven into the warp of all colonial situations, and an examination of the record of these processes provides problematic insights into the operations of hegemonic authority. Unsurprisingly, therefore, contemporary cultural translation studies are tightly bound to postcolonial studies. Translation theorists such as Niranjana, Cheyfitz, Rafael and, in the Irish context, Cronin, have, in different ways, historically contextualised translation acts as a way of illustrating the articulation of power relations in (post-)colonial societies and, by unveiling the ambivalences and contradictions embedded in these translations, have problematised the seemingly static nature of these relations.[22] Few activities in a colonial context are as rooted in ambivalence as translation: it can express, in the same moment, both a willingness to engage with the other and an indifference to the other. However, these ambivalences do not suggest that the relationships structured around and through the act of translation are based on forms of equality. On the contrary, many of these ambivalences are generated by the power differentials in this relationship; by the ways in which the 'weaker' element must accommodate this difference. Translation, as Bassnett and Trivedi phrased it, 'is not an innocent, transparent activity' and 'rarely, if ever, involves a relationship of equality between texts, authors or systems'.[23] Yet, translation has most frequently been presented as a benign and self-evident enterprise, and so analyses of translations that register these power imbalances, 'challenge and defeat both an imposed silence and the normalized quiet of unseen power', which is what Said saw as the function of the engaged intellectual.[24] In like manner, Field Day explored, through lenses supplied by the idea and act of translation, the history and ramifications of the language shift in Ireland, in order to uncover a series of contradictions that had been resolved into pillars that upheld the different constructions of identity that informed the politics on the island.

In this light, the contention here is that the contradictory dynamics of translation have underwritten the ethics of Field Day's intervention into the Northern Irish 'Troubles' and the wider discourse of Irish identities, and that this is the basis of what Roche has called Field Day's 'politics of creative contradiction'.[25] For this reason, little distinction is made between the translations and the other plays; indeed, the fact that only Mahon's play is a translation as such – the other translations are more precisely versions or adaptations – further indicates that Field Day's interest is in the idea of translation and its cultural ramifications. Instead, what is explored in the following chapters is how these dramas perform the dynamics of translation in a variety of different metaphorical guises: mapping the past; the politics of hospitality; the interaction of loyalty to the spirit and the letter in unionist identities; the necessity and impossibility of utopian impulses; and the ways in which the colonial relationship between Ireland and Britain has formed identities.

Movement and emancipation from stereotypical performances of 'Irishness' through theatrical excellence constituted the substance of Stephen Rea's formal proposal for setting up a theatre company (4 April 1979). These principles would remain at the enterprise's core as it gathered momentum and evolved through the subsequent decade.

> I have no particular axe to grind with regard to community drama, except that I believe 'knees-ups' and 'come-all-ye's' are often patronising. I believe that communities are entitled to a fine play by a fine Irish writer performed by the best actors available. The freedom of movement between traditional theatres and the more unpredictable community centres should be a liberating experience for the company and therefore for the audiences.[26]

There was no larger master-plan behind the creation of the company; rather, it was an essentially improvised occurrence. Having heard that Arts Council funds were available to produce a play in Northern Ireland, Rea approached Brian Friel, with whom he had worked in the early 1970s on *The Freedom of the City*, to see if he would be interested in writing a play that they might produce together and bring on tour around Ireland. It was a fortuitous moment, as Friel was working on *Translations*, and was also interested in exploring a mode of production that would allow him to venture beyond established theatrical centres. Needing to give the company a name, they came up with a rhyming amalgam of their surnames – Friel Rea: Field Day.

The unprecedented success of *Translations* in 1980 encouraged Friel and Rea to initiate a policy of producing and touring a new play each year, and the following decade saw the company bring its dramas to large and small towns throughout the island, playing in local theatres and other community venues, often before audiences that had limited access to professional theatre. In order to claim charitable status, the enterprise had to be set on a more formal footing, and invitations to become directors of the company were sent to four friends of the co-founders who were known to be interested in the idea of a Northern Irish cultural group. On 4 August 1981, letters welcoming these friends to Field Day stressed that while their advice would be welcome, their involvement with Field Day would involve 'no work'.[27] With the benefit of hindsight, it might appear that Seamus Heaney, Tom Paulin, Seamus Deane and David Hammond were being duped into joining the enterprise with this ultimately hollow promise. But this indicates both the extent of the ambitions of the company at this juncture – it was purely a touring theatre company – and how the involvement of these new directors changed the nature of the venture. Shortly after Field Day produced his 1986 play, *Double Cross*, Thomas Kilroy also joined the board, becoming its only director from the Republic of Ireland.

Despite the seeming formality of a board, Field Day, according to Heaney, was essentially a continuation of the conversations the directors had been having with each other throughout the 1970s,[28] and this is reflected in the witty ribbing that sets the tone of Deane's minutes of the directors' meetings (the August and December 1988 meetings are even rendered in rhyming couplets).[29] At the same time, these friends were also some of the most important names in Irish cultural life, and this context encouraged them to experiment and to explore the possibilities that the cross-fertilisation of their different talents might produce. Heaney, for instance, has recounted that Friel hoped 'that Field Day would induce the poets to have a go at work for the stage'.[30] Notably, Friel also acknowledged that Deane, who is most renowned as a critic, was 'the best touch-stone' for his work in the group.[31] It would have been difficult to gather together a more perspicacious group of readers in Ireland, and a commitment to critical rigour at times overrode the bonds of friendship; amongst other examples, the archive reveals that Friel and Paulin were not happy that plays they submitted were not produced.[32]

While the directors shared, as Deane phrased it, 'a set of cultural and political preoccupations' – in particular, a rejection of the inherently sectarian politics that had shaped Northern Ireland – these were never expressed in a doctrinaire or univocal manner and the company never issued an official manifesto.[33] Rather, Field Day was, for Friel, 'a forum where a more generous and noble notion of Irishness than the inherited one can be discussed'.[34] Differences of political opinion and approach were acknowledged and debated: according to Kilroy, views that were pro-nationalist, suspicious of nationalism, or that cleaved to eighteenth-century republicanism were aired. These designations, however, were starting points and the effort was to go beyond them; as Kilroy saw it, the driving force of Field Day was 'a belief in the imagination and in the capacity of writers to imaginatively investigate Irish life in a way which would […] reflect on current politics, but not to drive politics to any particular conclusion'.[35] A privileging of the imagination – seeing it as prior to and, in some senses, constitutive of politics – particularly distinguishes Field Day's activities: as Friel had it, the enterprise was working towards a 'cultural state', out of which 'a possibility of a political state follows'.[36] This 'cultural state' is never given a fixed form; rather, as is argued here, it embodied the processes of translation, literature and performance, and this dynamic is fundamental to its relationship with the political.

While the potentially enabling aspects of this focus on culture are brought to the fore in this study, this emphasis also placed severe limits on Field Day's interventions into political debates. The frustrations this engendered find expression through Deane, who, on occasion, sought to instil more organisational and theoretical rigour into the enterprise. In a confidential commentary on the first series of pamphlets in 1983, he took exception to Heaney's use of verse in *An Open Letter*: Deane felt this rendered Heaney's objection to being included in *The Penguin Book of Contemporary British Poetry* an entirely personal matter, and this perspective set *An Open Letter* apart from the other two pamphlets in the series, making them appear more strident in the process. He urged his fellow directors to work towards a more consistent conception of their activities:

> we need to decide what kind of group we are – one to which the doctrine of collective responsibility should apply – one which is involved in a series of concerted actions – one which is emblematic of a kind of excellence meant to be exemplary to others? The looser the framework, the less force this comment has; the tighter the framework, the more force it has.[37]

Yet, in a subsequent letter to Friel (6 April 1984), he sounded a different note, one that acknowledged, albeit with clear exasperation, the limits of Field Day's scope for concerted action: '[b]ut can we propose? The moment we do, we begin to sound hectoring, unconvincing even to ourselves. [...] we can't propose too directly without preaching or too indirectly without evading. We're not Young Ireland or the Frankfurt School.'[38] Unlike those two ventures, Field Day remained at heart an artistic initiative; class, economics and gender make but fleeting or belated appearances in its explorations of Irish life. Kirkland has noted that while Deane discussed the role of the intellectual in terms of class in the 1970s, this form of Marxian scrutiny was shed in his Field Day years.[39] In fact, class featured in a remarkably conservative fashion in a lecture Deane delivered in 1985 at Magee College in which he elaborated on 'what Field Day thinks it is about'. Working from the teleological assumption that the two polities in Ireland had, at best, 'twenty-five years to go', Deane did not greet their demise as a moment of revolutionary opportunity, but instead with trepidation: 'some species of anarchy' might be unleashed which would create 'zones in Ireland where nobody can travel safely'. Writers and intellectuals had 'a special kind of responsibility' to try to prevent such a disintegration from occurring,

> because middle class society in the last century and a half has given to them a freedom, a space in which to move and think which is almost unique to them and which they regard as a pre-condition of their being able to work at all. [...] if having being granted that privilege [...] they do not use it for the sake of the community as a whole then the intellectuals and the writers will be swept away with everybody else.[40]

Field Day's role, from this perspective, is essentially conservative: it is participating in a 'rescue act' for the good of the community.

This undertaking operated under an overarching paradox, as the conservative task of recuperation involved a radical deconstruction of the prevailing ideas of Irish cultural identity, precisely because they had generated the disabling and doomed political structures.

> Field Day is engaged upon a total cultural reappraisal of the Irish situation. Now the phrase is glib but the key word within the phrase is culture. In some ways we assume this, that seeing in cultural terms is believing, and if we see differently we may believe differently. [...] Our notion of culture is that all cultures [...] are politically governed. [...] there is a transaction, a negotiation, always taking place between that which is cultural and that which is political and the drastic nature of our situation, we believe, has something to do with the fact that we do not have a cultural vision of ourselves, that Ireland does not have a cultural vision of itself, which is sufficient to enable the possibility of some kind of political cohesion and social cohesion to be obtained within the island as a whole.[41]

There is an unresolved tension here between the claims of culture and politics: while cultural life is 'politically governed', ultimately a 'cultural vision' determines the shape of politics. The circularity of this logic is a symptom of Deane's effort to account for the disjunction inherent in Field Day's task of simultaneous conservation and deconstruction, and this speaks directly to the group's relationship with nationalism. In focusing on the issue of cultural identity and seeing it as constitutive of politics, Field Day co-opted the outlook of nationalism; at the same time, it rejected the conservative political forms nationalism had taken in Ireland: as Deane saw it, the two modes of political nationalism upon which the two states in Ireland were based had left them destined for history's junkyard. Richards has understood this fundamental ambivalence in Field Day's project in terms of Spivak's 'strategic essentialism': 'an appreciation of the need to mobilize identity as a site of resistance to its erasure while remaining constantly conscious of the provisional nature of that which was to be defended'.[42] This describes a point at which postcolonial thinking crosses with, and must exceed, nationalism. It is an ever-shifting location that has to be negotiated in a nimble and sure-footed manner in order to avoid falling into the essentialisms that are being disavowed. In particular, there is the constant threat that an analysis staged on these grounds might settle into a revised version of the binaries it explores. To avoid this, the relationship between the political and the cultural must be kept in play, perhaps in a manner not unlike that described by Deane above; the fact that this appeared to strain logic speaks to the difficulties of constructing such an argument. To use the distinction that is employed throughout this study, what is at stake is the relationship between the post-colonial and the postcolonial: 'post-colonial' here refers to a temporal, historical period – after colonialism – while 'postcolonial' indicates those theoretical elaborations that are informed by the power imbalances inscribed in the colonial experience, but which also have a salience beyond this immediate context and have the potential to disrupt understandings of the (post-)colonial moment. The Field Day pamphlets played a central role in introducing postcolonial thinking into scholarship on the Irish experience; as Deane put it: 'Field Day's analysis of the situation derives from the conviction that it is, above all, a colonial crisis.'[43] The postcolonial analyses offered in most, though not all, of the pamphlets involve subjecting this politico-historical crisis to readings that unveil aspects of its cultural foundations. While this is salutary, it is also a somewhat attenuated postcolonialism, as most of these polemics do not allow for a supplementary doubling-back that would allow the political to be re-viewed through the cultural disjunctions that have been brought to light. The plays, on the other hand, move according to the ambivalent dynamics manifested in these findings and, it is argued here, are ultimately more successful at articulating, or performing, an enabling postcolonial politics than the pamphlets.

While Said's signature is on one of these pamphlets, his influence can be traced in many of the others. In turn, as is testified to in the 'Afterword' to the second edition of *Orientalism*, Said viewed Field Day as an archetypal postcolonial enterprise, and he was a good friend of several people associated with it.[44] Culture and the question of cultural authority lie at the core of Said's work because, 'opposition and resistance to imperialism are articulated together on a largely common though disputed terrain provided by culture'.[45] But culture also plays what Said termed a 'contrapuntal' role, as it has equally been involved 'in consolidating imperial feeling', especially through forms of stereotyping.[46] Just as *Orientalism* laid bare the history of stereotyping that constructed the figure of the Oriental, the Field Day pamphlets look to provide an analytical dismantling of colonial stereotypes in Ireland, which in Deane's words, had 'become both a symptom and a cause of the current situation'.[47] This is most explicitly articulated in Deane's *Civilians and Barbarians*, which is a polemical reading of what he argues is the main stereotype that has obtained between Ireland and Britain: the 'four hundred-year-old distinction between barbarians and civilians'.[48] Exposing, in Said's words, 'the impression that interpretation of other cultures, texts, and peoples [...] occurs in a timeless vacuum, so forgiving and permissive as to deliver the interpretation directly to a universalism free from attachment, inhibition, and interest' has been a central objective of postcolonial thinking.[49] Deane, therefore, historically and geographically locates the 'universal' claims of British empiricists that had fostered the stultifying stereotype. By contextualising, for instance, Locke's assertion that '[l]aw makes men free', he demonstrates its colonial function: it simply casts the colonial lawmakers as civilians and those who resist them as barbarian others.[50]

That this stereotype still had traction in the early 1980s was manifestly evident. Margaret Thatcher had swept to power in London in 1979 and her political intransigence was quickly revealed in Northern Ireland, particularly with regard to the H-Block hunger strikes, her government's handling of which 'was inept to the point of criminality', according to Joe Lee.[51] For Deane, this was a situation in which 'both sides [...] played out their self-appointed roles to [...] a literal end', as they were locked into the logic of the 'civilian and barbarian' stereotype.[52] Furthermore, as David Lloyd notes, this stereotype has also framed subsequent representations of the hunger strikes in academic and media accounts both in Ireland and Britain, which have sought to contain the significance of the strikes by emphasising 'the atavistic or religious nature of the rhetoric and imagery' with which they were associated, with the result that 'the intervention of the state and its apparatuses is seen as secondary to the irrational and potentially violent forces of premodernity; [...] if not contained by the state, such forces will overwhelm civil society'.[53] As this indicates, unveiling the operations of this colonial stereotype is an essential move. However, Deane's pamphlet does not suggest how such an understanding might inform a more enabling politics; the discourse has been exposed but it has not been reimagined in a way that might displace its binary logic, and Donoghue's worry that Deane 'may be, in some appalling respects, right, and that his being right won't help the situation' is apposite.[54] Paulin's pamphlet, on the other hand, also offers a Saidian exploration of the imbrications of language and cultural authority in Ireland, but opens up a route towards the style of analysis that is performed in the plays, as, in contrast to Deane's pamphlet, it emphasises the possibilities inherent in the contradictions inscribed within Irish (post-) colonial experience.

Paulin's disruption of the 'civilian and barbarian' stereotype rejects dualist perspectives: he emphatically contends that an approach to the linguistic geography of Ireland that takes an unambiguous view of 'the English language [...] as an imposed colonial tongue and Irish as the autochthonous language of the island' is not tenable.[55] Rather, he subjects the cultural history that has engendered the stereotype to the movement of translation, by focusing, for instance, on the insecurities English writers had about their language in the late seventeenth and early eighteenth centuries. Dryden, as Paulin notes, considered English to be 'in a manner barbarous', and hardly suitable for the growing British empire.[56] To remedy this, Dryden turned to the example of the ways in which the Roman empire sought to consolidate its cultural authority through the translation of Greek culture and civilisation, and strove to link England's burgeoning imperial might with that of the *imperium Romanum* through the same process: translations of the classics. If the English language could capture the import of the classics, it was a fit vehicle for an imperial power. This moment can be phrased within the ancient theory of *translatio studii et imperii* (the 'translation/movement of learning and empire'): a semi-mystical theory that holds that knowledge and empire follow the sun, in other words, they travel from east to west.

It is, as Robinson points out, an a-historical idealisation of the inevitability of empire, which assumes an almost metaphysical form.[57] However, as a concept it is also constantly threatened by its governing tension between stability and movement; in other words, its efforts to construe an appearance of immutability are compromised by the very tactic it employs to effect this: translation. Paulin's pamphlet reinscribes the movement of translation into the relationship between Britain and the English language and, in doing so, undermines the stereotypical assumptions of cultural authority that became attached to the language.

In other words, these assumptions were fostered by translation. Cronin's account of this period supplements Paulin's work as he focuses on how the translations of the classics incorporated an implicit, and often explicit, assertion that the conquering language was now superior to that of the original:

> praise of foreign genius becomes the exaltation of native domesticity. [...] Translation is a form of homage to England's cultural superiority, and readers in English will soon find themselves granting the favour of poetic improvement to classical authors rather than remaining in mimetic thrall to their accomplishments.[58]

This outlines the motions of what Venuti has termed the domesticating mode of translation. It is the classical form of imperial translation, as it takes what is foreign home and appropriates it for itself. In terms of the practices of translation proper it is characterised by an excessively fluent rendering of the foreign-language text in the target language that seeks to remove any traces of foreignness. It is thus contrasted with a foreignising mode, which strives to register the otherness of the foreign in its translation practice.[59] In his pamphlet, Paulin not only exposes the history of imperial domestication that created an idea of the English language, but also calls for a postcolonial domestication of Hiberno-English through the compilation of a dictionary, which is conceived of in terms of Webster's project in the USA. As this indicates, domestication is not a unidirectional, unequivocal process and, time and again, its contradictory motion is played out in the dramas via the entirely ambivalent scene of hospitality, which is directly invoked by the terms domestication and foreignising. Indeed, it is frequently through performances of the translational dynamic of hospitality that the plays problematise the construction of national narratives.

Translation is disruptive of hegemonic constructions because it unveils the disruption inherent in language. For Steiner, '*human communication equals translation*', and so '[a] study of translation is a study of language'.[60] Lojek has argued that Steiner's 'broader notion of translation as interpretation [...] provides the most general and useful way of approaching [...] the whole body of Friel's works', and this holds true for all the Field Day dramas.[61] Language needs to be interpreted because of its inherent potential for metaphor, and translation provides a model of this process, because, in Derrida's words, it 'practices the difference between signified and signifier'.[62] Performing the movement of meaning in metaphor, translation can be conceptualised as a metaphor of metaphor, which, as de Man points out in his reading of Benjamin, does not mean that they are the same thing, or that a mimetic relationship links them.[63] Rather, a properly metaphorical relationship obtains between the trope of metaphor and translation: a translation both is and is not a metaphor, and *vice versa*. Metaphor and translation are not mimetic acts: in working towards forms of resemblance, they register differences and therefore animate the potential for otherness within language. They reveal that language is never sufficient unto itself, and that a resistance to one, set definition is nothing other than the fate of language.[64] This implacable movement sets the scene within which Field Day's literary, imaginative, translational, politics operate.

Heaney has argued that the task of literature is to search out and articulate '[i]mages and symbols adequate to our predicament'.[65] Implicit in this is a process of disruption based on literature's ability to bring to the fore the metaphorical nature of language. As Spivak put it, literature involves 'training the imagination – the great inbuilt instrument of othering': in disclosing the otherness that lies within words, literature reinvigorates language and reorganises one's assumed reality.[66] In a similar manner, Hallward locates the political import of literature in its problematising of the relationship between the signifier and its conventional signified: in 'its capacity to invent new ways of using words (new in either form or practice, or both), at a disruptive distance from inherited norms and expectations – in other words, its capacity to provoke people to *think*, rather than merely recognise, represent or consume'.[67] Such a view of literature is not predicated on the assumption that literature might have a direct relationship with the world, instead it is based precisely on the manner in which literature appears to turn in on itself. In this motion, rather than being simply relinquished, the relationship with the world has been displaced, perhaps in a way that produces a more powerful, imaginative, connection. As Ricoeur has posited, in literature 'the deferral of reference (as defined by the norms of descriptive discourse) is the negative condition for a more fundamental mode of reference to be brought out.' This describes 'the hermeneutic moment', which is,

> the work of thought by which the world of the text faces what we conventionally call reality in order to redescribe it. This confrontation may range from denial or even destruction (which is still a relation to the world) to the metamorphosis and the transfiguration of the real. [...] The impact between the world of the text and, simply, the world, within the space of reading, is the ultimate stake of the productive imagination. It engenders what I would dare to call the productive reference proper to fiction.[68]

Fundamentally this moment is a coming to terms through literature with the wide-ranging implications of the metaphorical nature of language. The open-ended act of interpretation that Ricoeur invokes chronicles the scene of translation and, as such, describes the very space of Field Day's political actions.

Lodged in this discussion of the act of interpretation is a crucial spatial aspect, the disjunctions of which speak to the dimensions of Field Day's notion of a cultural state. Metaphors, for Steiner, are 'new mappings of the world, they reorganise our habitation in reality'.[69] As summed up by A. T. Q. Stewart, the Northern Irish 'Troubles' were sustained by deterministic concepts of geography and demography: '[t]opography is the key to the Ulster conflict. Unless you know exactly who lives where, and why, much of it does not make sense.'[70] Shifting these apparently fixed *topoi* through the motions of metaphor and translation lay at the heart of the dramas' politics of translation. Both within their own terms and through the manner in which they were produced, the plays highlight the metaphorical movement that was occluded by the seemingly static political maps. They chart a 'cultural state' that, rather than being moored in the known, tells us 'something about displacement, disorientation, and agency in the contemporary world', which Appadurai believes is a function of literature.[71]

Similar to the manner in which Chakrabarty re-viewed the colonial metropole and periphery in *Provincializing Europe*, Field Day's alternative map placed Belfast and Dublin in a non-metropolitan perspective, which was more radical and unpredictable, because, as Friel told O'Toole in 1982, provincial audiences 'haven't been indoctrinated in the way a metropolitan audience is'.[72] For this reason, the company saw the provincial as a source of cultural revitalisation: according to Rea, '[t]he Abbey was just a dead space. It was a dead theatrical space. There was no live theatrical space in Belfast that I can think of.'[73] Whereas the Abbey project of the early twentieth century aspired towards a monumentalism – through the creation of a national theatre, both physically and conceptually – that would in time come to stand for more fixed, canonical notions of Irish identity, Field Day did not want to deliver diktats 'from headquarters' as Rea put it.[74] At the same time, while the tours were a way of linking audiences with these plays – suggesting even that they have emerged, in some sense, from impulses within these communities – the transitory nature of a touring company incorporates a distance into this relationship, as it disavows a too-easy identification between the audience and the performers: they are from elsewhere and will move on. Essentially, this expresses the spatial contradiction inherent in translation between proximity and distance: the plays both sought to engage with their audiences' sense of identity and, at the same time, inserted an unavoidable element of otherness into these expectations; what was considered distant is brought closer, while what was thought of as one's own is made somewhat foreign. Again this echoed Said, who, as Deane noted, held that, 'to be committed involves distance and […] to be distanced involves commitment'.[75] Such a dynamic is also in evidence in the plays, which take place in the doubled location of translation: both in specific locations – 1830s Ireland, apartheid South Africa, the pre-Revolutionary Russian provinces, and so on – and in an elsewhere that is, in some ways, contemporary Ireland. Bearing in mind that this enterprise came into being to form a cultural intervention into the Northern Irish crisis, it is immediately striking that only one play – Parker's *Pentecost* – is set in the context of the 'Troubles', and even this drama, as is shown in Chapter 3, seeks to transport this context to an elsewhere.

Smith and Katz have pointed out that, '[i]n social theory and literary criticism, spatial metaphors have become the predominant means by which social life is understood'.[76] They are referring to how these tropes are invoked to ground abstract theories; Field Day, however, employed spatial metaphors to re-envision the 'Troubles' in a more problematising fashion. Boltwood has posited that, for Friel, 'the Northern thing' came to describe a form of subaltern, critical, position,[77] while Kirkland has claimed that sites such as Derry and the 'fifth province' expressed for Field Day a 'scepticism towards history' rather than fixed places.[78] In short, places do not simply describe geographical locations, but become metaphors for the disruptive, political, operations of literature in the world. The most important of these artistic 'notions' and 'inspirations', especially in the company's initial years, was the 'fifth province', which served as the model for Field Day's imaginative rearrangement of the topography of Irish experience.[79]

The 'fifth province', according to Friel, 'may well be a province of the mind, through which we hope to devise another way of looking at Ireland, or another possible Ireland, and this really is the pursuit of the company'.[80] Phrased thus, the 'fifth province' clearly operates as a metaphor: it does not define a known place, but is a medium through which other ways of perceiving and thinking can be imagined. In so doing, it might be said to provincialise the provincial, to adapt Chakrabarty's concept. Provincialism has been a staple in the characterisation of (post-)colonial Ireland, which has been frequently cast, in Deane's words, 'as provincialism incarnate',[81] genetically, as it were, 'cut off from progress, modernity, cosmopolitanism and even revolution'.[82] Moreover, the boundaries of the term 'province' are contested in Northern Ireland by the two traditions: the province of Ulster contains nine counties, whereas Northern Ireland is composed of only six of these, but is frequently called Ulster, particularly by unionists, and Derry/Londonderry lies precisely on this fault line. The 'fifth province', because it lacks a basic 'grounding', defies and questions these understandings by keeping the tension between motion and stasis in play.

The 'fifth province' entered contemporary Irish cultural discourse with the announcement of its 'rediscovery' in the editorial of the first number of the journal *The Crane Bag* in 1977: '[m]odern Ireland is made up of four provinces, whose origins lie beyond the beginning of recorded history. And yet, the Irish word for province is *cóiced* which means a fifth. This fivefold division is as old as Ireland itself, yet there is disagreement about the identity of the fifth fifth.'[83] Linguistic and historical authorities such as O'Rahilly and Rees and Rees have disputed the whereabouts of this province: for Rees and Rees, the term came about through 'a subdivision of Munster' into East and West Munster, while O'Rahilly contended that it originally referred to a midland province that was ruled by the king of Tara; as he became the most important king, his domain was no longer reckoned to be a mere fifth part or province.[84] In any case, Kearney and Hederman, the *Crane Bag* editors, presented the 'fifth province' as a physically deracinated site, one 'marked by the absence of any particular political and geographical delineation, something more like a dis-position'. It became a quintessence, a utopian site: an equally important 'second centre', which provided a 'necessary balance' to the political world.[85] According to Hederman, the site's ability to contain contradiction afforded it the possibility of acting as a point of resolution: '1. Continuity of tradition is the basic condition for the existence of culture. 2. Breaking continuity, rupture of tradition, is the necessary condition for the formation of a new kind of culture. I would claim that "the fifth province" is an attempt to maintain the paradox of both these truths.'[86] Pivotally, the 'fifth province's' capacity to envelop these contradictions is directly linked to the way in which it is located in what Foucault has called the 'non-place of language', as its presence, or rather its cryptic (non-)being, is marked in Old Irish.[87] *Cóiced* is the Old Irish form of the Modern Irish *cúige*, and the *Crane Bag* version of the 'fifth province' is essentially a translation of *cóiced* in the fullest sense of the term, as it is presented as literally coming into being through the act of translation: *cóiced* signifies both 'a province' and 'a fifth', but not 'a fifth province', though that might be heard within this term. The seemingly innocuous shift from considering 'a fifth' solely as a noun to framing it, at the same time, as a complementary adjective, releases the energies that were latent within the term *cóiced*, and unveils a new possibility. It is a translation that manifestly creates its own original. Existing entirely as a possibility that might be glimpsed through translation, it remains an-other place; indeed, its very 'non-presence' is the source of its utopian qualities.[88] It might be recalled that Thomas More animated Utopia in a similar translational fashion, coining the location by bringing together 'the Greek adverb *ou* – "not" – and the noun *topos* – "place" – and giving the resulting compound a Latin ending', in such a manner that 'the word puns on another Greek compound, *eutopia* – "happy" or "fortunate" place'.[89] Just like Utopia, the 'fifth province' thematises its status as a translation and, as such, its existence, or rather its possibility, in language. Considered as a model for Field Day's 'cultural state', it represented an ideal translational process rather than a static location.

However, there was no unanimity about how the 'fifth province' constituted such a process. Even the two editors of *The Crane Bag* understood the relationship of this artistic idea to the world in a divergent manner. For Hederman, the 'fifth province' was a second, compensatory, centre that could only be accessed through art: it was a purely aesthetic realm that provided 'access to the transcendent', and could only be approached by 'the artist, and especially the poet'.[90] How 'the transcendent' is to be understood is unclear, and this opacity led Edna Longley to suspect that 'Hederman's terminology in invoking "the fifth province" owes more to transformed Nationalism, and transformed Catholicism, than perhaps he realises.'[91] This is of a piece with the tenor of the many critiques that the 'fifth province' has attracted in Ireland; in general, these have focused on the tenebrous nature of the site and have sought to outline the potentially negative politics that could issue from, or be hidden by, such obscurity.[92] Considered in isolation, Hederman's version of the 'fifth province' does invite this criticism; however, when it is read with Kearney's version, the transcendent assumes another, more challenging, role.

For Kearney, the 'fifth province' 'is always a question of thinking *otherwise*'.[93] In Hederman's vision, otherness is located in an implicitly hierarchical structure, beyond human possibility. Thought of in terms of translation, the transcendent would be a realm of perfect translation: an ideal place in which translation was not required, as it would be perfectly beyond translation.[94] In contrast, for Kearney, the other is encountered through the experience of communication: '[t]he creation of this fifth province calls for the creation of a new vocabulary, a new mode of communication which will acknowledge, and perhaps ultimately mediate between the sundered cultural identities of this island.'[95] Understood thus as a dynamic location of communication rather than as an access point to the transcendent, the 'fifth province' assumes the form of what Bhabha called an 'interstitial' site: a 'liminal space [...] between fixed identifications [which] opens up the possibility of a cultural hybridity that entertains difference without an assumed or imposed hierarchy'.[96] It describes an area wherein 'the processes that are produced in the articulation of cultural differences' are elaborated in such a way that they 'initiate new signs of identity, and innovative sites of collaboration, and contestation, in the act of defining the idea of society itself'.[97] In like manner, in Kearney's 'fifth province', identity is always being reimagined through a translational movement towards the other that brings one closer to one's self. In other words (those of Ricoeur), it would be the state of 'oneself-as-another', which Kearney has glossed in this manner: '[t]he shortest route from self to self is through the other. [...] the self is never sufficient unto itself, but constantly seeks out signs and signals of meaning in the other.'[98] While Hederman's conception of the artist as a cultural translator who negotiates the disjuncture between the continuity and rupture of tradition clearly speaks to Field Day's conception of itself, its productions and the forms of communication enacted in its tours placed their faith in the processes of translation itself.

For McGrath, Field Day's language policy was 'an aggressive attempt to appropriate a tongue that was originally forced upon the Irish': a call 'for a culture and a language in which the Irish, after centuries of billeting an alien other, feel at home with themselves'.[99] Like the 'fifth province', however, this form of accommodation was always elsewhere and to be worked towards; the translational dynamic driving Field Day's productions meant that, even as they attempted to chart a space that might stand as a home, the plays constantly questioned the possibility that such a place could be construed. Heaney observed that Field Day 'had less a position than a disposition, and that disposition came from a sense of displacement'.[100] In 1982, Friel also located Field Day's home in a disruptive, metaphorical, fashion: '[w]e think of ourselves', he told Comiskey, 'as Derry-based, northern-focused, but absolutely in terms of the whole island.'[101] Even as he appeared to claim Derry as home, Friel situated Field Day in other places. Like Sweeney in Heaney's translation of *Sweeney Astray* (which Field Day published in 1983), the company's travels through Ireland, and its dramatic journeys to other contexts and times, expressed a permanent exile from the home-place, which nonetheless continued to exert a pull. While the Field Day directors had strong filiative links with the North, they had no affiliative connection with the Northern Irish state; hence, rather than name a stable, comforting domestic site, Derry expressed the difficulties of constructing a home, or a condition of homesickness, or the complexities of what Said called 'problematic filiation', a theme that runs through the dramas.[102]

Derry also names a site of treason. For Friel, the 'fifth province' was a 'place for dissenters, traitors to the prevailing mythologies of the other four provinces', and running through the Field Day dramas are the interwoven themes of (in)authenticity, fidelity and treason.[103] These are key themes in Irish cultural history precisely because of the translated nature of Irish cultural identity. Instead of offering resolutions to this situation, the dramas provide models of identity structured on the contradictions inherent in the notion of *traduttore traditore*. Treachery, Deane held, has been 'one of the few available forms of freedom' available to someone who has not subscribed to the different hegemonic discourses of Irish authenticity.[104] Fidelity, as is movingly illustrated in his novel *Reading in the Dark*, inevitably took the form of treason in the context of the vicious, sectarian politics of mid-twentieth-century Derry.[105] 'As a place which exists between the North and the South but isn't entirely either', Derry, in Deane's words, expresses in its very location and in its two names the split and doubled identities that are to be found in Northern Ireland, and reflects the enmeshed consciousness within both communities of simultaneous belonging and isolation.[106] It is a city of borders, as Deane's 1983 portrait of the city (tellingly drawn from the perspective of a 'visitor-native') underlines, and in such an environment any movement will be an act of treason.

> The people of Derry are, in effect, locked into their city and its history. Ringed by the border, they are cut off from Donegal which the

> majority Catholic population looks to as its natural hinterland. Ringed by the walls and the zones of their housing estates, they are cut off from one another. The county of Derry beyond is looked to by the minority Protestant population as its natural hinterland. The geographical location and the historical background of the city confirm this oppressive sense of siege, one shared by both communities. Each regards itself as surrounding and being surrounded by enemies. [...] Derry, then, manages still to achieve its peculiar symbolic significance in Northern society. Its two names – Derry/Londonderry; its two communities; its history of siege; its absorption of the effects of violence and economic demoralisation along with economic subsidy; the contrast between its beautiful setting and its ruined cityscape; and above all its isolation, allow a visitor-native like myself to see it as a place, small enough to be understood and big enough to be typical of others, as the North's, maybe the whole island's exemplary town.[107]

If Derry has this representative status it is because it is the 'unhappy home', in Richards's resonant description, 'of significant events whose anniversaries have resounded bloodily across centuries of the city's history'.[108] In particular, this fractured home has nurtured the siege mentality that has been a determining feature of both communities in Northern Ireland.[109] The 1689 siege of Derry and the myths in which this event became encrusted – fermented, in Foster's words, through 'images of siege and delivery, treachery and self-reliance' – particularly define the identity of the Northern Irish unionist population.[110] The anniversary of this event is celebrated with the annual 12 August parade of the Apprentice Boys, who restate both the defiance of the Protestant community and its claim over Derry by beating drums around the city walls. This performance of a form of divinely predestined history was necessitated by the fact that the Catholic population of Derry long outnumbered the Protestant, and hence justification for the persistence of unionist domination of the city's political life could not be sought in purely democratic terms. While the Protestant community cannot but have felt a precarious sense of proprietorship, the city's Catholics were acutely aware of the blatant injustice of the long history of gerrymandering that kept the minority Protestant population in power. As Alasdair MacIntyre noted, the clash of 'justifications of legitimate authority' is encapsulated in the two names of the city: if Londonderry is central to unionist identity, Derry has been one of the clearest symbols of Catholic disaffection with Northern Ireland as a political entity.[111] This alienation reached a crisis point in the 1960s, as the first generation of Catholics to benefit from the 1944 Butler Education Act (which introduced free secondary education) began to find a more confident voice. When the Northern Ireland administration decided to locate a second university in stolid, unionist Coleraine rather than in Derry, this voice was roused. Derry was not only 'the second largest city in Northern Ireland', as McCann notes, but 'it already had, in Magee University College, the nucleus of a new university'.[112] As the Field Day directors were from this generation, it is possible to think of the company as a belated response to this decision: as a homeless, peripatetic university of sorts.

Reactions to this political degradation also took other, violent forms, and Derry was the site of some of the most important incidents in the early phases of the 'Troubles': the 'Battle of the Bogside' in August 1969 and, most notoriously, 'Bloody Sunday', 30 January 1972, when fourteen civilians on a civil rights march were killed by British paratroopers, who were subsequently exonerated by the discredited Widgery report.[113] The spark that ignited the 'Troubles' was the issue of housing, which brought to a head the political, historical and symbolic definitions and significances of space, accommodation and home in Northern Ireland. As Foster describes it:

> [i]n 1968 occupations, squattings, marches, might appear the simple mechanics of the international student protest; but within Ulster they represented symbolic invasion of ancient territory [...] And the forces of the Protestant state reacted accordingly. Violence began at a Derry march in October 1968, notably on the part of the police.[114]

The unionist grip on Derry's political life had led to notable inequities in the provision of social services, particularly in the area of the allocation of public housing to Catholic families. As the provision of housing was entirely in the gift of the gerrymandered local council, and as having a house entitled one to vote in local elections, the issue was at the heart of the injustices in Derry.[115] The problematic location of home and the laws of hospitality have therefore long been fundamental to quotidian experience in Northern Ireland, and the conflict can be phrased in terms of the problems of finding accommodation with and for the other. These difficulties are at the heart of the ethics of translation and the problematised, translational versions of home that are constructed in the dramas emerge from the intractable linking of hospitality and hostility that so acutely defined Northern Ireland.

Morash notes that the 1980 production of *Translations* 'was the first time in almost two centuries that a professional company had rehearsed a play in Derry'.[116] The founding of Field Day and the fact that this production was being staged in the Guildhall reflected an amelioration in the political life of Derry. The Guildhall, Derry's town hall, represented the unionist domination of the city, and its inhospitality to the non-unionist population of the city is encapsulated in Skinner's declaration in Friel's 1973 *The Freedom of the City*: 'this is theirs, boy, and your very presence here is a sacrilege'.[117] According to Rea, the Guildhall 'was a dead political space. We made it a live theatrical space. That meant that everything we did had a meaning, or a vibrancy, because of where we did it.'[118] The ever-changing domain of the theatre – a place where other places are imaginatively brought into possibility – is Field Day's home, and the transformation of a political site into a theatrical space stands not merely as a symbol of Field Day's political intervention, but as its enaction. To ab-use somewhat a term from Clifford Geertz, Northern Ireland had assumed the aspect of a 'theatre-state': while all states have a performative element, the lack of a properly democratic political structure in Northern Ireland meant that expressions of power and identification were more forcibly articulated there through symbolic manifestations, such as parades and funerals, than in most other Western contexts.[119] In a commensurate fashion, the hunger strikes, which formed the political backdrop to the founding of Field Day, were potently described by Deane as 'some kind of terrifying public theatre'.[120] Field Day's theatrical intervention sought to reconfigure the moribund, stereotyped performances through which the political manifested itself in Northern Ireland, and to propose more animated, more inherently problematised, alternatives to these: 'Field Day', as Andrews put it, 'is based on the belief that [...] imaginative strategies do "make things happen", however indirectly, and that the public ritual of theatre is one of the most potent, as well as one of the most sophisticated, ways we have of exploring our social, cultural and emotional world.'[121]

In other words, Field Day's political actions were intrinsically performative, in the sense that the plays brought to the fore the ways in which identities are performatively constructed. The politics of the performative that informs this book has much in common with Judith Butler's understanding that the '"*performative*" suggests a dramatic and contingent construction of meaning'.[122] Gender identities, Butler argues, are 'an effect of a corporeal signification':

> acts, gestures, and desire produce the effect of an internal core or substance, but produces this *on the surface* of the body, through the play of signifying absences that suggest, but never reveal, the organizing principle of identity as cause. Such acts, gestures, enactments, generally construed, are *performative* in the sense that the essence or identity that they otherwise purport to express are *fabrications* manufactured and sustained through corporeal signs and other discursive means. That the gendered body is performative suggests that it has no ontological status apart from the various acts which constitute its reality.[123]

Maintaining and regulating a rigid conception of an integral identity has been the project of hegemonic politics, which have structured this in binary terms – for instance, man/woman. When identities are understood as being performative, any idea that they express an authentic, original self is deracinated. As Butler illustrates, this performativity is exposed through performance: in its imitation of gender, drag '*implicitly reveals the imitative structure of gender itself – as well as its contingency*'.[124] Framed in terms of translation, drag is a reproduction that displaces the original: it is a 'parody [...] *of* the very notion of an original; [...] gender parody reveals that the original identity after which gender fashions itself is an imitation without an origin'.[125] By performing aspects of the construction of Irish identities, the Field Day dramas undermine the notion that an essential 'Irishness' exists. This is how they embody the processes of the 'fifth province', which, as has been shown, came into being through such a performative, translational process; indeed, it is only when it is understood as a translation without an original that the 'fifth province' can function as an enabling political model. In postcolonial terms, as McGrath has shown, the performative expresses a subaltern positioning. Employing Bhabha, McGrath has situated Friel's dramas, in particular his Field Day output, in 'the articulatory moment at the boundary of conflicting cultures [that] constitutes the hybrid space and the historical present occupied by the colonized and the colonizer alike'.[126] Such an ambivalent positioning eschews what McGrath sees as the limiting binaries of much postmodern thinking, and replaces them 'with performative enunciations from the boundaries between cultures'.[127] In this light, the performative marks the difference between the postcolonial articulations of the Field Day pamphlets and the plays. While the pamphlets unearth colonial binaries, the dramas call attention to the ambivalence that, as Bhabha has argued, gives colonial discourse its durability; the ways in which it, in particular, constructs otherness as 'at once an object of desire and derision'.[128] In manifesting this, they perform a disjunctive, subaltern (dis-)positioning, which, as Lloyd puts it, 'registers the radical non-identity of the colonized subject'.[129] Such a construction stands opposed to the shared objective of empire and nationalism to create, through forms of stereotyping, a unified subject with a unified identity.

Translation, as Bhabha put it, enacts 'the performative nature of cultural communication',[130] and not only do the dramas embody the motions of translation, but the dramatic performance of this performativity – the *mise en scène* – is also a 'translation', as Vitez has argued. Understood in this manner, the operations of performance reflect the ethics of translation that informed Field Day's intervention: precisely because it is impossible, a theatrical performance must be repeated.

> Works of art are enigmas to which we must constantly endeavour to find an answer. [...] The object remains forever insoluble, the trace of the text remains and we must translate it constantly. [...] And it is exactly because it is impossible to translate that I maintain: the production of a play is a translation. One cannot, but one has to try.[131]

Such uncertainty and possibility describe the place of Field Day, and this study seeks to register this contingency by viewing the plays through the optic of the never-finished processes of translation. In doing so, these text-based readings also lodge an implicit acknowledgement that a theatre text must be, as Bassnett notes, 'read as something *incomplete*', because it is 'impossible to separate *text* from *performance*, since theatre consists of the dialectical relationship between' the two.[132]

In the mid-1990s the Northern Irish 'Troubles' slowly began to peter out and Field Day drastically reined in its activities, calling a halt to its annual tours. The very gradual cooling of the political climate has also taken some of the heat out of the debates surrounding Field Day, and has led to a more discerning critical engagement with the enterprise. Histories of Irish drama, such as those written by Roche, Murray, Morash, Grene and Pilkington, which, from different perspectives, take seriously the political function of theatre, acknowledge Field Day's key role in late twentieth-century Irish theatre, even if they differ in their assessments of the company's activities and achievements.[133] In common with the style employed in important surveys of the relationship of culture and politics in Ireland by Cairns and Richards and Kirkland, studies of Friel's *oeuvre* – by Andrews, Pine, McGrath and Boltwood – subject this relationship in his work to a sophisticated theoretical scrutiny.[134] Despite egregious errors (for example, *Three Sisters* is referred to throughout as *The Three Sisters*, and the 1798 rebellion is given as having taken place in 1789 and, a few pages later, in 1793), Andrews provides a useful account of Friel's Field Day dramas that positions them in a Heideggerian postmodern tradition.[135] Pine, McGrath and Boltwood pursue this line of thought through to its articulation in Homi Bhabha's work, and lay particular emphases on how the ideas of liminality and hybridity help to unveil the contradictions embedded in Friel's political outlook. These studies, which have informed the approach taken in this book, reflect a shift from the more Manichaean reception that was accorded to the enterprise in its heyday in the 1980s and early 1990s, when Field Day was viewed by quite a number of commentators as a cell of militant cultural ideologues; a strain of criticism that reaches an egregious nadir with Colm Tóibín's tabloid-like description of the company as 'the literary wing of the IRA'.[136] As this indicates, a tendency to fix the company in a rigid conception of the political economy of Northern Ireland has marked Field Day's critical heritage. Pre-eminent in this regard is Edna Longley, the company's most prominent and consistent critic, 'always', as Deane quipped, 'preternaturally alert to the next step in the Field Day conspiracy'.[137] Field Day, for her, expressed an unreconstructed essentialism: it 'enacted the process whereby political fixity shuts off imaginative possibility, the ideological tail wags the creative dog'.[138] Extolling the virtues of '"English" empiricism', Longley distrusts any mixing of the social and the aesthetic spheres, seeing such a blend as deleterious to the arts and a mystification of politics.[139] However, a remarkably uneven politicisation of culture marks her criticism, because while she rejects Field Day's exploration of the social and the political through the literary, she nonetheless reads the literary through the political. As Kirkland noted, she 'has continually forced the poem as aesthetic object into the bear-pit of the social arena while consistently denying that it should have to function in such a manner'.[140] This inconsistency arises because, at bottom, her critical battle with Field Day is ultimately a defence of the Northern Irish state, and culture is viewed as a Trojan Horse for those who oppose this state. To a considerable extent, as Field Day illustrated, cultural authority had been seized by artists and scholars who were hostile to the politics of Northern Ireland, and in this context, strains of the siege mentality resonate in Longley's work, which figures the unionist community, more or less, as an embattled minority. Moreover, by defending unionism on these cultural grounds, the debilitating imbalances embedded in Northern Irish politics are elided: '[u]nionism does not linger like bog-mist in unsuspected crannies. For better or worse, you generally know it's there. Unionism exposes its contradictions in public: in the gap between its interior monologues and what it can get the rest of the world to believe.'[141] Being so transparent and, in any case, intrinsically better than opaque nationalism, unionism almost transcends analysis. Such a viewpoint ultimately perpetuates the sectarian politics that it ostensibly sets its face against, as can be heard in Longley's allusion in 2006 to the 'user-friendly Prods' who worked with Field Day.[142] So although, as McCarthy has suggested, the priority she accords to the literary imagination has 'a great deal in common' with Field Day's perspective, what sets them apart are their clashing outlooks on the politics of Northern Ireland.[143] For this reason, her contributions rarely generated a productive critical dialectic with the company; instead, what emerged from their fiery exchanges was a polarised critical atmosphere that had the unfortunate corollary effect of either sidelining other criticism of Field Day or marshalling it within the confines of the binary it created.[144]

Traces of this binary can be detected in the two monographs that have been published on Field Day: Richtarik's 1994 *Acting Between the Lines* and Szabó's 2007 *"Clearing the Ground"*.[145] Richtarik gives a very thorough account of the founding of the company and its initial five years, furnishing important insights into the practical challenges Field Day faced in getting its show on the road. Important as this is, the limits of this cultural-historical approach are felt in her readings of the plays, which are too dependent on contemporary newspaper reviews. While these give an idea of the reception of the dramas, very few of them offer any real explorations of the plays. Moreover, at the start of her book Richtarik provides a list of these newspapers aligned according to their political allegiances in the North. Implicitly, therefore, the book's critical approach revolves around locating Field Day in terms of the unionist/nationalist binary, and so it is not too surprising that she ultimately critiques the company for failing to break its logic. Szabó's book covers all the dramas and pays attention to significant aspects of set design and production. In common with this present study, it does not consider the dramas chronologically, but thematically. While there are correspondences between the themes examined in this book and hers, Szabó offers readings of the dramas based on 'the Field Day principles of identity discussed in the pamphlets'.[146] Not only is the relationship between the pamphlets and the dramas by no means straightforward, but the characterisation of these 'principles' is problematic: 'Field Day's intention was to "clear the ground" of colonial binary oppositions and embrace the pluralism opened up by James Joyce and Samuel Beckett with their definitions of a heterogeneous and at the same time European Irish identity.'[147] It is, as Szabó's book illustrates, very difficult to substantiate the claim that the concepts of 'pluralism' and 'modern Europe' constitute central elements in a Field Day 'ideology' (terms that recur regularly in her study). No doubt, some of the directors would have embraced versions of these ideals and, indeed, aspects of the enterprise's output speak to them: an invocation of Europe is to be found in Kearney's pamphlet, for instance, and the translations express an engagement with Europe. Fundamentally, though, in *"Clearing the Ground"* these concepts form one – intrinsically positive – side of an overarching binary, on the other side of which is Irish history, which remains obstinately 'gluey' and stuck 'in a loop' throughout the analysis, implicitly unlike the histories to be found elsewhere in the unproblematised 'modern Europe'. As is the case in Richtarik's study, Field Day is ultimately considered to have failed to liberate itself from a structure that has informed the approach taken by the critic to the company. A clear objective of the following readings is to offer an alternative perspective on Field Day's achievements; one that takes the company at its word, all the while insisting that this word is open to other, possibly contradictory, meanings.

1
Translating the Irish Past

Introduction

> To put it bluntly, all that FD has done is an exercise in inter[p]retation. We have a double text – history and the here and now. Each is read in terms of the other. There is no escape from that.
>
> — Seamus Deane[1]

'The owl of Minerva', as Hegel had it, 'takes its flight only when the shades of night are gathering'.[2] Set in moments of cultural twilight, these two key Field Day plays – *Translations* and *Making History* – unveil how the tensions manifested in these periods in Irish history speak to the construction of Irish cultural identities in the late twentieth century. These moments are crucial milestones in the decline of what might be thought of as authentic Irish culture: the nineteenth-century language shift and the late sixteenth-century deracination of the remnants of Gaelic civilisation. A crucial truth about Gaelic culture and the Irish language for Friel is that it is precisely because they have been lost that they have acquired a nostalgic, nationalist, patina of authenticity, which has obscured a critical appreciation of their significances. In different ways both plays (and *The Communication Cord*) chip away at this nationalist veneer, and in doing so offer reimagined versions of Ireland's colonial history. While both plays register the loss and dispossession involved in these episodes, their focus is on exploring the processes through which these moments have come to be understood. Grene noted that 'interpretation' is the 'main theme' of *Translations*, and this is even more emphatically the case with *Making History*.[3] More precisely, these dramas enact the method that, as Frazier noted, informed the first three volumes of the *Anthology*: of 'teach[ing] not your position on the past as if it were the past, but teach[ing] the conflicts of interpretation as stereoscopic views upon an irrecoverable and unsettled history'.[4] These 'conflicts of interpretation' are dramas of language, and the plays are concerned with how the apprehension of the past through language constructs identities. Both plays enact the postcolonial interplay of translation and history that was noted in the Introduction: translation is located in history, while historical concepts are problematised by being thought about in terms of translation, and it is through this doubled motion that the historical events in these plays are brought into dialogue with the present.

The fact that the first four dramas to be read here are the Friel Field Day plays is an acknowledgement of his centrality to the enterprise. Although Field Day did not set out with an explicit dramatic agenda, a Field Day play, to a large extent, bears the hallmarks of a Friel play: dramas that reimagine the political significances of the times and places in which they are set by focusing on issues of language and split (post-)colonial identities. Richard Pine in *The Diviner* – the finest study of Friel's *oeuvre* – encapsulates Friel's work in terms that speak to Field Day's project:

> Friel's genius has created metaphors to make vital connections between past and present, between myth and narrative, creating meanings which have both public and private resonance. But it is also a temporary genius, only valid at the transitory moments in which the metaphor inheres in the imagination.[5]

Framing Friel as a generator of metaphors, Pine indicates how Friel's awareness of the paradox of the possibilities and inadequacy of language produces an overriding sense of impermanence: a constant shuttling between places and times, the significances of which are created and recreated through this movement. The key location to undergo these metaphorical shifts is home. As Pine notes, Friel is never entirely comfortably at home; rather, he is, '[i]n some senses [...] permanently away from home, not even knowing what or where "home" might be', and so each 'homecoming' is 'followed inexorably by a new leavetaking, and the stories that demand to be told on each crossing of the threshold, constitute a permanent self-examination and redefinition'.[6] His Field Day plays project this process of personal redefinition on to a national canvas in precisely the same way: through narratives that speak of a movement to and from wherever stands as an Irish 'home' at a particular time. In *Translations* and *Making History*, Ireland, like a typical Friel character, is in search of a usable identity, imaginable versions of which emerge into language for a time from a translational movement between different temporal and physical locations: 'occupying a liminal, between-space', this Ireland is always coming-into-being through language.[7]

Charting dispositions: *Translations*

> I draw the line
> At being robbed of what is mine,

> My *patria*, my deep design
> To be at home
> In my own place and dwell within
> Its proper name
>
> — Seamus Heaney[8]

Heaney has claimed that the first productions of *Translations*, in Yeats's terms, 'engross[ed] the present and dominate[d] memory' in such a way that they 'allowed inchoate recognitions, both cultural and historical, to be clarified and comprehended'.[9] Field Day's project, it was noted in the Introduction, involved charting an alternative cultural map of Ireland that would place under question political configurations on the island, and the company embarked on this task with a play that portrayed how a politically sponsored map redrew Irish cultural arrangements. Two different types of lines clash in *Translations* – those on maps and those that plot the forms of local agency implicit in the line that Heaney draws in the stanza above from *An Open Letter*. Included in the *Penguin Book of Contemporary British Poetry* by his friends Blake Morrison and Andrew Motion, Heaney asserts his right not to be called British and takes a stand against the facile hegemonic assumptions that inform this erasing of his own name for his home.

Postcolonial thinkers have charted an apoptotic congruence between the activities of colonising map-makers and the disappearance of native languages; as Spivak notes, 'countless indigenous languages in the world [...] were programmed to vanish when the maps were made' – a seemingly automatic effect engendered by the would-be mechanical motives underlying the map-making.[10] While this describes the scenario of *Translations*, the play nonetheless privileges the glitches in this programme, the instants when it unwinds and reveals its contradictions, as it is in these short circuits that the possibility exists for reimagining Ireland's past and its contemporary ramifications. *Translations* establishes a dialogue between different temporal planes: between the Anglophone modernisation being ushered in by the Ordnance Survey and the Gaelic-speaking world of Baile Beag in the drama, and between this historical dialectic and late twentieth-century Ireland.[11] It is a history play in the doubled sense of having as its scenario a historical act of map-making that also stands as a metaphor for the act of making history. The map embodies a drama of representation and language, and how it is understood is therefore a central crux of the play. As Friel put it in early notes for the play: '[w]hat in fact is a map an image of? A state of mind? An area of recognisable consciousness? A political loyalty? A spiritual hope? An aesthetic? This is what the play is about: a conflict over a piece of map-making which means different things to the various characters.'[12] These different significances speak to different perceptions of the function of history, and the play explores these interpretations of the map in such a way that contemporary Irish identities are seen as emerging from a complex, fluid, contingent criss-crossing of these perspectives, rather than out of a secure relationship to a place and to a language. What had appeared to be fixed, mapped, positions are unveiled as dispositions. Reflected through this drama's disruptive prism, the sense of what constitutes one's 'proper name', to return to Heaney's verse, is splintered to the extent that what is one's own (*propre*) struggles to coincide with what is appropriate or usable.

Ricoeur has suggested that a 'secret dream of emulating the cartographer [...] animates the historical enterprise'.[13] This is the fantasy of an ideal form of objectivity and of a concomitant ability to piece historical experiences together into a seamless, unified, comprehensive map of a period. In the Renaissance and Enlightenment, cartography, as Zilsel has argued, was to the forefront of the evolving 'ideal of scientific progress'.[14] Maps seem to exist in a purer world, one somehow distant from man and his attendant baggage of history and language and, as such, are, in many respects, archetypal positivist documents. Related to empiricism, positivist epistemology dismisses metaphysical considerations and accords priority to experience and observation in the creation of meaning.[15] The link between what maps represent and how it is represented would seem to be straightforward: maps appear not to interfere with what they describe, rather they simply provide the perspective of a passive, disembodied gaze. In short, what you see is what there is; there is neither any need to dig deeper than the surface, nor can there be any argument about what is before your eyes, as this sensory apprehension exceeds interpretation. This profoundly a-historical conception of surveying is crucial to imperialism, which is, as Derek Gregory argues, 'a discourse which privileges vision'.[16] Dismissing the idea of the apprehending gaze being in any manner informed, it allows maps, in Klein's words, 'to pass off for evident truth what is hard won, culturally acquired knowledge about the world we inhabit'.[17] Fanon noted how '[f]or the native objectivity is always directed against him',[18] and, in the Irish context, from the time of Gerald of Wales there have been any number of British 'topographies' and 'geographies' of Ireland that have done little more than transmit stereotyping tropes through the ages.[19] Abraham Ortelius called geography 'the eye of history' in the 1606 English version of his noted atlas, *Theatrum Orbis Terrarum* ('Theatre of the Whole World'), and *Translations* suggests an analogous relationship between history and geography.[20] As noted in the Introduction, restoring a dynamic link between history and geography has been a central objective of postcolonial scholarship. Through its staging of debates between various understandings of the status of the map, *Translations* shows that history is situated knowledge; situated, that is, in locations that are themselves narrative constructions.

'A map is a representation on paper [...] a paper picture [...] showing your country in miniature' (*T*, p. 405): with this Lancey attempts to communicate his positivist vision of his job to the hedge school in Act I. Faced with the actual presence of an uncomprehending audience, Lancey stumbles over this description: the sheer banality of his perspective is manifested in the attempt to transmit it to the other. Owen, his interpreter, helps him regain his composure by advising him to take recourse in positivism's necessary deceit: '*assume* they understand you' (*T*, p. 406). As Cheyfitz put it, 'at the heart of every imperial fiction [...] there is a fiction of translation': the colonial other is translated into the terms of the imperial self, with the net result of alienation for the colonised and a fiction of understanding for the coloniser.[21] Taking on board this fiction, Lancey comfortably outlines both the operations of the Survey and the legal motivation informing it by literally citing the recommendations of the Spring Rice Committee, the body created by the Westminster parliament in 1824 that proposed the survey. Lancey and the Survey merely embody the letter of the law, which is clear and unambiguous, and so their acts of translation will be equally transparent. This is classic positivism, as Kearney has shown in an essay first published in 1983 that still stands as the fundamental text for understanding the positivist and ontological conceptions of language in *Translations*.

> Positivism maintains that words are mechanically given (*positum*) – no more mysterious than objects in a world of similar objects. They are eminently unmysterious entities to be used as instruments for the representation, mapping or classification of reality. And the reductionist goal of positivism is to produce an exact decoding of the world by establishing a one-to-one correlation between *words* and the *facts* of empirical experience. Language is thus reduced to a utilitarian weapon for the colonization of Being.[22]

In this view, the landscape and its inhabitants are reduced to a collection of inert signifiers in order to facilitate a 'transparent' translation from 'thing' to 'sign'. Such a translation is an entirely one-way process: an act that seeks to impose meaning on, rather than come to terms with, the other. Moreover, as Lancey's dependency on Owen's highly disruptive translations to gain the community's acquiescence illustrates, inscribed within this mode are forms of delusion. Clarity and transparency are employed, wittingly and unwittingly, to conceal the political import of the project. Lancey, for instance, cites the Spring Rice Committee's recommendation that the survey is being conducted in order to provide for a more equitable system of taxation in the country; but, as Mary Hamer has pointed out, the beneficiaries of this apparently just objective were two to three thousand Anglican Protestant families.[23]

In a remark that Friel cites in his programme notes for the first productions of *Translations*, Lord Salisbury noted that '[t]he most disagreeable part of the three kingdoms is Ireland; and therefore Ireland has a splendid map'; yet, the fictions of understanding that wrought this document were more ambivalent (both in terms of motives and effects) than this suggests.[24] The Survey was entrusted to Colonel Colby, the director of the Ordnance Survey in Britain, and he recruited Lieutenant Larcom as his representative in Ireland. These two men, on whom the characters of Lancey and Yolland are based, were given a relatively free hand, and took advantage of this to enlarge substantially the scope of their project. They envisaged producing a Domesday Book of the country, which Larcom spoke of as painting 'a full face portrait of Ireland'.[25] To this end, a 'memoir' was to be published to supplement the maps' literally superficial transcription of the landscape with an in-depth survey of those who lived on the land. Around 1828 Larcom set up a Historical Department to coordinate this information. This 'peripatetic university' was staffed by some of the most important Irish scholars of the day: George Petrie, John O'Donovan, Eugene O'Curry and, briefly, James Clarence Mangan.[26] Leerssen has hailed the group as 'the rescue team of Irish antiquarianism, the men who set the investigation of Gaelic antiquity on a new, scientific and critical footing, and whose enormous labours laid the groundwork for all subsequent work in the field'.[27] As is detailed in Larcom's instructions to his field officers, they were particularly interested in seeking out and cataloguing domestic knowledge.

> Habits of the people. Note the general style of the cottages, as stone, mud, slated, glass windows, one story or two, number of rooms, comfort and cleanliness. Food; fuel; dress; longevity; usual number in a family; early marriages; any remarkable instances on either of those heads? What are their amusements and recreation? Patrons and patrons' days; and traditions respecting them? What local customs prevail, Beal tinne, or fire on St John's Eve? Driving the cattle through fire, and through water? Peculiar games? Any legendary tales or poems recited around the fireside? Any ancient music, as clan marches or funeral cries? They differ in different districts, collect them if you can. Any peculiarity of costume? Nothing more indicates the state of civilisation and intercourse.[28]

This shopping list brings out the tensions and contradictions in the enterprise; it is an implicit acknowledgement of the insufficiency of a purely positivist apprehension of place, but it nevertheless sets out to remedy this in a positivist manner, through a process that reinforces the imperial gaze in its position of, in Pratt's terms, 'the-monarch-of-all-I-survey'.[29] Underpinning these enquiries is the colonial ambivalence that Bhabha noted: colonial fantasy and desire are registered in the litany of peculiarities that interpellates the native, calling her/him into being precisely as the other.[30]

At the same time, the Survey accumulated a huge amount of knowledge that might otherwise have been lost, and which moreover would play an important role in Irish cultural nationalism. Thomas Davis advocated studying the output of the Survey as a part of a strategy of reclaiming 'Ireland for the Irish', as 'to be able to keep [Ireland], and use it, and govern it, the men of Ireland must know what it is, what it was, and what it can be made'.[31] Through Davis it is possible to trace how cultural nationalism involved recasting the knowledge accumulated through the positivist research activities of groups such as the Survey in an ontological light. In *Our National Language*, for instance, the philological interest in the Irish language that formed an integral part of these studies is re-read by Davis as an insight into the Irish soul.

> The language, which grows up with a people, is conformed to their organs, descriptive of their climate, constitution, and manners, mingled inseparably with their history and their soil, fitted beyond any other language to express their prevalent thoughts in the most natural and efficient way.

For these autochthonous reasons, Davis, writing in English, evinces an implacable opposition to translation.

> To impose another language on such a people is to send their history adrift among the accidents of translation – 'tis to tear their identity from all places – 'tis to substitute arbitrary signs for picturesque and suggestive names – 'tis to cut off the entail of feeling, and separate the people from their forefathers by a deep gulf – 'tis to corrupt their very organs, and abridge their power of expression.[32]

As Thiele notes, via Nietzsche and Foucault, '[p]olitical engagement [...] is primarily about the preservation and enhancement of an essentially conflictual and ongoing process of self-definition and social definition.'[33] In its pursuit of origins and authenticity, a nationalist, ontological, view of language such as this seeks to stabilise identities in one form: as much as an imperial, positivist mode, it halts the processes of self-definition. While Kearney has shown that *Translations* presents an ontological mode of translation as an alternative to the positivist, the play neither allows this to be seen as a completely enabling disposition, nor does it entirely permit the two attitudes to language to define the political situation, with the ontologically self-sufficient Irish facing the positivist soldiers.[34] Rather, the play sets up this dualism, and then transgresses the boundaries between the two dispositions; a fraught business, as Jimmy Jack is well aware: 'you don't cross those borders casually – both sides get very angry' (*T*, p. 446), as to do this sets identity 'adrift among the accidents of translation'.

In Act II scene 1, Yolland reflects on the discoveries he has made in the course of his work: '[a]nd your place names – what was the one we came across this morning? – Termon, from Terminus, the god of boundaries' (*T*, p. 418). Termon is the Anglicised version of An Tearmann, a townland in north-west Donegal; the translation of a name that names a border describes a zone where the ontological and the positivist dispositions meet and reveal their differences, as they speak to radically divergent conceptions of boundaries. An extended discussion of the etymology of 'Termon' features in the only 'memoir' to be published within the lifetime of the Survey – *Memoir of the City and North Western Liberties of Londonderry, Parish of Templemore* (1837)[35] – as the 'City of Londonderry' was previously known as 'Termonderry'.[36] This analysis of the border inscribed within Field Day's 'home' – Derry/Londonderry, which would come to define the border between Northern Ireland and the Republic – opens up a way of coming to terms with the differing understandings of language and translation enacted in the play. 'Termon' names a defined, privileged area that functioned as an asylum or sanctuary, and so the Irish word for 'sanctuary' (*tearmann*) is related to the linguistic, temporal and spatial unit a 'term'.[37] A crucial difference between the ontological and positivist dispositions lies in the way they conceive of how language defines the sanctuary it offers identity. The ontological understanding is drawn out in Heidegger's etymological examination of the Greek word for boundary, *peras*: '[a] boundary is not that at which something stops but, as the Greeks recognized, the boundary is that from which something *begins its presencing*.'[38] Such a boundary cannot delimit a stable site in which identity finds meaning, because it is implicitly determined through interaction with the other. If, therefore, the ontological disposition, in Kearney's understanding, registers 'the very way in which language determines our innermost being-in-the-world', such a conception of the boundaries of the self would suggest that it does this in a manner that acknowledges the necessity and impossibility of ever fully grasping the other.[39] Positivism, on the other hand, sees boundaries as setting definitions (of time, of space, of the self) that stand more or less distinct from language, and that provide a sealed environment within which identity is enveloped. In such a structure, translational interactions result in meaning being imposed on the other, even if that operation is conducted in the belief that it does not alter the identity of the other.

The most sustained ontological/positivist debate in *Translations* concerns the place name Tobair Vree. According to Owen, this place would be literally 'Brian's Well' in English, but as this name finds no apparent resonance in the landscape – all physical traces of Brian and the well have been eradicated – Owen suggests abandoning it and finding a more suitable English equivalent for their name book. Lacking an unequivocal, 'original' source of authority, the identity of this place is, as Owen shows, a palimpsest of fragmentary narratives, and it is this that Tobair Vree names. To get to this well, one has no choice but to go through language, and in this way Tobair Vree gestures to the example Heidegger called on in his description of how language defines a place where being might dwell.

> The nature of language does not exhaust itself in signifying, nor is it merely something that has the character of sign or cipher. It is because language is the house of Being, that we reach what is by constantly going

> through this house. When we go to the well, when we go through the woods, we are always already going through the word 'well', through the word 'woods', even if we do not speak the words and do not think of anything relating to language. [...] all beings, each in its own way, are *qua* beings in the precinct of language.[40]

A topography constituted by language, and consequently in constant movement, is beyond the grasp of positivism. It charts a space that encompasses not only physical, visible features, but also the interiority of the self that engages with this landscape, and it marks out both in the same way: as manifestations of language. In Gaelic Ireland this attitude was embodied in the onomastic verses of the *Dindshenchas*, in which each place name was thought of as a delicate filigree of the myths and history of its location as well as its physical characteristics. While they could make no claims to normative historical and geographical accuracy, these verses nevertheless reflected a tradition of deep bonds between the people and the land.[41] Irrelevant to daily life, this way of thinking about place lies dormant and forgotten (if it was ever known) until these links are disrupted and put under threat by a renaming process such as that initiated by the Survey. In this situation, *Translations* suggests, a vague and imprecise memory of a tradition like this returns to haunt not only members of the community but also the colonising renamers.

It is Yolland who draws Owen's attention to the fact that, as he remembers the story behind Tobair Vree, the name still has traction. This marks the moment at which Yolland's 'traitorous' voyage from positivism towards an ontological understanding intersects with, and confronts, Owen's 'betrayal' of his community. Pine has argued that language is 'the vital link between public and private' and that this 'is the central tension of *Translations*'.[42] More precisely, the tension in the play is produced by the betrayals that language generates as it creates a link between the world and the self. These public and private treasons in *Translations* are principally performed through different forms of 'problematic filiation', to recall Said's terms from the Introduction.

Fathers and Sons, the title of a Friel translation, also denotes one of the key tropes in Irish theatre: for instance, in Revival plays, such as Yeats's *On Baile's Strand* and Synge's *The Playboy of the Western World*, the failures of this relationship betoken forms of broken authority in colonial Ireland.[43] However, regardless of the political context, there are no untroubled lines of communication between fathers and sons. In *Translations*, Yolland's relationship with his father is marked by (avoided) conflict: he is only in Ireland because he missed his boat to India and took the next available posting so as not to have to face his tough-minded father. His fear of his father, though, is shot through with a respect for him, especially for his energy and ability to shape the world to his own ends. He imagines the determination of his father's character is related to the fact that he was born the day that the Bastille fell, which brought in its wake a positivist universalism and an infectious sense of control over the future, similar to that which was experienced in the British empire. Having such an inheritance to call upon initially allows Yolland to feel equally 'at home' in Bombay or Baile Beag which, being interchangeable, can be renamed at will. But his father has also bequeathed him a privileged space in which he can reconsider his inheritance. In a sense, their relationship re-enacts the connection between the empire and Romanticism: while Romanticism was made possible by the economic and cultural benefits of the empire, it was a reaction against the machine of empire. Yolland's betrayal of his father's ideology follows the journey taken by Wordsworth, his erstwhile neighbour, who retreated from the universalisms of the French Revolution and the empire to write, amongst other things, *Poems on the Naming of Places*, and explore, as Heaney put it, 'the nurture that becomes available to the feelings through dwelling in one dear perpetual place'.[44] At the sharp end of the empire, however, such a desire to preserve or nurture an intimate relationship with local places cannot be construed as an avoidance of politics.

Hugh's relationship with his sons reflects aspects of the deformed structures of filiative authority that is moulded by this pressurised environment. In contrast to Yolland's fearful respect for his father, Manus and Owen are forced to act as their father's father on practical levels, while going through the charade of deferring to his pomposity as a source of authority. As Owen tells Manus, 'in a way' they 'complement each other' (*T*, p. 408): while Owen has taken the route of betrayal by leaving his family and community to join the Survey, Manus has remained and stayed faithful to his father and to his language. He tends to Hugh's domestic needs and takes his place when Hugh is too drunk to give lessons. Although he was literally crippled by his father when he was a child, he feels responsible for his father, and this fidelity handicaps him anew every day. He will not apply for the post at the national school as Hugh has put in for it, and so he will lose Maire who wants to be with a man who possesses a more enabling sense of himself. In being faithful to his father, he betrays himself. Similarly, his refusal to interact in English, despite his knowledge of the language, speaks of an adherence to a notion of authentic authority that he refuses to acknowledge has been fatally undermined. Manus personifies how a native helplessness in the face of the overwhelming might of the empire can cling to forms of fidelity that offer no way of coming to terms with the world, and which instead contribute to the deprivation of agency. The most this culture can muster is 'a gesture' that might, at least, 'indicate ... a presence' (*T*, p. 391). Manus even fluffs this: feeling himself to be in the wrong place, his shout at Yolland was, as he is painfully aware, '[t]he wrong gesture in the wrong language' (*T*, p. 432). It was also uttered at the wrong time – at the last sighting of Yolland – and so implicates him in a crime he didn't commit; rather than indicating his presence, his gesture requires him to erase his presence, to flee Baile Beag.

Departing, he leaves a set of instructions for Owen on how to take care of their father. Owen, though, will not replicate Manus's slavish fidelity: he is too aware of the chasm between his father's blustering charade of authority and the fact of authority. Precisely for this reason, his final exchange with his father gives evidence of a mature, compassionate and competent personality. He has acquired this character because he has put himself through a process of treason. Comments Friel wrote in two notes to Rea while he was composing the play testify to the fact that this can never be an easy form of education with which to empathise. It is evident that Friel was learning about the complexities of translational fidelity as his understanding of the character he was creating evolved. In the first note, which accompanied a draft of the first act, Friel asked Rea to consider taking on the role of Owen (which he would play).

> Yes, I did say he's a P.R. man and that if he had been alive today he could have been a good host of a TV chat show. BUT – he is a very interesting and complex character (LOOK OUT FOR EPISODE TWO!!!) and will need subtle and delicate playing. Indeed it seems to me that he becomes more and more central. As the play unfolds he won't cease to be – except for a few minutes – the smiling, charming, quick-talking guy; but the reasons for his assumption of this mask are, I think, important and compelling for social and historical reasons.[45]

These sentiments are strongly reaffirmed when he sends Rea the first two acts – 'Owen becomes more interesting. I regret the easy tag I gave him when you were here.'[46] Friel has also admitted to having had to overcome an aversion to John O'Donovan, upon whom the character of Owen is based. O'Donovan was the foremost scholar of the Irish language of the age, who was employed (on a pittance) by the Ordnance Survey to ascertain the most correct version each place's name, and to advise on the orthography to be used on the maps.[47] While the Survey, and Larcom in particular, was sympathetic to O'Donovan's painstaking etymological investigations, he was nevertheless instructed that the place names on the 'maps should be recognizable and "pronounceable" by the English map-reader',[48] and in order to learn the correct orthography of a name, he was further enjoined to consult with 'the best authorities'[49] within an area: 'local landowners, agents, clergy, and schoolmasters, county records such as grand-jury presentments and lists of freeholders, and [...] existing maps both printed and in manuscript'.[50] This preserves the structures of colonial and hierarchical interpellation: the power of naming expresses authority in the fullest sense, particularly in this context where it links political and economic power with the power of the written word; written, that is, in a language that was not spoken by the majority of people in the country at the time. O'Donovan therefore played a crucial role in what Friel called 'the naming-taming process', and he initially viewed O'Donovan's work with the Survey as an unsavoury act of complicity:[51]

> I read into O'Donovan's exemplary career as a scholar and orthographer the actions and perfidy of a quisling. (The only excuse I can offer for this short-lived delusion is that the political situation in the North was particularly tense at that time.) Thankfully that absurd and cruel reading of O'Donovan was short-lived. But it soured a full tasting of the man.[52]

Through writing the play, Friel appeared to have revalued the performance of contradictions that is treason as a more enabling, yet always double-edged, response to the Northern Irish crisis.

As this play is all in English, Owen's translations negotiate between the positivist and ontological dispositions rather than between languages. He is conversant in both approaches, slipping quickly from the positivist – the names are 'just going to be standardized' – to the ontological – '[u]ncertainty in meaning is incipient poetry' (*T*, p. 408). Crucially, he does not place complete trust in either. While Yolland looks to give himself over to the idea that a condition of ontological wholeness existed in Baile Beag prior to the arrival of the Survey, Owen at no point holds such an idea. However, Yolland's growing estrangement from positivist logic – his belief that their activities are 'an eviction of sorts' and that '[s]omething is being eroded' (*T*, p. 420) – forces Owen to attend to the appeal of language: this is the precise point at which he gives the account of Tobair Vree. Furthermore, by pointing out to Owen the simple fact that he remembers his grandfather's story about the origin of this place name, Yolland brings home to him an alternative, familial, narrative-based authority that stands apart from the sanctioned authorities he has been using for his renaming. Having felt that there were no usable traditions of authority in the native community, Owen is reminded by Yolland that he has inherited one that resides in the vagaries of language. His immediate reaction to this is to reclaim the name given to him by his family; in the longer term, this perspective enables him to engage with his father.

This engagement never resolves itself into an unproblematic return of the prodigal son, as the very nature of the home to which he is returning is subjected to constant questioning. Even Owen's reclaiming of his name cannot be read as an unequivocal recovery of agency. When Sarah finally articulates her own name at the start of the play, Manus greets this with the promise/threat: '[s]oon you'll be telling me all the secrets that have been in that head of yours all these years' (*T*, p. 385). Withholding one's name is a way of protecting what Yolland (via Steiner) called, '[t]he private core' (*T*, p. 416) of an identity, as to pretend to be someone else is, in Steiner's words, 'to employ the "alternative" powers of language in the most thorough, ontologically liberating way'.[53] Owen, however, was using a name that was imposed on him, not one that he had withheld. At the same time, his retaking of his name does not betoken a cut-and-dried process of substitution, with 'Roland' standing as an unambiguous symbol of his colonial submission and 'Owen' restoring him to a state of pre-colonial wholeness. Pilkington has noted that Tobair Vree is an Anglicisation rather that an original Irish name; the same, of course, is true of Owen.[54] Once again, the colonial name-changing is not bringing about the destruction of some sort of an original and authentic 'Irishness'; rather, it is enforcing a halting of a translational struggle for self-definition and identity. Struggle assumes an other of some form, as does naming; in fact, it was the fantasy of self-naming that drove and ultimately sealed the fate of those who constructed Babel: they wanted to make a name for themselves.[55] Insisting on the inevitable interaction with the other, *Translations* portrays how the ontological view of language must undergo a translation, through the other – the positivist – in order for it to be incorporated into the world and how, through this translational interaction, it calls the positivist disposition back to the duty it owes to language. At the same time, *Translations* registers the ways in which the political context limits translational possibilities, as the play proposes this dynamic through a context in which one disposition has overwhelmed the other. But, as Friel learnt from Steiner's model of translation in *After Babel*, these disabling impulses are latent within every act of translation.

On 22 November 1978 Friel made some preparatory notes for the as-yet unnamed play, which concluded with this preliminary, capitalised, summary:

> It is, in fact, a play about
> DEFINITION——N.B.
> CONQUEST
> ABSORPTION
> INTER-RELATIONSHIP
> &
> AFFECTION[56]

These informing concepts all describe crucial facets of the act of translation, and echo three of the four parts of the hermeneutic cycle into which Steiner has codified the act. The motion that is not accounted for is the first one, namely the 'initiative trust' in the other, which is the springboard for all acts of translation. Following this, there is 'aggression', which is a form of 'conquest'. Friel's concept of 'absorption' finds a parallel in the third, 'incorporative' motion into the target language. Finally and crucially, the most difficult motion, the one that completes the hermeneutic cycle, 'restitution', can be reached towards through forms of affectionate interrelationships.[57] As can be seen by the terms in which the different motions are cast, the processes of translation involve negotiating a series of ethical decisions. Even a complete translation – one that embodies a form of restitution – must also incorporate an aggressive encounter with the other, and Friel looks to capture in *Translations* the problematic ethical struggle that the hermeneutic cycle describes.

Friel's debt to Steiner is very well known and has been commented upon by almost every critic who has examined *Translations*; however, his translation of *After Babel* into the context of pre-Famine Ireland also enacts a form of restitution. Steiner has been critiqued for his blindness to the situation of writers in (post-)colonial contexts: Samia Mehrez detects in *After Babel* the absence of 'the political context and power relations within which language acquisition takes place', and an exclusion of 'questions of colonialism and cultural hegemony'.[58] Niranjana reiterates this criticism and locates the source of this lack in the idea of restitution, which she sees as nothing less than a 'lapse into [...] classical ahistoricism'.[59] At the core of Niranjana's critique is an understanding that a complete translation is the only mode of translation allowed for in *After Babel*. However, Steiner is explicit that a '"perfect" translation is no more than a formal ideal'.[60] The idealism embodied in the final motion of restitution can be the motive for re-translation and so, rather than express an a-historicism, it injects the need for history – for change and renewal – into the cycle as a whole. *Translations* illustrates how such a conception of the final motion of Steiner's model might have a problematising, postcolonial salience. At precisely the point at which the drama reaches towards an ideal translation – the love scene between Yolland and Maire – the a-historical idea of a permanent, perfect translation is dismissed; so, even though the inevitability of submission to the authority of English is inscribed throughout the play in the fact that it is all in English, at the same time, the a-historical ideas of inevitability and immutability are being continually deconstructed.

The dynamics of communication in sex, as Steiner noted, embody those of translation, and he turns to a metaphysical image of the communication that takes place in love to capture the fourth movement of the hermeneutic cycle, borrowing the term 'interanimates' from Donne's poem 'The Extasie' in order to describe this action of restitution.[61] Restitution is, as Smith succinctly put it, 'made under the sign of the "other"'.[62] It is premised on an interaction of the self and the other in a dynamic that goes beyond forms of appropriation, and that instead looks to add something to both; an action Heidegger sought to encapsulate in the adjective *ereigende*, which, according to his translator, speaks 'of an activity or process by which nothing "selfish" occurs, but rather by which the different members of the world are brought into belonging to and with one another and are helped to realize themselves and each other in realizing this belonging'.[63] The love scene in *Translations* is made possible by Owen who, as the go-between, introduces Maire and Yolland and helps interpret for them. In assisting them to set up their meeting, Owen unwittingly betrays his brother: as ever, translation involves treason. However, as is signalled by Yolland's signature uncomprehending plea, '[s]orry-sorry?' (*T*, pp. 424–5), Owen's interventions do not aid communication. It is only in the subsequent scene, when they have dispensed with the translator, that communication begins to flow through the only piece of Irish dialogue in the drama. Maire first reaches across via her other languages – Latin and her sole, useless, line of English – and in response, Yolland names the towns he has been renaming:

> YOLLAND: Maire.
> (*She still moves away.*)
> Maire Cathach.
> (*She still moves away.*)
> Bun na hAbhann? (*He says the name softly, almost privately, very tentatively, as if he were searching for a sound she might respond to. He tries again.*) Druim Dubh?
> (MAIRE *stops. She is listening.* YOLLAND *is encouraged.*)
> Poll na gCaorach. Lis Maol.
> (MAIRE *turns towards him.*)
> Lis na Gall.
> MAIRE: Lis na nGradh.
> (*They are now facing each other and begin moving – almost imperceptibly – towards one another.*)
> MAIRE: Carraig an Phoil.
> YOLLAND: Carraig na Ri. Loch na nEan.
> MAIRE: Loch an Iubhair. Machaire Buidhe.
> YOLLAND: Machaire Mor. Cnoc na Mona.
> MAIRE: Cnoc na nGabhar. (*T*, pp. 428–9)

This call-and-response becomes a form of mirroring increment rather than simple repetition. It is an act of translation that opens up the inner resonances of these names, as Maire and Yolland locate themselves and each other in the topography these names weave. At the same time, the sheer fragility of this site is always present: the names have just about lost their grip on actual places, and their significance is coming to reside entirely in a form of evocative poetic incantation. Maire and Yolland come together and communicate under a sign of incipient silence, and there can be no doubt that this spell cannot last. It is broken by nothing less than the English word 'always':

> YOLLAND: I would tell you how I want to be here—to live here—always—with you—always, always.
> MAIRE: 'Always'? What is that word—'always'?
> YOLLAND: Yes—yes; always. (*T*, p. 429)

'Always' names the impossible wish of lovers; it gestures towards a realm of full comprehension, a place beyond the clumsy intervention of translators such as Owen. Yet, despite linking the pair through its repetition, it is a point of incomprehension. The dialogue ends with an almost exact repetition of the formula above, but with 'always' now in 'Irish'. Playing on the theatrical conceit of having all the characters speak in English, it appears that the word has been so drained of meaning that Yolland does not recognise it in his own tongue.

> MAIRE: I want to live with you—anywhere—anywhere at all—always—always.
> YOLLAND: 'Always'? What is that word—'always'? (*T*, p. 430)

Yolland's almost immediate disappearance after the love scene shows how 'always' is at the mercy of historical, social and political factors. In this categorical manner, Friel resists presenting the love scene as a fully consummated, ideal translation. There are perhaps few other words in any language that are as anachronistic and a-historical as 'always', and it is in this zone of fantasy that 'always' describes that the limits of the positivist and ontological dispositions intersect.

It becomes apparent that both the ontological and positivist dispositions are insufficient on their own for the same reason: both tend towards an a-historical idealism. Earlier in the play, Hugh employed other phrases from *After Babel* to point to the ontological resistance incorporated in Irish to precisely the inexorable tide of history: 'it is a rich language, Lieutenant, full of the mythologies of fantasy and hope and self-deception – a syntax opulent with tomorrows. It is our response to mud cabins and a diet of potatoes; our only method of replying to ... inevitabilities' (*T*, pp. 418–19). While this might be read as a celebration of the imaginative power of language to refuse to accept the world as it is, it can also be understood as a description of an ontological sclerosis: a rigid, unchanging focus on tomorrow or on impossible alternative times, rather than on the present. Positivism can be similarly limited, but, in its case, by an engrossment in the ever-present. It is only when the two dispositions enter into a translational dialogue that a language remains vigorous. Hugh's statement, though, was an act of cultural bravado in front of the English soldiers and not one that defines his relationship with language; in fact, he is the character who reinserts history, in the sense of change through time, into the apprehension of the cultural changes that are taking place. 'To remember everything is a form of madness' (*T*, p. 445), he avers, and so he agrees to teach Maire English, though he will not start with the word 'always', as it is 'a silly word' (*T*, p. 446). Hugh brings translation into play with history, in such a way that history is recalled to the duty it owes to language, and historical awareness is incorporated into an onto-linguistic apprehension of the world. Instead of taking refuge in the idyllic space 'always' describes, Hugh will provide her, 'with the available words and the available grammar. But will that help you to interpret between privacies? I have no idea. But it's all we have' (*T*, p. 446). For Pine, '[t]he fact that English is the language that Maire needs to know is the central fact of *Translations*', and the way that Hugh strives to fashion a space in which he might teach her English between, and containing, the two dispositions marks him out as a subaltern pedagogue for the late twentieth century.[64]

Friel acknowledged that his vision of the hedge school was informed, in large part, by Dowling's *The Hedge Schools of Ireland*.[65] In this text, the hedge schoolmaster is portrayed as an unofficial source of authority and, according to Dowling, was viewed by colonial officials as a potentially subversive force.[66] Bereft of a fixed position in the official hierarchy, Hugh constructs his own alternative, subaltern form of authority through his use of Greek and Latin, which sets him at a remove from the community he is deeply embedded in. While these languages are the classic sources of authority, they are rendered expressions of disjuncture in this context: even Hugh's 'need for [his] own' is phrased in another language, Latin, as his '*desiderium nostrorum*' (*T*, p. 445). Hugh's inability to get beyond translation and chart a univocal authority is the lesson he teaches the audiences of *Translations*. This is most emphatically underlined in his concluding perspective on the newly Anglicised place names: '[w]e must learn those new names. [...] We must learn where we live. We must learn to make them our own. We must make them our new home' (*T*, p. 444). Home is not a stable or an inherently known site: it must be translated into being. *Translations*, in short, stages Heidegger's notion that 'language is the house of Being' in a way that sets this dwelling in a translational flux and places it under the constant threat of destruction by the forces of history.[67] Such a 'house of Being' can never be construed as something pure, original or authentic, nor, indeed, exclusively one's own. Instead, it expresses the translational contradictions that for Steiner define a '[g]ood translation': 'that in which the dialectic of impenetrability and ingress, of intractable alienness and felt "at-homeness" remains unresolved, but expressive'.[68] The postcolonial import of *Translations* resides in the ways in which it illustrates that such unresolved tensions describe the dimensions of an enabling home in late twentieth-century Ireland. While it is impossible to chart such a home in any empirical fashion, this does not render it an a-historical ideal; rather, the constant struggle to construct such a dwelling is, as the play shows, an ethical response to historical circumstances.

The impossible possession of the past: *Making History*

> In some ways the inherited images of 1916, or 1690, control and rule our lives much more profoundly than the

> historical truth of what happened on those two occasions. The complication of that problem is how do we come to terms with it using an English language. For example, is our understanding of the Siege of Derry going to be determined by Macaulay's history of it, or is our understanding of Parnell going to be determined by [F. S. L.] Lyons's portrait of Parnell? This is a matter which will require a type of eternal linguistic vigilance.
>
> — Brian Friel[69]

Nietzsche averred that, '[o]ne can estimate the degree of an age's historical sense from the way it makes *translations* and seeks to absorb past ages and books'.[70] *Making History* is a meta-historical play in which the protagonists essentially review and critique how contemporary Ireland has translated them and their epoch: it is not so much about Hugh O'Neill and the Battle of Kinsale as it is about 'the historical sense' of late twentieth-century Ireland. In this respect, *Making History* can, like *The Communication Cord,* be regarded as an echoing response to Field Day's inaugural drama: in very different fashions, both reiterate the links between attitudes to language and the representation of the Irish past that were central to *Translations.* As with *The Communication Cord, Making History* does this by blending an explicit academic element into the drama, which some critics have found off-putting, reading it as Friel's most doctrinaire play, and as evidence of how his involvement in Field Day was harming his art.[71] There is no doubt that episodes resemble academic debates (indeed they are the principal focus of this academic study); at the same time, as a play about the difficulties of accounting for public and private pasts, *Making History* speaks to concerns that have been at the centre of Friel's work both before and after his time with Field Day. Whereas Friel has most often located this drama in the vagaries of individual memory and identity, in *Making History* he risks transferring this debate on to the dramatically unpromising terrain of historiography.[72] If the experiment does not entirely come off on a dramatic level – it is a 'failure as theatre', according to McGrath[73] – the manner in which personal memory is brought into play with public narrative in order to question the claims (to truth, to authenticity, to importance) of both, echoes the structures of undecidability in plays that operate in very different contexts, such as *Philadelphia, Here I Come!* and *Faith Healer.*[74]

Specifically *Making History* stages the debate that had been running in Ireland since the late 1960s between revisionist and nationalist historians. Revisionism presented itself as an empirical, archive-based corrective to what it perceived as the dominant nationalist discourse in Ireland, which it believed was not history: according to Fanning, revisionism involved 'revising [...] not history but myth and legend'.[75] Properly speaking, this was second-wave revisionism because, since at least 1938 and the founding of

Irish Historical Studies by Moody and Edwards, academic history in Ireland has been revisionist, in the sense that it has been rigorously archive-based.[76] Hence, the examples of nationalist discourse cited in a polemically revisionist text such as Foster's *The Irish Story* are nineteenth- or early twentieth-century histories, literary texts, the *Catholic Bulletin*, and history textbooks for schools.[77] But, while revisionist discourse was pitched on a methodological level – exposing the mythical historiography that informed these texts – its objective was highly political: it was a reaction to the 'Troubles' that sought to undermine any narratives that might be employed by militant nationalists.

Translations attracted revisionist-style criticism from commentators such as Longley, Connolly, Fenton and Andrews, who held that it was more mythic than realistically historical.[78] Informing these critiques is an aversion to the idea that colonialism had a deleterious effect on Irish society, and so any portrayal that hints that an educated, sophisticated world existed apart from, and in distinction to, the modernisation brought by colonialism is therefore inherently historically inaccurate: 'an artificial contrast between the hopelessly idealised and the hopelessly debased', as Connolly put it.[79] In like fashion, for Andrews, Baile Beag is a prelapsarian site, rather than a location in the grip of history. Yet, not inconsequentially, Andrews portrays what he calls the 'fallen world' beyond Baile Beag – the world of history – as an almost Platonic realm, which is revealed through the way in which he equates the loss of meaning of the English place names mentioned in *Translations* with the Irish experience: 'you don't need a foreign tyrant to transform your language: ordinary peaceful historical processes will do the job just as well – but ordinary peaceful historical processes are too complicated to be put on stage'.[80] Only a mythical understanding of history proceeding 'through', in Benjamin's words, 'a homogenous, empty time', could permit a comparison of the processes of language change in Ireland with those that took place in England.[81] This distorted comparison reflects a trend in Irish revisionist historiography that (for the most part implicitly) evaluates the Irish past in light of England's history. The clear political implications of this perspective are renounced by flattening out historical experiences in the same way a map-maker translates a landscape. As is evident from Andrews, the comparison has clear political ramifications: his account implicitly frames English colonial activities as 'ordinary peaceful historical processes', which is perhaps why he so disliked the sight of British soldiers with bayonets onstage.[82] Moreover, inscribed within the logic of this comparison is the assertion that English historical experience represents something of a 'norm' and that English historiography embodies objective rationality. Even, or most particularly, on a methodological level, this is a highly inappropriate analogy: England's exceptional past – the stability and longevity of its parliamentary structures, for instance – renders it a highly problematic state with which to construct comparisons.

As Deane noted, '[o]ne of the outstanding features of historical revisionism is its philosophical innocence, perhaps the basic requirement for its peculiar brand of political incrimination that pretends to be a rational form of "detachment"'.[83] This antagonism to thinking about the Irish past in theoretical terms is most thoroughly articulated in Howe's wide-ranging study *Ireland and Empire*, which, amongst many other things, takes aim at Field Day's introduction of postcolonial discourse into Irish Studies.[84] Howe considers thinking about Ireland as a colony inappropriate for the remarkably anodyne reason that Ireland's relationship with Britain reflects nothing other than 'the inevitable asymmetry of relations between a relatively large (and formerly globally dominant) state and a very small one'.[85] Colonial power imbalances are dismissed as 'natural' expressions of geography and consequently unworthy of any particular scrutiny. While Howe's text would stake its claim to be a 'realistic' approach to history through what Lloyd called a selective 'shift of focus from heroes to bureaus',[86] it is, as Guha had it, an 'unhistorical historiography' precisely because it leaves out 'the *politics of the people*'.[87] Furthermore, such a perspective casts the processes of modernisation, the gift of colonialism, as an inevitable, 'natural' aspect of the progress of history; in the context of the strategic power-vacuum it constructs, the reluctance to reap the unquestioned benefits of modernisation can only be a symptom of a cultural perversion, a deliberate going against the natural order of things. Hence, returning to the critiques of *Translations*, Maire's wish to learn English in *Translations* expresses nothing more for Connolly than one of the 'forces for change [...] within the Irish-speaking community': the economic, social, political and cultural reasons why she needed to learn the language are elided, and in the process colonialism is figured as a vehicle for beneficent progress and modernisation.[88] However, *Translations* has also been critiqued for not challenging the discourse of modernisation: Pilkington has argued that the play evinces an acceptance of 'the inevitability of the loss of Irish as a contemporary spoken vernacular', and reads the representation of the Irish past in *Translations* as articulating a revisionist, anti-republican attitude that fails to challenge the political status quo of post-colonial Ireland.[89] Like Andrews, he takes issue with the ways in which the drama translates history into politics or politics into history.[90] But, whereas the revisionist historian takes exception to the 'spirit' of the translation, Pilkington focuses on the 'letter' of the translation: he provides a very literal reading of the politics of language in the play through the terms of Sinn Féin's attitude to language in the late 1970s, a position he fails to interrogate.[91]

Like *Translations, Making History* is set in a moment in which the forces of colonial modernisation encounter and defeat native ways of life: 'the emergence of "Modern Ireland"', as Cleary noted, 'is conventionally ascribed in Irish historiography to the sixteenth and seventeenth centuries: the inception of that modernity is attributed [...] to the Tudor and Stuart colonisations of the islands'.[92] In *Making History*, modernisation, rather than being seen as a form of natural progression, is historically and discursively located in the context of this colonial activity, as is evident in the exchange between Mabel and her sister Mary in Act I scene 2. Mary regards the Irish as 'a savage people' (*MH*, p. 278) who have 'no right' to their land as they do not cultivate it, and whose way of life is doomed, 'because civility is God's way [...] and because superstition must yield before reason' (*MH*, p. 279). Mabel, however, is, in some respects, from the same family as Yolland: appreciative of the benefits of modernisation, but having empathised with the situation of those who have had no say in its operations, she has a confused, but deep, sense of loyalty to a departing way of life, which she struggles to comprehend. She problematises the portrait of the cultivated civility of the Baganel house that Mary had painted by wondering if 'the Cistercian monks in Newry didn't think our grandfather an agent of civilization when he routed them out of their monastery and took it over as our home' (*MH*, p. 279). The lack of civility that the coloniser sees in Gaelic culture is not an expression of an atavistic reaction against the light of modernisation, but has been created, she has realised, by the dislocations of colonialism. Moreover, in this exchange, it is apparent that the maintenance of this 'civilian and barbarian' stereotype is an essential condition of the claims of modernisation, which evicts alternative narratives, casting them 'outside history', to use Lloyd's phrase.[93] This process has continued in post-colonial Ireland where, particularly since the Lemass years of the 1960s, modernisation has been, as McCarthy put it, the 'narrative in terms of which the "imagined community" of the Republic [has] understood itself, and envisioned its future': the inherent conservatism of modernisation theory – its positivist tendency 'to de-politicise social and economic issues' – invariably suits the purposes of the ruling classes.[94] As Benjamin noted: '[h]istoricism' (his term for the Rankean method favoured by revisionism) sympathises 'with the victor. And all the rulers are the heirs of prior conquerors.'[95] Hence, precisely because of their joint espousal of the logic of modernisation, it would be wrong to think of revisionism as being inherently at odds with the ideology of the post-colonial Republic of Ireland, despite that state's occasional need to pay homage to nationalist pieties.

In *Making History*, the disjunctions wrought by modernisation are refracted through Hugh O'Neill, who was, in Hiram Morgan's description, 'one of the most enigmatic figures in Irish history'.[96] He was the crucial character in the massive cultural shift that took place in Ireland in the late sixteenth and early seventeenth centuries: the decline and defeat of the old Gaelic order and the decisive victory of the 'New English' colonists. O'Neill looked to straddle the ever-widening rift between an older way of life and the urgent demands of colonisation and modernity: before organising an army to oppose the English, the resistance of which was defeated at the Battle of Kinsale in 1601, he had worked and fought with the colonists while simultaneously searching for a form of compromise that would secure a future of some sort for the Gaelic order. These ambivalences render O'Neill an important figure in revisionist historical accounts, where he is read as embodying the contradictions of 'Irishness' that a nationalist history must elide in order to create the myth of the hero. In Foster's account, he personifies 'the Janus-face of Ireland': brought up between Ireland and Britain, he is caught between these two identifications.[97] But, rather than being explored in a productive manner, these tensions merely assume fixed roles in a revisionist allegorical drama in which modernisation bravely forges ahead, despite the reactionary forces that threaten to drag it down. Foster's brief biographical account of O'Neill in *Modern Ireland, 1600–1972*, for instance, neglects important aspects of his engagement with Gaelic culture – his assumption of the title 'The O'Neill', for example – but does not fail to note that he 'lamented his countrymen's unwillingness to accept English ways', which are implicitly progressive.[98] In place of the unproblematised nationalist tale of O'Neill as national hero, this effects an equally fluent story of him as a martyr in the cause of modernisation.

Making History, on the other hand, presents an O'Neill ensnared between these narratives. While the emotional thrust of the piece favours the revisionist position, it does not, in Hohenleitner's words, allow the audience 'the comfort of a single interpretation of Hugh O'Neill'.[99] Chakrabarty has pointed out that the 'problem of capitalist modernity' is, on one level, 'a problem of translation', and,

> what translation produces out of seeming 'incommensurabilities' is neither an absence of relationship between dominant and dominating forms of knowledge nor equivalents that successfully mediate between differences, but precisely the partly opaque relationship we call 'difference'.[100]

O'Neill's doubled and split identity is employed to articulate this zone of difference, which, more generally, is where both *Making History* and *Translations* stage the drama of the colonial encounter. In this location the past is never fully past nor is it fully graspable in the present, but is continually in the process of being recreated. Through O'Neill, *Making History* foregrounds this process by exposing the limits of nationalism's and revisionism's rhetorical constructions of temporality.

Never too far from O'Neill's side in *Making History* is Archbishop Lombard, who is writing a history of O'Neill. Or, rather, he is taking O'Neill's life as the basis for a national parable, as for Lombard the function of a history is to provide an uplifting national allegory: a perspective that casts him as an archetypal nationalist historian.

> People think they just want to know the 'facts'; they think they believe in some sort of empirical truth, but what they really want is a story. And that is what this will be: the events of your life categorized and classified and then structured as you would structure any story. [...] I'm not talking

> about falsifying, about lying [...]. I'm simply talking about [...] offering a cohesion to that random catalogue of deliberate achievement and sheer accident that constitutes your life. And that cohesion will be a narrative that people will read and be satisfied by. [...] Think of this [*book*] as an act of *pietas*. Ireland is reduced as it has never been reduced before – we are talking about a colonized people on the brink of extinction. This isn't the time for a critical assessment of your 'ploys' and your 'disgraces' and your 'betrayal' – that's the stuff for another history for another time. Now is the time for a hero. Now is the time for a heroic literature. So I am offering Gaelic Ireland two things. I'm offering them this narrative that has the elements of myth. And I'm offering Hugh O'Neill as a national hero. A hero and the story of a hero. (*MH*, pp. 334–5)

These arguments uncannily pre-empt those of another priest, Brendan Bradshaw, in the most noted exposition of the motives behind Irish nationalist historiography – his 1989 essay, 'Nationalism and Historical Scholarship in Modern Ireland'. To a large extent, this was a critical response to an earlier revisionist article by Steven Ellis,[101] and the main charges it presents against revisionism are: it produces a dull form of writing; it cannot live up to the strictures of its own interpretative strategies and so disguises its bias; it filters out the trauma of history; it removes the role of the historian as mediator between the past and the public; it is marked by 'an invincible scepticism'; and it denies continuity in Irish history.[102] Bradshaw traces this mode to Butterfield's 1931 *The Whig Interpretation of History* and, in the second part of the essay, looks at how Butterfield himself repudiated such a methodology in his 1944 *The Englishman and His History*, when the Second World War made it imperative in his eyes to produce a different type of history of England.[103] Bradshaw would follow this later move of Butterfield's, and calls for a re-evaluation of the notions of empathy and imagination in historical accounts and for a form of 'purposeful unhistoricity'. Imitating the way in which Butterfield sought to consolidate English national sentiment, this would similarly produce, in the Irish context, a story of liberty and stability that would build up the idea of the nation. Thought of in terms of translation, a drastic act of domestication is described here, whereby the past is thoroughly colonised, 'to accommodate [...] the needs of the present', as Bradshaw put it. He expresses a fundamental distrust in the past, believing that a coherent narrative, 'its wrongness notwithstanding', is preferable to one that struggles to come to terms with the events of the past.[104] This is, to cite Lloyd, a history 'written [...] with the aim of producing a non-contradictory subject'.[105] And, as the example of Butterfield makes clear, Bradshaw essentially advocates an explicit repetition of the motions of imperial history: a positivist translation (nothing more really than a mimicking) of an inherently conservative story of modernisation that would merely involve changing the names, as it were, into Irish ones.

While Lombard's duty is to his conception of the public present and future, Hugh O'Neill insists upon an account that bears witness to his personal past and present – a 'warts-and-all' portrait that encompasses his contradictions: '[t]he schemer, the leader, the liar, the statesman, the lecher, the patriot, the drunk, the soured, bitter émigré – [...] the *whole* life' (*MH*, pp. 329–30). By taking this stance, O'Neill, as Canny pointed out, 'takes the place of the contemporary revisionist'.[106] In Foster's terms, the ambition of revisionism is Rankean: 'to get behind hindsight', which involved producing interpretations 'in terms of what people knew or assumed at the time, rather than what we know (or assume) to have happened subsequently'.[107] In the present tense of *Making History*, O'Neill is a man caught between various loyalties and it is evident that his decisions are made for reasons that are not consistent with the broad and glorious brush strokes that nationalist historians have employed in representing them. These rhetorical strategies are analysed by Foster in *The Irish Story*, where he argues that the Irish nationalist rendition of O'Neill 'uncannily conformed to the classical structure of the fairy tale' as outlined by the Russian structuralist Propp.[108] Nationalist history in this telling resorts to narrative strategies and rhetorical devices that are more proper to fiction. As Hayden White has argued, '[i]t is often thought that history's principal enemy is the lie, but actually it has two enemies considered to be more deadly to its mission to tell the truth and nothing but the truth about the past: rhetoric and fiction.'[109] Lombard is explicit in his use of such devices, and even offers the interweaving of fiction in history as a justification for his ideological agenda. His responsibility is,

> to tell the best possible narrative. Isn't that what history is, a kind of storytelling? [...] Imposing a pattern on events that were mostly casual and haphazard and shaping them into a narrative that is logical and interesting. [...] I'm not sure that 'truth' is a primary ingredient [...] Maybe when the time comes, imagination will be as important as information. (*MH*, p. 257)

As Lombard is the villain of the piece, these literary methods are a description of his perfidy. Yet, the inherent irony of a history play staging an argument against the intertwining of history and literature cannot be discounted; McGrath, indeed, has suggested that Lombard is, to some extent, 'the spokesman for [a] poststructuralist historiography' that Friel espouses.[110] More generally, the play illustrates how the act of translating the past involves negotiating the problems created by the inherently metaphorical nature of language.

Since the rise of Romanticism and its privileging of the imagination, history has fashioned for itself a place apart from literature, with which it had been closely aligned.[111] Yet, as de Certeau noted, the tension between literature and history has never been resolved, and is moreover inscribed within the very term 'historiography' '(that is "history" and "writing")' which speaks to 'the paradox – almost an oxymoron – of a relation established between two antinomic terms, between the real and discourse'.[112] Ricoeur has argued that this split has never been as thoroughgoing as might be imagined, and has posited that there is more fiction in history than empirically minded historians would claim, and that the essentially mimetic quality of narrative fiction means that there is more history in fiction than is usually admitted by structuralists. Working from this position, he proposes 'that the references of "true history" and "fictional history" *cross* upon the basic historicity of human experience'.[113] In other words, there is a fictional element in history precisely because identity is itself historical.

> A meditation on historicity of a more ontological than epistemological character [...] would bring out features in the historical condition itself which demand that the historicity of human experience can be brought to language only as narrativity, and moreover that this narrativity itself can be articulated only by the crossed interplay of two narrative modes. For historicity comes to language only so far as we tell stories or tell history. In a word, if our historical condition requires nothing less than the conjunction of two narrative genres, it is because of the very nature of our experience of *being* historical. [...] *We belong to history before telling stories or writing history*. The game of telling is included in the reality told.[114]

While the comprehensive nature of this claim and the vision of narrative it relies upon are open to criticism, Ricoeur nonetheless recalibrates the relationship between history and fiction in a way that poses fundamental, challenging questions about the roles of history and the imagination in the formation of identities.[115] Recalling Aristotle's claim that poetry is 'more scientific and serious than history, because poetry tends to give general truths while history gives particular facts',[116] Ricoeur suggests that the ontological role of history comes to fruition through fiction: '*the world of fiction leads us to the heart of the real world of action*'.[117] This speaks directly to Field Day's practice and to *Making History*, more specifically.

Even when he was alive, O'Neill's identity would appear to have been, to a large extent, a fiction. 'To trust this traitor upon oath is to trust a devil upon his religion', is what Elizabeth I is reported to have said of O'Neill;[118] as a traitor, at different times, to both sides, O'Neill is, to redeploy Mary's description in the play, a 'slippery' historical subject (*MH*, p. 279). Morgan recounts that O'Neill 'was a consummate liar. [...] Thereby he made the enigma he is today because the elaborate campaign of disinformation he waged was so successful that it confounded subsequent historians as well.'[119] Friel reflected in 1972 that 'autobiographical fact can be pure fiction and no less true or reliable for that',[120] and he responded to O'Neill's opacity by forging an O'Neill out of the interplay of historical events and fiction; a procedure he signalled in a note on his historical methods in the programme for the first productions:

> MAKING HISTORY is a dramatic fiction that uses some actual and some imagined events in the life of Hugh O'Neill to make a story. [...] when there was a tension between historical 'fact' and the imperatives of fiction, I'm glad to say I kept faith with the narrative. For example, even though Mabel, Hugh's wife, died in 1591, it suited my story to keep her alive for another ten years. Part of me regrets taking these occasional liberties. But then I remind myself that history and fiction are related and comparable forms of discourse and that an historical text is a kind of literary artifact.[121]

Clearly O'Neill's version of his own life is just as fictional, albeit in different ways, as Lombard's putative history.[122] Essentially, he brings to light the conceit informing an ideal literal translation of the past, by showing that the past cannot be entirely possessed, not even by the person who has lived it. As the oral historian Passerini reflected: '[a]t the roots of our memory [...] I find a rupture. Our identity is constructed on contradictions.'[123] Indeed, the dialogical processes of psychoanalysis point to the essential otherness of one's past, and of the past in general.[124]

Friel thus explicitly charts out his *'positionality'* as a historian, to use Koselleck's term.[125] In doing this, he compromises O'Neill's position as a revisionist historian, or rather he brings out the essentially fictional location from which such histories are composed. A number of the dramatic devices that sustain the representation of O'Neill in the play as a revisionist are historically flawed: the historical archbishop Lombard did not write such a history of O'Neill and, besides her extended lifespan, Mabel's role in O'Neill's life would appear to have been very different from the way she is portrayed in *Making History*.[126] Morgan portrays the marriage in an entirely different and less romantic light, as 'a matter of power politics, not love'.[127] Friel's version of O'Neill emanates from O'Faoláin's *The Great O'Neill*, a source that explicitly makes no claim to historical accuracy, but which is, as Deane has argued, an 'important work[ ] in the genesis of the revisionist school in Irish historical writing'.[128] It is clearly also the source for Foster's brief account of O'Neill in *Modern Ireland, 1600–1972*, as, like *Making History*, it picks up on O'Neill's youthful sojourn in Sir Philip Sidney's home at Penshurst, a key episode in O'Faoláin's argument that O'Neill cannot be simply read in Irish terms.[129] Morgan, though, is emphatic in his debunking of this event, and the presumption of an identity crisis that it engenders.[130] So while O'Neill's revisionist position takes on a battle-scarred heroism in the play, the processes required to render him in this light have a lot in common with those practised by Lombard; as Kiberd has observed: 'Friel's plays are implicit critiques of the value-free approach to history.'[131]

There is, as White averred, 'an irreducible ideological component in every historical account of reality',[132] and so the choices facing the historian are, in Deane's words, 'always moral and/or aesthetic'.[133] O'Neill, in classic revisionist style, locates himself in the normative and non-ideological position of history itself; as he put it: 'the slow sure tide of history is with me [...] All I have to do is ... just sit – and – wait' (*MH*, p. 283). But the claim to have history on one's side is, as Koselleck points out, 'an ideological fiction [...] [which] thrives on the category of *necessity*'.[134] As such it is a tautological construction:

> [i]dentifying an event as necessary amounts to a double statement about that event. Whether I say that something happened or whether I say that it happened by necessity is identical from an ex post facto perspective. Something did not happen more so only because it had to happen. By making a statement about an event and by adding that it had to have happened, I vindicate for this event a necessary causal chain – a necessity that in the end derives from the omnipotence of God, in whose place the historian is acting.[135]

Historical time is not, as White notes, the same as 'natural' time, and the conflation of the two in O'Neill's argument is nothing other than rhetorical manipulation. Both revisionist and nationalist histories are made by conforming to predetermined notions of time that claim to be natural and beyond question, and so both produce 'a historical knowledge true to its "concept"', be it of modernisation or the inevitability of the nation.[136]

Making History brings these two ideological constructions into conflict, and points to the essential artifice of both, and of history in general. Historical dramas operate between tenses, in 'the time lag between the *now* of the performance and the *then* of the historical events themselves', to employ Rokem's words.[137] By self-consciously playing on the 'permanent anachronism of history', as Deane has it,[138] what Bhabha called 'the disjunctive temporality of translation' takes the place of nationalist and revisionist teleologies in *Making History*.[139] In other words, thinking of history as an act of translation exposes how 'natural' concepts of progress are constructed. As de Man points out in his reading of Benjamin's 'The Task of the Translator', translation and history should not

> be understood by analogy with any kind of natural process. We are not supposed to think of history as ripening, as organic growth, or even as a dialectic, as anything that resembles a natural process of growth and of movement. We are to think of history rather in the reverse way: we are to understand natural changes from the perspective of history, rather than understand history from the perspective of natural changes. [...] In the same way, the relationship between the translation and the original is not to be understood by analogy with natural processes such as resemblance or derivation by formal analogy; rather we are to understand the original from the perspective of the translation.[140]

Not only does this posit a decisive rupture between past events and their historical representation, but it suggests that the understanding of time itself is gained through history, not the other way around. Historical time is fundamentally textual and is the site of interpretation: as Koselleck intimated, it is precisely as a result of '[t]he impossibility of intuiting pure time' that history 'as a discipline lives by metaphorical expression'.[141] The basic act of history – the carrying over of meaning from the past to the present – is strictly speaking metaphorical and, as Gadamer has elaborated it, this movement is constitutive of historical interpretation:

> time is not a chasm which we could bridge over in order to recover the past; in reality, it is the ground which supports the arrival of the past and where the present takes its roots. Temporal distance is not a distance in the sense of a distance to be bridged or overcome. This was the naive prejudice of historicism. It believed it could reach the solid terrain of historical objectivity by striving to place itself within the vantage point of a past age and think with the concepts and representations particular to that epoch. Actually, it is rather a matter of considering temporal distance as a fundament of positive and productive possibilities for understanding.[142]

Understanding of the past takes place in a flux, and so, rather than imagining that only one reading exists that might be uncovered by an impossible return to the conditions of the past, the act of writing history takes the form of a dialogue between the present and the past that changes over time. Gadamer emphasises that this is an essentially critical dialogue, which precludes any pretensions to complete objectivity. In short, it involves an act of translation: '[t]he dialogue we enter into with the past confronts us with a fundamentally different situation from our own – we will say a "foreign" situation – and consequently it demands an interpretative approach.'[143] Not only are the preconceptions that are brought to bear on a historical question and the ways in which they condition understanding unveiled by thinking of history as an act of translation, but, as Gadamer argues, translation also speaks to the effort required to communicate the otherness of the past.[144] History, as such, is an act of interpretation and communication, and the ethical obligation and crux are once again those of translation: the necessity and impossibility of transmitting this other – the past – into one's own terms in a way that also acknowledges the otherness of the other. As Koselleck put it: '[a] historical statement is only meaningful when it speaks to its addressees in such a way that the otherness of past or foreign experiences can be integrated into their own experience.'[145] In *Making History*, Friel offers an O'Neill who is, to cite Derrida, 'the crossing between [...] two phantoms of witnesses who will never come down to the same'; in this light, truth is, as Lombard suggested, perhaps not a useful historical category at all, as it betokens an unchanging, hypostasised state.[146] Rather, criteria relevant to translation such as fidelity and trust, which speak to acts, and in which there is an implicit horizon of truth, are more relevant, as the task of historical studies is fundamentally to establish, demonstrate and maintain trust with its sources.

As was noted at the outset of this chapter, *Translations* and *Making History* are set in moments of cultural twilight, and it is by performing the crisis of historical representation that Friel indirectly speaks of the cultural traumas that underpin both scenarios: the loss of the Irish language and intimations of the Famine in *Translations*, and the decline of the system of Gaelic civilisation in *Making History*. Trauma is that which cannot be easily grasped, even by those who have directly endured it and, for this reason, literature, which 'is interested in the complex relation between knowing and not knowing', as Caruth put it, plays a crucial, ethical role in the representation of historical trauma.[147] Rather than speak directly about these traumatic events, both plays speak of the predicament of trauma: 'the way that its very unassimilated nature – the way it was precisely *not known* in the first instance – returns to haunt the survivor later on'.[148] By enacting a form of what LaCapra has called 'empathic unsettlement', the plays look to dislodge these pasts from the essentially repetitive roles they have held in Irish histories, and to set them off in different, more varied, trajectories.[149] Repetition is nothing other than a symptom of trauma, and instead of this, these two dramas seek to instantiate a process of 'working through', which is, in LaCapra's terms, an act of mourning through which 'one is able to distinguish between past and present and to recognize something as having happened to one (or to one's people) back then which is related to, but not identical with, here and now'.[150] The past cannot be fully possessed or claimed, nor is it, for that matter, to be disclaimed; rather, it assumes the role of the other that calls for yet another translation: this is the entirely ambivalent political message of both plays.[151]

2
Translation, Home and Hospitality

Introduction

> The [translation] model to be preferred is the more modest one proposed by von Humboldt, i.e. that of raising the distinctive spirit of his own language to the level of that of the foreign language [...]. It really is a matter of living with the other in order to take that other to one's home as a guest. [...] In this sense we can speak of a *translation ethos* whose goal would be to repeat at the cultural and spiritual level [this] gesture of linguistic hospitality.
>
> — Paul Ricoeur[1]

When, in *Making History*, O'Donnell learns the identity of O'Neill's new wife, his advice is brutal: 'keep her for a month and then kick her out. [...] Give her some clothes and a few shillings and kick her back home to Staffordshire' (*MH*, p. 264). O'Neill emphatically rejects this, informing O'Donnell and Lombard that '[h]er home is Newry' (*MH*, p. 264) and moreover insists that they show her hospitality: 'I particularly ask you two to welcome her here' (*MH*, p. 265). For her part, Mabel soon discovers that marrying O'Neill does not mean that she will feel at home in his house: '[e]verything's so different here. I knew it would be strange [...] But I didn't think it would be so ... foreign. I'm only fifty miles from home but I feel very far away from everything I know' (*MH*, p. 271). Home and the processes of hospitality form a metaphor through which this colonial encounter is performed. Likewise, in *Translations*, the English soldiers are both welcomed as guests and treated as hostile invaders. At the same time, their map-making activities can be thought of as charting out a 'house' within which they might stake a claim to the role of host. The definition of home is the site of colonial conflict: it is precisely what is at stake. Friel, as was noted in the last chapter, reframed this in terms of the Heideggerian notion of language as 'the house of Being': substituting, as it were, a fixed conception of home with the dynamics of linguistic hospitality that are dictated by the shifting, translational notions of trust, fidelity and betrayal. Translation is an act of hospitality, and while, as Ricoeur suggests above, it can serve as a possible imaginative model of political and cultural integration, it can only do so in an enabling fashion if the equivocal nature of its hospitality is taken into account: translation not only undermines the notion of a fixed home within which hospitality can be enacted, but it also exposes the hostility in hospitality, as these dramas illustrate.

In the Introduction it was noted that the politics of home and hospitality have been particularly fraught in Northern Ireland. By controlling the allocation of public housing, unionists were able to back up their claim that Northern Ireland was, in Craig's notorious terms, a Protestant state for a Protestant people.[2] For its part, the Catholic community has claimed the island of Ireland as its home – an assertion based on a sense of prior ownership – in order to cast unionists as outsiders. Unionists, in response, have asserted, like Mary in *Making History*, that their superior 'house-keeping' – their 'civilising' of the country – entitles them to claim the place as their possession. Moreover, should one community extend hospitality to the other, it is also making a claim to be the host, with the result that the offer is clearly also a hostile act. In light of these vicious circles, Field Day's diagnosis suggests that the types of hospitality tendered in Northern Ireland have produced nothing other than a shared experience of homelessness; the company's plays, therefore, problematise the notion of home, and try to disentangle it from unquestioned ideas of origin and belonging.

In a 1982 interview, Friel pointed out that these Northern Irish conditions had bequeathed a troubled, thoroughly ambivalent sense of home to the Field Day directors:

> the contradiction [...] is that we are trying to make a home. [...] we aspire to a home condition in some way. We don't think that exile is practical. We think that exile is miserable in fact. And that's what's constantly being offered to us, particularly in the North, and this is one of the problems for us, is that we are constantly being offered the English home; we have been educated by the English home and we have been pigmented by an English home. [...] And the rejection of all that, and the rejection into what, is the big problem.[3]

Since the time of O'Neill's failure at Kinsale and the collapse of the Gaelic world, the experience of displacement has been a keynote in Irish literature, and Friel turned to perhaps the most famous twentieth-century example of the tensions induced by this sense of cultural and linguistic fracture: Stephen Dedalus's conversation with the English Dean of Studies in *A Portrait of the Artist as a Young Man*.[4] Home – the burdensome location Dedalus ultimately sets out to re-forge – is the first word to sound different to him in the Dean's

English accent.[5] Even as he experiences a downtrodden pride in the fact that the word 'tundish' was more Anglo-Saxon than the Dean's Latinate 'funnel', this induces in him a mood of irreconcilable, fragmented resentment, as he knows that by thinking about his word as more original, he is merely operating according to the dictates of the hegemony of English, in which his 'home' will always sound different. On the other hand, his reaction later in the novel to Mulrennan's account of meeting with an old Irish-speaking man in the West of Ireland articulates nothing other than a vehement Oedipal confrontation, in which he recognises, disowns, fights against and ultimately feels he cannot escape from what is, and is not, his past – his 'origin' or 'home', which is also his other.[6]

Hiberno-English expresses the colonial history of possession and dispossession in Ireland, and in a 1980 interview, Friel is explicit about how this has shaped the political analysis of Field Day.[7]

> There will be no solution until the British leave this island, but even when they have gone, the residue of their presence will still be with us. This is an area that we still have to resolve, and that brings us back to the question of language for this is one of the big inheritances we have received from the British. In fact twenty miles from where we are sitting, you can hear very strong elements of Elizabethan English being spoken every day. [...] We must continually look at ourselves, recognize and identify ourselves. We must make English identifiably our own language.[8]

Echoing Hugh's pronouncement in *Translations* that the residents of Baile Beag would have to learn the names of their places anew in order to make them their home, Friel is unequivocal about the fact that it is nothing other than a fantasy to imagine that the disappearance of British rule from the island of Ireland would result in a restoration of some form of renewed, authentic Irish home. While Friel and Field Day undoubtedly believed that a British withdrawal from Northern Ireland was a political imperative, this is never directly articulated in the dramas. In fact, while the three dramas examined in this chapter address the issues of displacement that Friel sees as central to his Northern Irish experience, not one is set there. The action of *The Communication Cord* takes place just over the border in Donegal, in a jerry-built version of an Irish post-colonial home that stands as a satirical symbol of the state of contemporary politics and mores in the Republic of Ireland. Although the other two plays are set in nineteenth-century Russia and apartheid South Africa, in some respects they translate more smoothly into the Northern Irish experience Friel described: *Three Sisters* performs the equivocal freedoms of being displaced in one's own house, while *Boesman and Lena* portrays the misery of a permanent exile in what should be home.

Re-*Translations*: *The Communication Cord* and the travesty of the post-colonial home

> Nationalism, cultural or political, is no more than the inverted image of the colonialism it seeks to replace. It too is an act of translation or even of retranslation. The assumption that it shares with colonialism is the existence of an original condition that must be transmitted, restored, recuperated, and which must replace the fallen condition which at present obtains.
>
> — Seamus Deane[9]

> The post-colonial translator must be wary of essentialist anti-colonial narratives; in fact, s/he must attempt to deconstruct them, to show their complicity with the master-narrative of imperialism.
>
> — Tejaswini Niranjana[10]

One of the most frequently aired objections to the employment of postcolonial theory in Irish Studies has been that Irish historical experience was not properly colonial and has nothing in common with non-European colonies, because Ireland was, from 1800, a part of the United Kingdom. Consequently, it is argued that movements such as negritude and Afro-Asian cultural nationalism have little relevance to an analysis of contemporary Ireland, and their use does nothing other than posit a uniform, homogenising, experience of colonialism across continents.[11] These critiques are informed by a strategically attenuated version of Irish history and an intellectual insularity that is encapsulated in the catch-phrase dismissals of Field Day's postcolonial analyses, such as Longley's linking of Derry with Derrida and Howe's 'Fanon comes to the Foyle', which betoken incongruous conjunctions rather than possible modes of rethinking Northern Irish experience.[12] These arguments also misrepresent Field Day's postcolonial project: rather than offering unproblematised comparisons between (post-)colonial contexts, Field Day was, as *The Communication Cord* illustrates, very wary of undifferentiated postcolonial analyses.

The Communication Cord is a response to *Translations* that, in essence, takes the form of a translation of the earlier play. It is a re-translation – a translation of a performance of acts of translation – that re-enacts in a post-colonial Irish frame the processes involved in the Ordnance Survey's Anglicisation of Irish place names. Friel felt the need to translate *Translations* precisely because reactions to the play had spoken to a generalising post-colonial experience. He believed that these responses occluded the postcolonial import of *Translations* as, instead of coming to terms with the difficult, ambivalent questions at the core of the play, they treated it 'much too respectfully':

> when you get notices especially from outside the island, saying, 'If you want to know what happened in Cuba, if you want to know what happened in Chile, if you want to know what happened in Vietnam, read *Translations*', that's nonsense. And I just can't accept that sort of pious rubbish.[13]

Translations inspired an immediate and intense critical reaction, and a lot of this was, in different ways, essentialising and pious. For a few romantically inclined nationalist critics, it was an evocation of a lost Irish idyll: Desmond Rushe, for instance, articulated a remarkably sentimental and inaccurate view of the play, seeing it as 'an authentic reflection of how things were', 'an extraordinary evocation of a time when the flashing-eyed Athene of Grecian saga was as familiar to Donegal peasants as the Grainne of Irish legend'.[14] Such thinking is little more than the flip side of a critical coin that is minted in conjunction with the critics discussed in the previous chapter who held *Translations* to be a dangerous play that refurbished old nationalist myths – that, in Henderson's words, 'bleat[s] plaintively of old wounds' – as both reactions read the drama in a politically essentialist manner.[15] However, perhaps the most insidiously pious critical reaction was the dominant one, which set the play on a pedestal as a classic.[16] Friel felt uncomfortable: 'I was being categorized in some sort of a way that I didn't feel easy about, and it seemed to me that a farce would disrupt that kind of categorizing.'[17] As a farce, *The Communication Cord* is a wholeheartedly anti-pious play, sending up post-colonial nationalist piety and the aura that had encrusted *Translations*. But, it also looks to recuperate a more enabling form of piety – which might be termed *pietas* – that might inform a constructive postcolonial approach to the past. It does this by exposing the post-colonial piety in the play as a form of debased re-translation.

With the exception of Mahon's *High Time*, all the Field Day translations are re-translations. Benjamin, though, warned that re-translation is a fraught procedure: translations are untranslatable, he maintained, 'because of the looseness with which meaning attaches to them'.[18] This looseness is the result of the travesty inherent in all translation; earlier in the same essay, Benjamin had described the relationship between what he called the untranslatable nucleus and the translatable words of the original in this manner:

> the relationship between content and language is quite different in the original and the translation. Whereas content and language form a certain unity in the original, like a fruit and its skin, the language of the translation envelops its content like a royal robe with ample folds. For it

> signifies a more exalted language than its own and thus remains unsuited to its content, overpowering and alien.[19]

In these terms, re-translation is nothing other than a travesty of a travesty: an entirely farcical act in which language becomes excessively unsuitable to the meaning. Friel employs this strategy in *The Communication Cord* to loosen the robes of meaning that had been draped on *Translations* and to re-clothe the play in the Emperor's new clothes – unveiling anew the absence of an unequivocal, authoritative original. *Translations* thus stands exposed as a translated original that performs the impossibility of a return to a whole, authentic original. *The Communication Cord* re-translates this very impossibility, which is central to the political significance of *Translations* and the Field Day plays in general. Cast in this light, *The Communication Cord* is not simply a case of, in Simmons's elegant description, Friel 'pissing on his own monument', rather it highlights the always already fragmented nature of the monument.[20]

Excessive travesty is a suitable dramatic and translational register for a play that satirises post-colonial Irish society: it underscores how, in paying homage to nationalist pieties, elements of this society have assumed mock-imperial clothing. In *The Communication Cord* the post-colonial, nationalist translation of colonialism is enacted through the ways in which it translates colonial translation strategies. Dr. Donovan, the blustering senator, most explicitly personifies this collusive disposition. He has the name of the 'namer' (translator/orthographer) upon whom the character of Owen in *Translations* is based – John O'Donovan – and he too fancies himself as someone who can unfold meanings within names. When the hapless lecturer in linguistics, Tim, invents a character called 'Jack the Cod', Donovan reacts excitedly: 'I love that. Call a man Jack the Cod and you tell me his name and his profession and that he's not very good at his profession. Concise, accurate and nicely malicious. Beautiful!' (*CC*, p. 46). He attempts to provide an ontological-style reading of this name – seeing it as an insight into the man's origins – however, its style is resolutely positivist. The effort to cast his interpretation in an ontological light is simply Donovan's obligatory gesture to a supposedly purer past. It is an act of piety, the absurdity of which is reinforced by the fact that it concerns a name invented on the spur of the moment.

As a farce, the play is weighed down by a plethora of invented names and mistaken identities; acts of travesty that set out to mislead and misrepresent, to gain short-term advantages. Donovan is Teddy to his French mistress, Evette, and Dr. Bollocks for the German who is, at various stages, called Barney the Banks, Barney Munich and Willie Hausenbach. In this manner, *The Communication Cord* re-enacts, in excess, the portrayal in *Translations* of the limits of a positivist translation of names. However, it also reposes a question about the nature of the ontological approach, as it mocks the idea that the untranslatability of proper names is a preservation of inner meaning. Just like *Translations*, this play is ultimately chary of the '[f]earsome authenticity' that, as Levinas has pointed out, plays a 'primordial' role in Heidegger's system of thought.[21] This may well have been obscured for the audiences of *Translations* by the historical and political context of that play: by the fact that it formed the basis of the resistance to an imposed, colonial linguistic model. Replaying this as farce in a post-colonial context exposes this critique, and *The Communication Cord* playfully questions the notion that there might be a pure and secret proper name to be found underneath the characters' aliases. Instead, the aporetic desire that Derrida submits is 'at work in every proper name' – 'translate me, don't translate me' – is suggested: rather than signifying an inalienable ontological wholeness, one's proper name denotes an inherent longing for a stable identity.[22]

On a political level, this dialectic between the positivist and ontological dispositions was resolved in the Republic of Ireland through the promotion of an official national identity based on an *ersatz* form of the ontological disposition; one that produced an illusion of difference between the colonial and post-colonial modernisation projects, and closed off any form of rigorous questioning of language and identity. As Lloyd put it, Irish nationalism appealed, 'to its premodernity as the site of significant cultural differences on which to found a distinct but no less *modern* state formation, equivalent to if not identical with that of Britain'.[23] Maintaining this illusion of difference-in-sameness involves, amongst other things, an incantatory repetition of pious formulae, which is an essential aspect of Senator Donovan's claim to authority in the post-colonial state. Ironically, his proper name – 'Patrick Mary Pious' (*CC*, p. 80) – indicates his character, yet through his actions he sends up and deconstructs the concept of piety that gives him his identity. Gerald Dawe has offered a devastating verdict on the insidious pervasiveness of this form of piety in Ireland, arguing that, 'the chief enemy of positive change has always been piety: the need we have to sanctify and ritualise ourselves by way of compensation for what may well be the impoverished reality of our politics and culture.'[24] However, there is a more enabling version of piety than the one that sustains Donovan: a *pietas* that, amongst other things, provided an initial impulse for Field Day's activities. At a press conference at the Guildhall on 3 June 1980, Friel launched the enterprise in these terms:

> [w]hen I talk about the specific reason that Stephen Rea and I came together for this enterprise, I'm talking about something that hasn't got an exact English word to describe it. The Irish word is ductas [*sic*]; the Latin word is pietas. I suppose the nearest English equivalent would be a sense of dutifulness to one's native place, a sense of loyalty to one's home.[25]

This would not be a form of fixed allegiance, as Kearney has indicated in his attempt to recuperate the relevance of this sense of *pietas* for contemporary Irish life in his Field Day pamphlet. Invoking the filial piety shown by Aeneas towards his father when he carried him out of burning Troy, he suggested that such *pietas* facilitates a form of enabling translation:

> this is what tradition means – *tradere*, carrying or transferring the past into the present and the present into the past. Myths of tradition defy the historical logic of non-contradiction (*either/or*); they lay claim to a supralogical order where something can be *both* what it is *and* what it is not – the past can be present, the human divine, and so on.[26]

In contrast to a *pietas* that informs a sense of tradition that opens up rather than closes off future possibilities, the post-colonial piety and tradition on display in *The Communication Cord* are nothing other than simulacra that entrap the self.

The inverted commas in the first stage directions, which describe the house within which this play takes place, draw attention to the fact that the piece will revolve on understandings of the terms 'traditional' and 'authentic'. The drama is located in the 'present' in a restored thatched cottage in Ballybeg, the location of many of Friel's plays including *Translations*, which bore witness to the change of its name from Baile Beag. In the shift from colony to post-colony, it has maintained its peripheral status: it is now a place 'away from it all' for the new Dublin-based elites.

> *The action takes place in a 'traditional' Irish cottage. [...] Every detail of the kitchen and its furnishings is accurate of its time (from 1900 to 1930). But one quickly senses something false about the place. It is too pat, too 'authentic'. It is in fact a restored house, a reproduction, an artefact of today making obeisance to a home of yesterday.* (*CC*, p. 11)

To employ Kearney's distinction, the cottage is not traditional, rather it has been fashioned to appear traditional, and the eagerness to replicate the traditional has produced an evident travesty – something that is 'too "authentic"'. Similarly, the piety exhibited in the play is 'an artefact of today making obeisance to a [...] yesterday', which it simultaneously reinvents as an authentic site of purity. Taken as a symbol of the Republic of Ireland, the house suggests that instead of invigorating an unfolding sense of tradition and an authentic post-colonial freedom to fashion for oneself ('*auto-hentes*: to fashion for one's self', as Kearney notes in a Heideggerian fashion), this state has merely reproduced colonial concepts of piety, tradition and authenticity.[27] Eighteenth- and nineteenth-century translation projects, such as the Ordnance Survey's memoirs, began a process of re-reading what was seen as Irish culture's attachment to the past as a source of authenticity that might serve to counterbalance, in Arnoldian terms, the loss of such depth in the industrialising metropole. These colonial constructions of Irish authenticity were, as Graham has noted, co-opted by the post-colonial state: '[s]omewhere between colonization and post-colonialism, domination and independence, the in/authenticity of the colonized is overturned. The role of authenticity alters from being a signifier of the colonized's cultural "incapacities", to being a marketable sign of value.'[28] In *Translations*, this process is pointed out by the subaltern pedagogue, Hugh, when he cautions Yolland against the lures of such a vision of Irish culture; in *The Communication Cord*, this colonial discourse is seen to have been reproduced as a post-colonial farce.

The play opens with this unlived-in holiday home being set up as a trap. It belongs to the family of Jack McNeilis, who is a barrister and so a member of the middle-class Irish who have prospered by serving the post-colonial state. He has hatched a plan to loan it for a couple of hours to his friend Tim Gallagher, a lecturer of linguistics (tellingly without tenure – the rhetoric of home and property is pervasive). By playing the role of host in such a cottage, Tim can impress Donovan – his girlfriend Susan's father – who is noted for his appreciation of such Irish antiquities. Donovan is indeed immediately enchanted by the place, finding himself very much at home there, and agreeing with Jack's deliberately parodic description of it: '[t]his is where we all come from. This is our first cathedral. This shaped all our souls. This determined our first pieties' (*CC*, p. 15). Donovan finds himself in communication with the place: '[t]his speaks to me, Tim. This whispers to me. [...] This is the touchstone. That landscape, that sea, this house – this is the apotheosis. [...] I suppose all I'm really saying is that for me this is the absolute verity' (*CC*, pp. 32–3). He gives a fulsome rendition of Adorno's description of the jargon of authenticity: '[w]hoever is versed in the jargon does not have to say what he thinks, does not even have to think it properly. The jargon takes over this task and devalues thought.'[29] Like Donovan's ontological disposition, the house is a piously translated construction that produces an effect of authenticity in order to deny its status as a translation. However, when they come into contact with each other, both reveal an underlying positivist-style literalism.

In common with positivism, the pious attitude ignores the otherness of the past, as is most vividly performed in Donovan's misadventure with the post and chains that were used to tether the cattle in the house. He is at first overjoyed to see them because they embody a literal translation: they are, 'incorporated [...] into the kitchen as of course it should be because that it is exactly as it was!' (*CC*, p. 34). He decides to act out a pious historical scenario to illustrate how these posts and chains functioned. In doing so he literally traps himself in this recreated, idealised past, and spends most of the rest of the play tied up in the posts and chains. The limits of the house's style of translation are exposed by the fact that the clasp of the post and chains is rusted and does not work properly: the whole contraption is literally of no use. As Lloyd has argued, the post-colonial Irish 'state is [...] founded upon a fetishization of invented traditions',[30] and Fanon offered a stark reading of this fixation: '[t]he desire to attach oneself to tradition or bring abandoned traditions to life again does not only mean going against the current of history but also opposing one's own people.'[31] Donovan incorporates this analysis: while he phrases his actions as a form of piety towards the past, their real objective is a colonisation of the past, an attempt to render the past inert and so open to modes of manipulation that might further his political objectives. Trapped in the realisation of his rhetoric, his real false piety is uncovered: '[t]his determined our first priorities! This is our native simplicity! Don't give me that shit! [...] This is the greatest dump' (*CC*, p. 75). His piety has travestied him, and he can only respond to the predicament of being caught in his positivist pretence of being a cow with a series of repeated, bovine 'oooghs' of discomfort.

If this scene is further read as a travestied translation of the Circe episode of *The Odyssey*, a diagnosis of this situation can be unearthed. Lojek has noted that Steiner repeatedly invokes 'Odysseus, the wily deceiver' as a symbol of the potentially liberating ability of language to conceal and to create alternative worlds. Jimmy Jack in *Translations*, as Lojek further points out, picks up on this theme and 'refers regularly to the Greek whose words transform reality' and, in this light, his disorientation serves as a warning of the perils of counter-factuality.[32] As the Field Day plays constantly explored the ways in which language creates realities, it is not surprising that Odysseus should feature in them; yet, what is notable is that he functions as negative example. In particular, in Heaney's *The Cure at Troy*, he is the arch-politician whose use of language is seen to be self-serving and crucially lacking an ethical dimension. In *The Odyssey*, the limits of his linguistic craft are most evident in the Circe episode, in which, like Donovan, his men are turned into animals, losing their language and their humanity. In order to continue his eccentric journey and release himself from Circe's clutches, Odysseus has to descend to the House of Death and consult with Tiresias and, crucially, confront his dead ancestors and friends. He is specifically enjoined to practise appropriate rites of piety on encountering the dead. In effect, this piety involves the difficult coming to terms with the fact that his loved ones are no more than shades; indeed, thrice he tries and fails to embrace his mother.[33] Not only does encountering the past bring no joy to Odysseus, he literally cannot grasp it. Donovan, on the other hand, assumed a full comprehension of his past, until he found himself ensnared in its grip. The difference between *pietas* and its false reflection can therefore be witnessed in the contrast between Odysseus's pious need to mourn, and so endure the experience of the lack of presence, and Donovan's celebratory recreation of the past, which produces nothing more than a pantomime of piety. LaCapra's linking of mourning and 'working through' historical trauma was noted in the last chapter, and the political and linguistic salience of this nexus can be developed here by examining Donovan's resistance to mourning that, amongst other things, echoes Lombard's attitude in *Making History*.

In Freudian terms, rather than mourning, Donovan displays a type of melancholia, which as Ricoeur summarises it, is 'the permanent claim of the pleasure principle'.[34] It is an immature and disabling response to the past: one that refuses to acknowledge that the past is irretrievably other. As such, it is marked by forms of repetition, a motion that will be witnessed presently in a more devastating fashion in *Boesman and Lena*. The repetition involved in *The Communication Cord* is, to a large extent, a parody of the impulse towards extreme literalism that informs a positivist-style translation: what could be more literal than repetition? Donovan's nationalism, in short, is a melancholic repetition of the colonial mode of translation. In contrast, mourning is, in Ricoeur's reading of Freud, an expression of 'the principle of reality', as it is a 'reconciliation [...] [w]ith the loss of some objects of love'; a reconciliation in the sense that it involves acknowledging that what was loved has gone, and can only be accessed through the difficult, unpredictable and perhaps contradictory 'work of memory'.[35] Mourning describes an attitude that is based on the acceptance of the task of seeing the past both as part of oneself and as precisely, irretrievably, other. It is a sign under which the past, rather than being assumed to live with one, must instead be considered an other in need of hospitality, through acts of memory and history. Mourning exists on the other side of intelligibility: it is the figure *par excellence* for non-comprehension, based as it is on the impossibility of comprehending the absence of the other. As de Man suggests in his reading of Rousseau, this describes a fundamental feature of language; mourning registers the decisive fissure between the signifier and the signified: the fact that language can invoke that which cannot be fully comprehended and which is irretrievably lost.[36] Mourning, in other words, speaks of the inherently metaphorical nature of language.

Accounting for the possibility of metaphor is at the heart of what is literally the thesis of *The Communication Cord*: the PhD dissertation on discourse analysis and response cries with which Tim is struggling.[37] The nature of his problems can be gauged from his description of this work to Jack. He proposes a rather strict, positivist-style codification of the functions and processes of discourse: '[l]anguage. An agreed code. I encode my message; I transmit it to you; you receive the message and decode it.' He goes on to draw out the broader relevance of his approach, noting that '[a]ll social behaviour, the entire social order, depends on [...] words mutually agreed on and mutually understood. Without that agreement, without that shared code, you have chaos' (*CC*, pp. 18–19). Such a linguistic and social system is devoid of ambiguity, as the relationship between signifier and signified is constant and unproblematic. However, his focus on the role of 'response cries' is deconstructing this system. In the midst of his lengthy descriptions, Jack interjects an exclamatory 'God', and Tim seizes on this:

> [a] response cry! And that's really the kernel of my thesis. A response cry blurted out as an involuntary reaction to what you've just heard. And what does it tell me? Does your 'God' say: I never knew that before? Does it say: This is fascinating—please continue? Does it say: Yes, I do desire to share your experience? Does it say: Tim you're boring me? Or is your expletive really involuntary? Maybe—because we're both playing roles, if we're both playing roles—maybe your 'God' is a *pretence* at surprise, at interest, at boredom. And if this is a pretence, why is it a pretence? (*CC*, pp. 19–20)

Evidently Tim is caught between the desire to codify language and the knowledge that this is not sufficient precisely because even a seemingly involuntary (one might even say 'originary'), cry needs to be interpreted; inscribed within it, in short, is the possibility of metaphor.[38] As *Translations* and Tim's struggles show, this is a possibility that cannot be accommodated in a purely positivist translation.

This linguistic crux induces yet another travesty at the conclusion of the play, with the house collapsing in a Babel-like fashion. Babel was brought down by an angry god who resented the fact that the Shems constructed this tower in order to make a name for themselves. In other words, the desire for a universal tongue that the tower was to embody was not going to be fulfilled through the creation of a new transparent language; rather, the tower was an aggressive symbol of the Shems' imposition of a linguistic hegemony.[39] While this episode has an obvious colonial relevance, its restaging within *The Communication Cord* renders it also as a post-colonial quandary: how to recuperate one's past and the ability to name oneself without reimposing a hegemony based on a pretence that some form of pure Adamic act of naming is available. Looked at in this manner, the collapse of the cottage repeats the lessons of the Tower of Babel and *Translations*: namely, the pretence that a prelapsarian, 'original', coincidence of sign and signified is possible underwrites hegemonic linguistic and political systems, be they colonial or post-colonial.

In *The Communication Cord*, the post-colonial house falls because Tim and Claire lean on the beam that was holding the house up. They had not noticed this as they were in the process of falling in love with each other. Their conversation – a form of linguistic foreplay that is also explicitly a deconstructive translation-as-analysis of Maire and Yolland's dialogue in *Translations* – farcically performs the issues that have been impeding Tim's thesis.

TIM: We're conversing now but we're not exchanging units, are we?
CLAIRE: I don't think so, are we?
TIM: I don't think we can because I'm not too sure what I'm saying.
CLAIRE: I don't know what you're saying either but I think I know what's implicit in it.
[...]
TIM: Even if what I'm saying is rubbish?
CLAIRE: Yes.
TIM: Like 'this is our first cathedral'?
CLAIRE: Like that.
TIM: Like 'this is the true centre'?
CLAIRE: I think I know what's implicit in that.
TIM: Maybe the message doesn't matter at all then.
[...]
CLAIRE: It's the occasion that matters.
TIM: And the reverberations that the occasion generates.
[...]
CLAIRE: And the desire to sustain the occasion.
TIM: And saying anything, anything at all, that keeps the occasion going.
[...]
CLAIRE: Maybe even saying nothing.
TIM: Maybe. Maybe silence is the perfect discourse. (*CC*, p. 92)

On one level, this suggests that some meaning might even be uncovered in the *ersatz* ontological phrases that had apparently been completely hollowed out in the course of the play. But, in a seeming paradox, in order for these phrases to have meaning, they must resist comprehension.[40] Language, in this view, does not say everything explicitly, as something is manifestly being suppressed as it is spoken. As Gentzler has pointed out, this metaphorical, precisely mournful, understanding of language informs Heidegger's thinking.

> Heidegger argues that we do not hear everything, for there is something essential which cannot be heard or read. Something is withheld as language speaks. Words not only reveal what is there [...] but language also holds back. If we let language speak for itself, what is revealed is something about the nature of language: words not only show what is there, but also show what is there and at the same time is not.[41]

Thinking about language as an act of (non-)revelation disavows the possibility that it is amenable to effortless, arbitrary reproduction. In this light, the house – a reification of a positivist travesty of an ontological approach – is brought down by a renewed confirmation of the Heideggerian notion of language as the 'house of Being'. By destroying this edifice, a blueprint for the construction of a more accommodating post-colonial home is suggested: one that resists the hegemonic lures of authenticity and the coterminous pretence of a full comprehension of the past. In this manner, the play re-performs and amplifies Hugh's conclusions in *Translations* that 'confusion is not an ignoble condition' (*T*, p. 446) and that '[t]o remember everything is a form of madness' (*T*, p. 445). In place of a post-colonial hegemony, *The Communication Cord* re-posits a difficult, non-essentialist, performative, counter-hegemonic postcolonial strategy involving simultaneous understanding and confusion, deconstruction and construction; a strategy of movement and resistance to stasis. One, in short, that attends primarily to the unpredictable, mournful claims of language, even though they can never be fully comprehended. Language, in brief, remains the master of Man, and it has been shown once again that '[w]hen this relationship of dominance is inverted, man has recourse to strange contrivances.'[42] As these Friel plays underline time and again, this mastery never assumes a fixed form, as language is not static, but rather is in constant dialogue with experience and history. Such a process, as Hugh puts it via Steiner, prevents us from petrifying: 'it is not the literal past, the "facts" of history, that shape us, but images of the past embodied in language [...] we must never cease renewing those images; because once we do we fossilize' (*T*, p. 445). For this reason, as Kearney concludes, 'we will go on telling stories, inventing and re-inventing myths, until we have brought history home to itself.'[43] And it is only on such an ever-shifting, problematising site that secure foundations for an accommodating post-colonial Irish home can be sunk.

Hosting Chekhov: *Three Sisters*

> Chekhov needs no special papers to take up Irish residence. He can be at home there in a standard translation, or in the totally Irish form of Thomas Kilroy's *The Seagull*. Friel's mistake is to offer a half-nationalized text, still Russian in its names and social rituals, but sprinkled with Edwardian songs and plonkingly underlined local references.
>
> — Irving Wardle[44]

Wardle's review of the 1981 Dublin Theatre Festival production of *Three Sisters* critiques the hospitality it offered Chekhov. Friel, he feels, failed in his role as host by offering the text a confused, uncomfortable form of accommodation: rather than confining this transaction either to Chekhov and the English language, or to Chekhov and Ireland, Friel has tried to accommodate all three terms, and has produced a hotchpotch – a 'coarsely reductive exercise in Irish Chekhov'.[45] But what the terms of this English review also make apparent are the array of cultural sensitivities that are aroused by Friel's translation, as built into it are unspoken hegemonic assumptions about the English language and about Chekhov's place in the English theatre. As such, it gestures to the processes of hospitality and domestication that operate between Chekhov, the English language and Ireland that Friel sought to negotiate. A commensurate dynamic of hospitality propels the various tripartite, intermeshing relationships within *Three Sisters* itself, and by concentrating on these a more nuanced reflection on the type of accommodation offered by Friel's act of translation is uncovered.

Although it was produced a year after *Translations*, Friel had begun work on *Three Sisters* before composing Field Day's inaugural drama; indeed, it was this act of translation that led him to Steiner and to reflect on the broader implications of translation.[46] As this indicates, translation is a complex, protean activity that disavows the possibility of conclusion; in Berman's terms, translation 'is not mere mediation: It is a process in which our entire relation to the Other is played out.'[47] Precisely because the other can never be fully comprehended, 'we must go through the experience [*épreuve*] of translation, incessantly'.[48] What this suggests is that the ethical imperative of fidelity assumes the form of an inexorable, Sisyphean process, as *épreuve* not only translates as 'experience' but also, at the same time, signifies something closer to an 'ordeal'.[49] As has been seen, Steiner opined that a faithful translation is one in which a precarious balance is achieved between domestic intelligibility and an accommodation of the otherness of the foreign text. Yet even this open-ended, stringently ambivalent and paradoxical definition offers a becalming, ideal, perspective on what is an endlessly fraught process between competing conceptions of domestication. Even when a translator strives to communicate the cultural differences intrinsic to the text, in practice s/he, as Venuti notes, has to resort to

> another set of differences, basically domestic, drawn from the receiving language and culture to enable the foreign to be received there. The foreign text, then, is not so much communicated as inscribed with domestic intelligibilities and interests. [...] Hence, the domesticating process is totalizing, even if never total, never seamless or final.[50]

In this light, the hospitality enacted in translation is, in many respects, a form of domestication: as Benveniste pointed out, hostility is always already inscribed within hospitality.[51] Derrida further glossed these etymological excavations to point out that a form of hostility is necessary to create the pact of hospitality, and this '*hostpitality*' (to employ Derrida's coinage) is in play in the domesticating hostility that generates a potentially hospitable translation.[52]

As is evident, translation regularly draws on vocabulary pertinent to the *topos* of the home, such as domestication. Thus far, the concept of domestication has been employed as it is used in translation studies – to indicate a style of translation that is noticeably biased in favour of the exigencies of the receiving language and culture.[53] This metaphor might be interrogated through *Three Sisters*, which is a drama about domestication, in which modes of domestication dictate the dynamic of hospitality. Reading the different ways in which the characters interpret the home within which the play is set opens up ways of understanding the ambivalence informing Friel's domesticating intentions towards Chekhov.

The three Prozorov sisters – Olga, Masha and Irina – live with their brother Andrey in their dead parents' house in the Russian provinces, having moved there from Moscow eleven years before the play opens. They feel themselves to be in exile from their spiritual home of Moscow and somewhat superior to the local provincials. The dull round of their existences is disrupted when Andrey marries a local girl, Natasha, and brings her into the family home, and this relationship becomes the axis around which the play revolves.[54] The sisters react to the first reports of this romance in a dismissive fashion: they can hardly credit it to be true and, in any case, appear confident of their ability either to domesticate Natasha or to frighten her off with their level of domestication. This is unambiguously registered on Natasha's first appearance. She has been invited by Andrey to Irina's birthday party and is nervous at the prospect of having to face the other guests. Olga greets her:

> OLGA: Ah, Natasha's here. How are you, Natasha?
> NATASHA: As my mother used to say, 'Never felt better and had less'—ha-ha-ha. (*They kiss*) God but that's a wild big crowd, Olga. I could never face that crowd.
> OLGA: Of course you can. They're our friends. (*Quietly*) And you're wearing a green sash. That's ... unusual.
> NATASHA: You mean it's unlucky, Olga? Because it's green? Is it a bad omen, Olga?
> OLGA: Not at all. It's just that against the pink it—it—it's quite ... distinctive.
> NATASHA: (*Tearfully*) You're right. It's wrong. (*TS*, p. 33)

In this scene of welcoming, the sisters' urge to exercise the power of domestication informs the way in which they perform their roles as hosts. To domesticate, as the *OED* notes, always brings into play contrary impulses, as it signifies 'to naturalize', '[t]o make to be or to feel "at home"' and 'to tame or bring under control; *transf.* to civilize'. In other words, it brings into play the terms of the 'civilian and barbarian' binary, the colonial logic of which is subverted, in a problematic fashion, in this play.

The hostility within hospitality can work both ways, and the guest's acceptance of hospitality can also be hostile. Rather then being cowed, once admitted into the house, Natasha proceeds to chase the sisters out of it and take it over. She could therefore be viewed solely as the outsider/enemy who transforms her hosts into hostages.[55] However, that is clearly not the full story, because the full effects of her entry into the house are more complex; understanding Natasha involves getting beyond, or encompassing simultaneously, terms such as 'guest' and 'enemy', which are both fitting and insufficient. To put it another way, rather than attempting to bypass the aporia of '*hostpitality*', it is necessary to rethink it in a way that might avoid the consolidation of rigid dichotomies. A way of doing this involves framing Natasha as a 'third' that participates in and disrupts assumed binaries – to think of her as a parasite rather than as an enemy, as the parasite operates precisely in the relationship between host and guest, assuming at times both of these roles.

Natasha displays many of the attributes ascribed to the parasite by Serres in *The Parasite.* Act II is set a year and a half after the end of the first act, during which time Andrey and Natasha have married and have had a son, Bobik. The scene opens with Natasha distrustfully checking up on the servants. Throughout the play, Natasha displays a hatred of the servants of the house, which as Serres points out is an expression of the inherent vulnerability of the parasite when it assumes the role of host: '[i]f someone knows all too well that I am not a master, it is the servant.'[56] This gives rise to a form of transference whereby the parasite casts the servants as parasites, and in this scene Natasha is checking to see if the maids have wastefully left a candle alight.[57] Indeed, it is precisely over the issue of servants as parasites that Natasha, in a rage, openly reveals her vision of her role as mistress of the house to the sisters in the cathartic third act. She wants to get rid of the sisters' old nanny, Anfisa, who does nothing but 'sleep or sit about' (*TS*, p. 71). Olga reacts to her treatment of Anfisa by employing her only weapon – reminding her of the differences in their levels of domestication: '[w]hat I'm trying to say is that—that—maybe the way we were brought up seems peculiar to you. But to hear a servant talked to like that upsets me terribly. I just can't—it just makes me feel actually ... physically sick' (*TS*, p. 70). But, by this point, Natasha is in a position to respond defiantly: 'I run this house. And when I make a decision about the servants that decision will be respected. [...] And don't you ever cross me again!' (*TS*, p. 71). This is, as Moss pointed out, 'the breaking point of the play'.[58] However, it is prefigured in the brief scene at the start of Act II, as taking on the role of the one who checks on the servants inevitably implies a claim to the role of the host. Moreover, another aspect of Natasha's behaviour in this earlier scene is a clear example of the dynamics of '*hostpitality*' as, in the space of a couple of minutes, the well-nigh maternal love she expresses for Olga and Irina's welfare becomes the motive that initiates the process of moving the sisters out of the house: she proposes to move her son into Irina's room, which she believes to be less damp. Her strategy is precisely one of domestication and, as her clash with Olga over the control of the domestics highlighted, the action of the play revolves around competing modes of domestication.

To reiterate, this does not simply take the form of an aggressive substitution; rather, it functions always within a three-way process. A parasite, in Serres's terms, 'has a relation with the relation and not with the station',[59] as to be positioned thus within a flux is the optimum location for the acquisition of potential power; in short, it is the indeterminable position of the politician.[60] Natasha fully exploits these possibilities by carefully interfering in the relationship between Andrey and his three sisters. For instance, when she prompts Andrey to perform as the host – asking him to 'say something' (*TS*, p. 39) to his sisters about the proposal to move Bobik into Irina's room – she is getting him to act as proxy for her, which at this juncture is a more effective way for her to operate as the host. She moves between the positions of parasite, host and guest, forcing the sisters and Andrey to play the other roles in this three-way dynamic at the times of her choosing. This unsettling motion is not entirely negative because, as Serres asserts, the parasite introduces essential change into a system.[61] Natasha's arrival in the house has at least helped to force the sisters into making decisions; they have to respond, to find an accommodation between the impossible nature of their earlier dreams and the necessity of reacting. Tellingly, in a letter to the actress Vera Komisarjevskaya, Chekhov described the play as having 'four heroines',[62] and Moss has even read the sisters' passivity as a form of invitation to Natasha.[63] Furthermore, her introduction has uncovered a series of parasitic dispositions in the play, based around complicated, overlapping, three-way relationships of dependence and exclusion, hospitality and hostility. Dynamics of this sort can be uncovered in the combinations, for example, of Andrey, Natasha and Protopopov; Masha, Kulygin and Vershinin; Vershinin, his wife and Masha; Kulygin, Olga and Masha; Irina, Baron Tusenbach and Solyony; and, most obviously, the three sisters themselves. In all these relationships the roles of parasite, host and guest could, in different contexts, be conceivably assigned to each of the three elements in turn.

As can be discerned in Friel's motives for producing a version of *Three Sisters*, his translation is equally parasitic and also creates a three-way relationship: '[t]his translation was undertaken primarily as an act of love and, since the only Chekhov translations available to the Irish theatre are American and English, in the hope that this translation may make the unique experience of Chekhov more accessible to Irish audiences' (*TS*, Jacket notes).[64] He elaborated on this in an interview published just before the first production of the play, in which he located the act of translation within Field Day's broader postcolonial agenda of rethinking Irish identities.

> I wrote this play in an Irish idiom because with English translations Irish actors become more and more remote. They have to pretend, first of all, that they're English and then that they're Russians. I'd like our audience to see Captains and Lieutenants who look as if they came from Finner

> or Tullamore. The decolonisation process of the imagination is very important if a new Irish personality is to emerge.[65]

There is a palpable tension in these statements between Friel's love for the unique experience of Chekhov and the wish to domesticate this into an Irish context. The explicit political inflection he gives to this dilemma sets out the terms within which the postcolonial resonance of Friel's translation can be understood. Friel's Irish domestication of Chekhov is, in short, a counter to the English domestication of Chekhov that implicitly informed Wardle's critical review of the play cited above. The English stage has subjected Chekhov to such a strong process of domestication that it could more properly be termed appropriation: as Kenneth Tynan noted with waspish sarcasm in 1958, having 'remade' Chekhov in their 'own image', the English 'then congratulate him on having achieved honorary English citizenship'.[66] So, even though Friel's version does not differ radically from British English, some critics used to hearing Chekhov pronounced in English tones found the performances disconcerting.[67] Most notable was a review of the opening night in the unionist *Londonderry Sentinel*, which bizarrely wondered: '[a]re we next to have the beauty of Shakespeare's language translated into Irish phrases and repeated in various blends of an Irish brogue?'[68] Leaving aside the potential irony of the fact that, as Friel noted, Elizabethan English might sound more like 'an Irish brogue' than received English pronunciation, this implies that Chekhov is as English as Shakespeare. In this light, Friel's version is an act of resistance against an imperial cultural hegemony; against the ways in which, as he put it in an interview in 1980, 'we are constantly overshadowed by the sound of English language, as well as by the printed word.'[69] His translation is therefore not only an engagement with the Chekhovian other, but it also deals with a third element, another other – the English-language versions of Chekhov.

At the same time, these English versions of Chekhov have also directly mediated Friel's engagement with Chekhov. As he speaks no Russian, his translation was produced with the assistance of six other English-language translations:

> I don't know a word of Russian, of course, and so I feel a wee bit diffident about using the word 'translation'. What I did was simply to put six texts in front of me and tackle each line at a time, to see first of all what was the meaning of it, then what was the tone and then eventually what was the sound. It took nine months in all.[70]

This act of re-translation involves a three-way relationship that reflects an aspect of the nature of Irish engagement with the concept of 'Europe'. Such an engagement has often been evoked as a potentially enabling and liberating strategy in the creation of a post-colonial identity. But, if this is considered as a move away from British influence, it also immediately uncovers the difficulties of disentangling Irish identity from the cultural and linguistic mediation and/or contamination of English, as the English language has been, as Dantanus points out, the major 'vehicle through which cultural influences have penetrated Ireland'.[71] This postcolonial dilemma is reflected in Friel's attempt to accommodate Chekhov in a form of English that speaks to Irish experience.

Located at the centre of a flow of information from one person to another, the translator is a classic parasite.[72] In the play, Andrey, who is a translator, is clearly a parasite. He has inherited the role of host, and while he invited Natasha into the house, he finds his filial inheritance problematic and fails entirely as an authority figure. Instead, as a parasitic translator, he radically undermines the family's home: in Act III, it is learnt that he has mortgaged the house without his sisters' consent in order to cover his gambling debts. However, even before Andrey's interventions, the sisters' conception of home had already been destabilised by the ability to translate. In Act I, Masha indicates the alienation they feel within their home-place by pointing to their ability with languages: 'you can imagine how useful it is to be able to speak three languages in a town like this! It's almost a necessity, isn't it? Like having eleven toes or a sixth finger. Hah! Nearly everything we know is ... useless' (*TS*, p. 26). The question of language(s) brings to the fore their sense of not being at-home at home: while speaking other languages is a mark of social achievement, in the provinces it is not appreciated and merely sets them apart.[73] It also reveals that a large part of their desire to return to Moscow and to that version of their home is the wish to return to a more cosmopolitan and multilingual society, not to an idealised, 'authentic' Russian homeland; as Vershinin hints, they are probably living closer to the heart of an idyllic Russia where they are. So, in yet another Field Day play, home is not a place of idealised purity, but is elsewhere and, to use the language of parasitism, in some way contaminated.

Nothing is known of Andrey's style of translation, but it might be suspected that it is a somewhat arid practice that embodies the uselessness that Masha alluded to above. This is betrayed by Andrey's generally ineffectual manner and by the fact that he only communicates with the deaf Ferapont: he is a translator who, in Szondi's words, 'can speak only when he knows he will not be understood'.[74] Once again, this must inflect upon any understanding of Natasha's actions and her achievement of taking control of the house; simply put, her parasitical intervention is at least vital and productive. As Bloom put it: '[u]nlike the sisters, whose vitality is thwarted, the uncultured Natasha is extending the life of the Prozorov family; she is peopling the house with babies.'[75] Tempting though it is, particularly as Natasha is given most of the idiomatic Irish expressions in Friel's version, a strict allegorical correspondence between the relationships that link Friel's version, the other English-language translations and the Russian Chekhov, on the one hand, and the relationships between Natasha, the three sisters and Andrey, on the other, is not being proposed here.[76] Instead, what is being suggested is that the introduction of the third term – the parasite – acts like a chemical reagent and highlights the parasite/host/guest system that was already operating between, and sustaining, Chekhov's Russian text and its English translations. Not only does it uncover the domesticating hospitality that Chekhov has been afforded by English translations, but it also casts another light on Chekhov's declared insistence on remaining emphatically and unequivocally as a host, beyond translation. Reviewed in this construction, Chekhov not only plays the role of host – as the 'original' – and guest – of the translations – but also that of a parasite.

While Mirsky believed that what he thought of as Chekhov's dull use of language made him particularly amenable to translation, and so 'of all Russian writers, he has least to fear from the treachery of translators', Chekhov nonetheless had a resolute antipathy to translation.[77] Scolnicov recounts that he expressed 'a total denial of the possibility of transferring a play from one culture to another', as he 'did not believe that non-Russian audiences could possibly understand the full meaning of the selling of the estate in *The Cherry Orchard*'.[78] The example is telling, as it suggests that it is the significance of home that is beyond translation. Yet, even as it proposes this conception of a home as an inviolable citadel whose meaning is entirely contained within its walls, it reveals the impossibility of this idea – the estate is being sold, is taking on new meaning and becoming something else. Likewise, Chekhov's resistance to translation is not only impossible but, having been translated, his texts have acquired many more and unexpected meanings. Even the rather extreme process of domestication he has undergone at the hands of the English stage has produced a new and different Chekhov. His translational afterlife is a form of parasitism, and reflects the inherent ambiguity in the figure of the parasite, which as Cronin argues, is 'a key to the ambivalence felt towards translation itself. The parasite feeds on another organism, breaks it down, destroys it. It is the symbol of aggression, rot, decay. The parasite also transforms, recycles and ensures the continuity of the cycle of life.'[79] In short, the hospitality/hostility to which Chekhov has been subjected by his hosts/parasites/guests has regenerated Chekhov.

This process is furthered by Friel's 'alternative' English translation, which he has spoken of 'as "re-freshing" Chekhov'.[80] It not only draws attention to the assumptions built into the English translations, it also opens up Chekhov to other translational possibilities and assists in restoring an essential element of otherness to his texts within the Anglophone tradition. Moreover, in an Irish context, Friel's project is a potentially destabilising postcolonial initiative: if his intention was to make turn-of-the-century Russia as familiar as Finner or Tullamore, this also suggests that rural Ireland might be as foreign as provincial Russia for the audiences of his version of

Three Sisters. The subtle disruptions hinted at both within the play and by the play's status as a translation might provoke a rethinking of the notion of home and a questioning of the defined, cosy and 'known' versions of that concept.

Friel's translation not only disrupts a nationalist construction of home, but it also silently deconstructs the foundations of another form of post-colonial Irish home, one that is even more explicitly tied to colonial times: the so-called 'Big House', the abode of the landlord class in colonial Ireland. Culturally deracinated with the advent of the Irish Free State, through the so-called 'Big House' literature these houses assumed a literary afterlife as symbols of a form of lost imperial civility that had been overtaken by the barbarous provincialism of Catholic, middle-class commercialism. The parallels between such a view on Irish political history and the action of *Three Sisters* are evident, and it is no surprise that Chekhov, or a version of Chekhov, became an important example for early revisionist writers such as O'Faoláin and O'Connor who viewed post-colonial Ireland, in Deane's words, as a site of 'stifling provincialism and slow-moving disintegration'.[81] But, while Friel has been justly described as 'a Chekhovian dramatist', his versions of the 'Big House' dramas cannot be read as threnodies for the passing of the landowning class; rather, the ambivalent effects of the shift in power from an older, more genteel family to a brash newcomer like Natasha are hinted at by the sense of possible release that it engenders.[82] This is most pronounced in his 1979 *Aristocrats*, Friel's most explicit rendition of an Irish-Chekhovian 'Big House' drama.[83] In a manner not entirely dissimilar to the ways in which *The Communication Cord* deconstructs the form of *Translations* to amplify both plays' postcolonial reading of Irish life, and *Making History* employs revisionist arguments to question their presumptions, *Aristocrats* uses the conventions of the 'Big House' genre to interrogate the myths that supported the landowning class, and ultimately gestures towards a potentially reinvigorating liberation from these myths at the end.[84]

In a commensurate fashion, the sluggish provincialism that forms the framework for *Three Sisters* is revealed as a possible site of a problematising and reviving 'fifth province' in Friel's Irish interpretation. Central to this is the performance of the intractable knot of '*hostpitality*' that does not attempt to untie and resolve it, with the result that the notion of home is deconstructed in a manner that holds out the possibility of another form of accommodation. The relevance of this translational dynamic to Field Day was underlined in Deane's essay in the programme for *Three Sisters*, in which he cast the enterprise's activities in terms of a complicated, parasitic, ambivalent scene of hospitality between the theatre company and its audience.

> [Field Day] goes to the people, not for their sake but for its own – but in the conviction that it will eventually be for their benefit if they are sufficiently to its benefit. Theatre finds its sense of origin in an audience

it has to invent. Field Day is inventing an audience. It is therefore also inventing an idea of origin. It finds itself in terms of its new audience. For the contemporary Irish theatrical audience determines contemporary Irish theatre. It does not find *otherness* in the theatre. It only finds the self it knows.

[...] [Field Day] needs a dialectic. It has to itself be liberated by the audience it liberates. At first most people will read the plays as though they were written in braille. But the plays may be so foreign and recognisable at the same time (Chekhov?) that they might eventually open their eyes. I think the idea of a theatre without a roof over its head is precisely right for now.[85]

Displacement and reiteration: *Boesman and Lena*

> Field Day has grown out of [...] [a] sense of impermanence, of people who feel themselves native to a province or certainly to an island but in some way feel that a disinheritance is offered to them.
>
> — Brian Friel[86]

Crucial moments in both *The Communication Cord* and *Three Sisters* were induced and symbolised by the collapse of houses. In *The Communication Cord*, the sham post-colonial attitudes found expression in an unstable house, which folded under the weight of a renewed attention to the claims and vicissitudes of language. Pivotal decisions to move on with their lives are made by the sisters in Act III of *Three Sisters*, while they watch from Olga's and Irina's bedroom – a site of dislocation for them – a fire displacing a group of people from their homes. *Boesman and Lena* is initiated by the destruction of an abode before the play opens. However, this is of a completely different order to the events in the other two dramas, as this play is emphatically an example of what Soyinka termed the 'theatre of the dispossessed'.[87]

Boesman and Lena – the first non-Friel play to be produced by Field Day – is not only set in a context that is far removed from that of Ireland, but it is also not an original Field Day production, having been first performed at the Rhodes University Little Theatre, Grahamstown, South Africa, on 10 July 1969. For Field Day, *Boesman and Lena* was a substitute production, or rather a doubled substitution. The company's original plan for its 1983 tour was to produce a play that had been commissioned from David Rudkin. However, on receipt of Rudkin's script of *The Saxon Shore*, the directors decided against putting it on. As the motives informing this decision concern, to some extent, the representation of the Northern Protestant community, they are discussed in the introduction to the next chapter. Having made this decision, the directors though did not immediately turn to *Boesman and Lena*, but first considered Fugard's *The Island*, which explores the plight of political prisoners on Robben Island through the framework of *Antigone*.[88] However, according to Richtarik, the Northern Ireland Arts Council intimated that that was too political a play, and in order to maintain their funding the directors turned to *Boesman and Lena*.[89] Yet, notwithstanding the fact that this production came about through a process that created a clear distance between Field Day's stage and Northern Ireland's political life, it and Parker's *Pentecost* are the two most immediately political plays in the Field Day *oeuvre* (in the sense that the action of both these dramas is inseparable from, and is entirely engendered by, a political situation that was ongoing in the 1980s).

The drama of commitment and distance that Deane noted in Said, and which was discussed in the Introduction, is explicit here: a fundamental aspect of Field Day's response to the 'Troubles' was a reluctance to perform them directly. In 1983, in answer to a question about the lack of a Field Day 'production dealing with the current troubles', Friel gave examples of plays that were relevant to the political situation: '"Boesman and Lena" with its references to the effects of apartheid on the individual and "Translations" with its theme which dealt with the passing of Irish place names, do have obvious modern references.'[90] As this intriguing response reveals, Friel's almost instinctive dramatic tendency is to displace, temporally and physically, the political situation in Northern Ireland. Surveying Friel's work, Pine points out that 'Friel's approach to the question of political theatre has been characteristically circumspect and discreet', and this has engendered 'a theatre which appears oblique rather than direct but which is concerned with the psychological basis of political action'.[91] While Friel's politics have shaped his dramas, he has striven to ensure that his plays are not read as direct commentaries on contemporary politics. On the occasions when the pressure of events has flooded his dramas – as in, for instance, *The Freedom of the City* – he was dissatisfied with the results. One reason for this is that, according to Friel in 1971, for an effective dramatisation of a conflict, 'you must have a conflict of equals or at least of near equals. There is no drama in Rhodesia or South Africa, and similarly there is no drama in Northern Ireland.'[92] The reality of political inequality quashes the possibility of drama, as it can only produce essentially predictable results. Therefore, the action must be displaced into a context where the protagonists have some latitude to act, which is where they are located not only in *Translations*, but also, albeit to a highly restricted degree, in the South African setting of *Boesman and Lena*.

Once again, it is a question of variable distance: as in Friel, the political in Fugard's play is performed through the personal in order to bring the political home and give it an immediate resonance. However, in this context, the political entirely invades the private, as the extreme conditions of apartheid completely erase the boundary between these states, and so the drama stages its account of South African politics through the devastation it has wreaked on the lives of its two protagonists. For Fugard, the fundamental political objective of his drama was 'to *witness*', which required him to 'throw in [his] lot [with the] "here and now"', and the protagonists of *Boesman and Lena* bear ample witness to the brutal logic of Fugard's equation '[p]overty = the violence of immediacy'.[93] Paradoxically, though, this aesthetic of immediacy has at its core forms of displacement that bring it closer to Friel's attitude to the representation of politics onstage.

The representation of immediacy, the witnessing in the play, is refracted through a series of dramatic conventions and styles derived, in particular, from Brecht and Beckett. But this displacement does not block the political import of the piece; rather, these inspirations are marshalled in a way that produces an imaginatively effective way of bearing witness to the subaltern world of apartheid South Africa. Gordimer has read Beckett's influence on Fugard as 'an interesting example of a writer's methodological response to his socio-political situation', and as a contradictory distancing that brings immediacy: 'a fascinating example of an esoteric mode, in which character is sacrificed to symbolic abstraction, and dialogue largely disembodied, returned to flesh and the individual involved rather than alienated'.[94] This translational route ultimately returns Fugard back to a refigured 'home' through a process that also allows the immediacy of this 'home' to be understood beyond its own confines.

If this has much in common with Field Day's aesthetic approach to the 'Troubles', on a practical level, Fugard's collaborative theatrical ventures also had correspondences with Field Day. Fugard was the most famous name in the tradition of the South African 'workshop' plays which, as Crow and Banfield describe it, were ventures in which both black and white artists worked together, and simply by doing so, 'challenge[d] the very basis of apartheid', as the very notion of blacks and whites even associating was repugnant to the apartheid regime.[95] These 'workshop' theatre groups did not merely defy and indict the state, they were also consciously enacting a future non-racial society, and in the process becoming, as Anne Fuchs comments, 'a metonymic image of a New South Africa'.[96] The 'workshop' groups represented a positive form of displacement that created an other space within the political realities of South Africa. As a cultural image of a more ideal state to come, this artistic, political, collaborative, theatrical project also echoes the logic of the 'fifth province' and the means through which Field Day sought to articulate it.

However, just as certain Field Day directors (most notably Heaney in the early 1970s) were accused by nationalists of having failed to support explicitly and unambiguously their struggle, it is telling that Fugard has been charged with displacing political responsibility in favour of artistic considerations. Lewis Nkosi, who collaborated with Fugard on the script of the 1958 *No-Good Friday*, has recalled that Fugard, 'gave no sign of being directly interested in politics[;] [...] [he] was interested in learning about how we lived and in practising his art. The politics of the South African situation touched him on these two levels.'[97] In fact, Fugard's declaration, cited above, about the need to witness on an immediate level was made while he was struggling with his response to the international artistic boycott of South Africa that was called for in 1968: the need to testify to the everyday devastation wrought by the South African regime was precisely the reason why Fugard opposed the boycott. As he recounts in his *Notebooks*, this was a massive moral dilemma for him; it brought into conflict his personal, artistic and political responses to the apartheid state. Ultimately, the manner in which he aligned these responses privileged the artistic and the personal over the political, because he was convinced that his plays articulated a voice against white complacency and so constituted his most effective opposition to the South African regime.[98] This attitude has correspondences with Friel's vision of his political role, and so it is ironic that, as Kadar Asmal notes, Friel backed this South African boycott. Even more ironically, Asmal (who in 1983 was the head of Ireland's anti-apartheid movement and has subsequently held key ministerial posts in ANC-led governments in South Africa) noted this in a short essay in the programme for the Field Day production of *Boesman and Lena*. In this, he firmly rebukes Fugard for his stance on the issue of the boycott; unequivocally calling him to account for having 'never been able to commit himself to the liberation struggle'.[99] To recall, Field Day turned to this play as a substitute production that would be political, but not too obviously so, and so it is possible to perceive an instructive double standard here: there can be little doubt that Asmal would not have said the same thing about Field Day's attitude to the Northern 'Troubles' as, amongst other reasons, Irish intellectual and artistic discourse has privileged the artist's freedom to respond in a manner of her/his own choosing to political matters. While Fugard is implicitly called upon to live up to expectations that are not considered appropriate in an Irish context, Friel nevertheless saw *Boesman and Lena* as speaking to the Northern Irish political situation. But, as Friel's support of the boycott indicates, South Africa was explicitly another case: a situation in which political and ethical decisions were considerably more clear-cut. So, despite Friel's linking of apartheid with the situation in Northern Ireland (a connection that sailed close to echoing the sort of undifferentiated postcolonial comparisons that Friel rejected when they were made about *Translations*), there is an implicit recognition that the North represented a considerably more entangled political situation. Asmal's essay points towards a confused, swirling dynamic of hostility and hospitality in Field Day's hosting of *Boesman and Lena* which is, in some senses, the most foreign drama that the company produced, as it is the only entirely unmediated (by translation), non-Irish play in its list of productions. Field Day seemed to struggle to domesticate the play – to locate it within an Irish context – a process that was either overtly or covertly accomplished through the translations of the other non-Irish dramas. But even as *Boesman and Lena* was being framed in an uneasy, displaced manner by Field Day, it brought to the company's stage crucial, different, aspects of the dilemma of domestication, and the imperative of witnessing.

The fundamental structure of *Boesman and Lena* is one of displacement and repetition. Boesman and Lena are a thoroughly dispossessed mixed-race couple, whose shack has been destroyed by the apartheid authorities in a 'clean-up' operation earlier in the day (a distorted echo might be heard here of Lancey's threatened clearance at the end of *Translations*). This is not the first time this has happened, and they do not imagine that it will be the last. They trudge around in circles in a bleakly desolate South African no-man's land with the tattered remnants of their lives on their backs. The stage is a stage on their unplanned, unending journey: a halting post in their wanderings. As the opening stage directions underline, the endless wear and tear they have endured has erased significant portions of their identities: '[*a*] *life of hardship and dissipation obscures their ages, but they are most probably in their fifties*' (*BL*, p. 193). They have been at the receiving end of repeated brutality for so long that their anger has been displaced inwards; the cruelty they have experienced finds an amplifying echo in their habitual self-destruction, and a mode of damaging repetition pervades the piece. As a result, there is no need to represent directly the oppressive apartheid state onstage, and it is kept offstage. Boesman and Lena take on its form by becoming that which it excreted: '[w]e're whiteman's rubbish. That's why he's so *beneukt* with us. He can't get rid of his rubbish. He throws it away, we pick it up. Wear it. Sleep in it. Eat it. We're made of it now. His rubbish is people' (*BL*, p. 231).[100] They repeat the structure of their apartheid situation just as Senator Donovan enacts a disabling post-colonial repetition of the structure of colonialism in *The Communication Cord*. But what is different here is the presence of violence: as Walder has commented, Fugard portrays the inevitability of violence 'in a society which has adopted violent methods to stifle opposition and prevent change'.[101] A vicious spiral operates in which the violence of the absent-but-ubiquitous state gathers redoubled momentum through the violence of repetition and the repetition of the violence it breeds. It is a condition of trauma, and Boesman is floundering in this maelstrom: the state's violence provides him with the only terms he has to make sense of his situation, which is what he fails to do, simply because he knows no other context. So, for instance, he declares that the flattening of their shack has made him happy—the 'whiteman [...] did me a favour. I was sick of it' (*BL*, p. 196). He attempts to welcome the state's repeated action and excoriates himself for having fallen into the habit of constructing another shack:

> I saw that piece of *sinkplaat* on the side of the road, I should have passed it. Gone on! Freedom's a long walk.

> [...]
> So I picked it up. Finish and *klaar*. Another *pondok*.
> [*Shouting violently*.]
> It's no use, *baas*. Boesman's done it again. Bring your bulldozer tomorrow and push it over! (*BL*, p. 230)[102]

He has been driven to a point where he endeavours to imagine his enforced trek as an emancipatory walk; indeed, '[f]reedom's a long walk' both echoes and anticipates the titles of Nelson Mandela's books.[103] Boesman blames himself for impeding his progress towards this objective by securing even the basest shelter, which indicates that he cannot even imagine what freedom might look like. Lena, on the other hand, does not try to fool herself either by reframing their wanderings as having a destination or by indulging in pointless acts of self-flagellation; she is acutely aware of the difference between their movements and freedom, such as that enjoyed by the birds she curses because, '[t]omorrow they'll hang up there in the wind and laugh. We'll be in the mud. I hate them' (*BL*, p. 194). Like the birds, she and Boesman will have 'the sky for a roof again' (*BL*, p. 195) but this boundless space is a limit to their freedom, as she well knows.

Their situation is defined by just such a paradox, as their endless movement is ultimately a form of fixity. It is an unremitting revisiting of places from where they have been repeatedly evicted, the names of which Lena regularly regurgitates: 'Redhouse – Swartkops – Veeplaas ...' (*BL*, p. 202). In Bhabha's terms they are acting out their roles as stereotypes, which are generated through an ambivalent process of simultaneous fixity and movement; a motion that is precisely one of repetition: 'the stereotype [...] is a form of knowledge and identification that vacillates between what is always "in place", already known, and something that must be anxiously repeated'.[104] It is a movement that goes nowhere, an illusion of progress that merely confirms the already-known. In this context, it is, as ever, a manifestation of the absent but all-controlling state, as it is obvious that the would-be paradox of their simultaneous movement and fixity can only be explained by the fact that Boesman and Lena go to places to which they are allowed/forced to go: though they are homeless and without place, they are very definitely placed by the state.

It was argued in the Introduction that Field Day employed the resources of translation as a way of deconstructing (post-)colonial Irish stereotypes: a process that put them in motion and that placed conceptual and physical borders under question. As the readings so far have illustrated, what is most radically destabilised in this procedure is the concept of home. Home though has an entirely different significance in *Boesman and Lena*: whereas there is a need to disrupt notions of home as a secure, given, location of selfhood in almost every other Field Day play, here home and a concept of the self are explicitly lacking and so constitute the objective. The gulf between

Irish experience and that of black South Africa is clearly demarcated here, and must be registered both in considering the role of translation in each and in the translation of South African experiences into Ireland. However, rather than disqualifying *Boesman and Lena* from having a salience in Ireland, this should caution against the privileging of migrancy for its own sake and a too-easy Western theoretical occlusion of the necessity of a sense of home. Bassnett and Trivedi have critiqued a commensurate tendency to conflate the conceptual with the actual in Bhabha's work, noting that, in it, 'the translated hybridity of the "unhomed" migrant now inhabits a "Third Space" [...] which presumably becomes accessible only after one has left the Third World.'[105] Considering that Field Day represents the generation of the Northern Irish civil rights campaign, which was precisely about the issue of housing, there can be little doubt that the company did not think of home merely as a theoretical abstraction; instead, this experience informed the company's need to theorise the concept of home. Indeed, even in the extreme context of *Boesman and Lena*, home cannot simply be a place of stasis. Enacted through Boesman and Lena's different responses to the other, a translational movement of sorts is shown to be required to imagine a version of home which is empowering. What is witnessed onstage is not just how fraught this process is, but also how actual political circumstances can expunge in a trice the possibility of imagining such a location.

A *Translations*-like translation policy of renaming, relocation and remapping played an essential role in the South African state's self-creation. As Loren Kruger noted '*African* cannot be translated into Afrikaans, since its place is already occupied by *Afrikaner*.'[106] Fugard underlined the colonial and political imperatives of this conceptual displacement:

> [t]he apartheid government went through all sorts of linguistic manoeuvres to avoid calling black people Africans, and thereby recognizing their prior claim to the land. The term 'Bantu' was used to refer to blacks; in one of the original black languages it simply means 'people'.[107]

Boesman and Lena is, amongst other things, a drama about language and translation, and this is most explicit in the way in which it is laced with the different languages and dialects of South Africa (which apparently caused some bewilderment at Field Day performances).[108] Boesman and Lena's deracination is evidenced in their language, which is a pidgin mixture of English and Afrikaans, which, according to Walder, is an act of translation on Fugard's part, as '[a]ll the characters [...] apart from [...] Outa [...] would normally speak Afrikaans; but Fugard has "translated" their speech into his own Eastern Cape idiom – thus reaching a world audience without sacrificing what he calls his local "specifics"'.[109] Such a hybridised, glocal linguistic practice speaks to Field Day's non-essentialising attitude to Hiberno-English. Fugard's background is both English and Afrikaans, and it was while he was writing *Boesman and Lena* that he came to understand that 'an important aspect of my writing is an element of "translation"': 'the whole problem of the form and content of one language as opposed to another'. Musing on this, he notes that in *Boesman and Lena*, '[t]o begin with phrases were all in Afrikaans – some are still, and still defy "translation" e.g. "nog 'n vrot huis vir die vrot mens".'[110] The phrase that thwarts translation could be translated as 'another rotten house for rotten people'. Just as Chekhov believed that the significance of the selling of the estate in *The Cherry Orchard* could not be translated, a very different abode here underlines how the concept of home, because it is intrinsic to one's sense of identity, defines the very limits of translatability, or, to put it another way, calls most urgently for translation.

The constant erosion of Boesman and Lena's sense of themselves is shown to be deeply linked to the impossibility they experience of constructing a domestic space. Even Lena's momentary aspiration that she might work towards a home by being taken on as a domestic in a white person's house is incredibly unrealistic: the white children simply laugh at her, as does Boesman. The repeated destruction of their shacks has reduced her *'to a dumb, animal-like submission'* (*BL*, p. 193) while Boesman expresses his servility by thanking the man who bulldozed their hovel for throwing him a cigarette butt and calling him his boss/master: *'my baas'* (*BL*, p. 228). They are impotent, posing no threat to those who command them; without the basis of a domestic space from which to articulate themselves, they are entirely domesticated. As Freud notes, a possible German term for this form of domestication is *'heimlich'*, and in 'The "Uncanny"' he demonstrates how the meanings of *'heimlich'* and its opposite *'unheimlich'* overlap.[111] This interplay has been charted in the previous two plays, in readings which have suggested that the metaphor of home as the centre of a secure, self-sufficient identity is constantly undermined by the uncanny need for the other to confirm one's identity. Boesman and Lena exemplify a different interweaving of these two terms, in that their homelessness has domesticated them, while their helpless repetitions are one of the most telling symptoms of the uncanny: 'the constant recurrence of the same thing – the repetition of the same features or character-traits or vicissitudes'.[112] Boesman repeats the brutality of the apartheid state through the regular beatings he gives Lena, who nevertheless follows him; like Boesman helping the 'whiteman' to burn their shack, she cannot imagine an alternative to her tormentor.

In 'Remembering, Repeating and Working-Through', Freud links repetition with repression: 'the patient does not *remember* anything of what he has forgotten and repressed, but *acts* it out. He reproduces it not as a memory but as an action; he *repeats* it, without, of course, knowing that he is repeating it.'[113] In contrast to the 'working through' of mourning, repetition, as has been noted, is the symptom of melancholia, and is, in Ricoeur's reading, 'an obstacle to remembering'.[114] Boesman and Lena's endless repetitions are a traumatised defence mechanism against having to recall their situation. But these repetitions are also literally getting them nowhere. Lena articulates the vulnerability and loss that this process entails when she makes an effort to remember. Declaring that 'I meet the memory of myself on the old roads' (*BL*, p. 219) she senses that her identity is inscribed in the landscape she has trod: '[t]hose little paths on the veld ... Boesman and Lena helped write them' (*BL*, p. 218). In order to know where she is, she needs to recall the order of the places through which she has passed, but she cannot get it straight – 'Veeplaas—Redhouse—Korsten? [*Pause*] Where's Swartkops?' She looks to Boesman for help, but sensing that she is trying to locate herself and being incapable of being constructive, he simply takes the opportunity to have some sadistic fun by furthering her disorientation.

> LENA: [...] Help me, Boesman!
> BOESMAN: What? Find yourself?
> [*Boesman launches into a grotesque pantomime of a search. Lena watches him with hatred.*]
> [...]
> Sorry, Auntie. Better go to Verplaas. Maybe you're there. (*BL*, pp. 205–6)

Remembering requires an other; Lena needs someone to help her in the difficult process of remembering and as Boesman has no interest in any form of progress, she needs another other, who emerges in the unlikely shape of the old man, Outa.

Rather than immediate hospitality, Outa receives hostility when he emerges and they see that he is an *'old African* [...] *an image of age and decrepitude'* (*BL*, p. 210). Their first word of greeting is *'Kaffer'* ('kaffir'): the racist structure of South African society automatically reproduces itself, and Boesman takes his place as a mixed-race, 'coloured' person before a black man. 'BOESMAN: He's not brown people, he's black people. LENA: They got feelings too. Not so, *Outa*?' (*BL*, p. 212). Race determined every aspect of a person's life in the apartheid regime, and this racist political structure underwrote the complete lack of responsibility for the other that is most simply expressed by Boesman's statement, '[h]e doesn't belong to us' (*BL*, p. 212). He refuses to let Outa enter what can only be described as their space, as Boesman has not yet put together their shack. Despite this, Boesman assumes the role of host that the power structure of apartheid society grants him, and warns that if Outa joins them, he will '[t]urn my place into a *kaffer nes*!' (*BL*, p. 213).[115] Once again, the scene of hospitality, as the primary encounter with the other, is the site of crucial ethical and political decisions.

Lena, after her initial disappointment, welcomes Outa as '[b]etter than nothing' (*BL*, p. 211). She is not in a position to grant unlimited hospitality. Instead, Boesman enforces an economy, whereby the opportunity to converse with Outa is bought at the price of two bottles of wine and exclusion from the rough shelter he eventually throws together.

> LENA: [*impulsively*]. You can have the wine. All of it. Next time as well.
> [...]
> BOESMAN: [*unbelievingly*]. For that!
> [...]
> BOESMAN: This is wine, Lena. That's a *kaffer*. He won't help you forget. You want to sit sober in this world? You know what it looks like then?
> LENA: I want him. (*BL*, p. 222)

Levinas, as Llewelyn has noted, held that 'the contentment of living monadically is a kind of drunkenness', and that an awakening 'sobering' of the 'subject' is brought about through an 'interruption' 'by the Other'.[116] It is at this moment, and in precisely these terms, that the crucial difference between Boesman and Lena is displayed. His relationship with the world is one of hatred and violence, which he unsuccessfully tries to keep in check through forgetful drunkenness: his mantra is '*Weg wêreld, kom brandewyn*' – 'go away world, come brandy' (*BL*, p. 213). He lapses into periods of sullen silence and refuses to speak to Lena, who has to plead with him: '[d]on't be like that tonight, man. This is a lonely place. Just us two. Talk to me. BOESMAN: I've nothing left to say to you. Talk to yourself. LENA: I'll go mad' (*BL*, p. 199). While his inability to communicate entraps him in what Kane called 'the hell of the self', Lena needs to recount her life, to engage with the task of remembering, of working through the repetitions that have blighted her life.[117] As Ricoeur's commentary on Freud has it: '[t]he way towards reconciliation with oneself [...] is called "working through" (*Durcharbeiten*). [...] memory for Freud is work, what we might call a *travail*.'[118] Lena has had to pay for this opportunity to place herself in such a potentially enabling situation, but that is only one of the obstacles that she must surmount in order to continue this '*travail*'. Another, far more substantial one, is created by the fact that she and Outa do not speak the same language. While, initially, this compounds the difficulty involved in her attempt to remember, it gradually becomes apparent that the effort to communicate with someone who does not speak her language provides her with a different, perhaps even richer, experience, as she has to imagine the life of this other person if she wants to communicate at all.

Lena's reactions to the fact that Outa speaks Xhosa mirror the dynamics of the act of hospitality that is simultaneously being played out. At first she is hostile, dismissing Xhosa as 'kaffir *taal*' ('Kaffir language') (*BL*, p. 211) and even 'baboon language' (*BL*, p. 212). But, as she relaxes and becomes more accustomed to Outa, she devises a style of communication. Just as in *Translations* and *The Communication Cord*, this produces another scene of communication across linguistic borders that comes about because of the establishment of an emotional attachment. Echoing the scene in *Translations*, this one also starts with naming, as Lena's teaches Outa her name.

> My name is Lena.
> [*She pats herself on the chest. Nothing happens. She tries again, but this time she pats him.*]
> Outa ... You ... [*patting herself*] ... Lena ... me.
> OLD MAN: Lena.
> LENA: [*excited*]. *Ewe!* Lena!
> OLD MAN: Lena.
> LENA: [*softly*]. My god! (*BL*, pp. 214–15)

Outa, as can be seen, is not the old man's name. Lena never learns it and instead calls him 'Outa', a respectful mode of address to an elder that could be translated as 'old father'. Inscribed within this name is a possible source of Boesman's hostility towards Outa, as there is a sense that in Outa he is faced with their 'original' language, from which they have become estranged, the proper name of which has even been lost. In other words, this is another Oedipal-like drama of 'problematic filiation', to return again to Said, as Outa may well represent for Boesman a symbol of their abandoned status; a lost father perhaps.

This scene of communication across linguistic barriers also discloses the connection and difference between translation and repetition, as the quiddity of these repetitions is considerably different to those that have been heard thus far. Translation is a different form of repetition: it is a process of repeating differently, as can be seen in the manner in which Lena extends hospitality to Outa.

> LENA: [*offering the bottle*]. Water. *Water! Manzi!*
> [*She helps him get it to his lips. He drinks. In between mouthfuls he murmurs away in Xhosa. Lena picks up the odd phrase and echoes it ...* 'Bhomboloza Outa, Bhomoloza' ... 'Mlomo, ewe mlomo' ... '*Yes*, Outa, dala' ... *as if she understood him.*
> *The whole of the monologue follows this pattern: the old man murmuring intermittently* [...] *and Lena surrendering herself more and more to the illusion of conversation.*]
> LENA: *Safa ... safa ...*
> [*Pause.*]
> What's all that mean?
> [*He hands her back the bottle.*]
> If *Outa*'s saying ...
> [*She stops, takes another quick look to make sure Boesman is out of sight, then returns to the old man's side. She speaks secretively and with intensity.*]

> It's true! You're right. [*He is still murmuring.*] Wait now. Listen to mine. (*BL*, p. 215)[119]

According to Bassnett and Trivedi, the Sanskrit the word for translation, *anuvad*, conceives of the act as a form of expository repetition: it 'etymologically and primarily means "saying after or again, repeating by way of explanation, explanatory repetition or reiteration with corroboration or illustration, explanatory reference to anything already said"'.[120] Rather than figuring the act as a form of unmediated repetition, this temporal understanding of translation involves an attempt to come to terms with the difference and the otherness inscribed within the language that is to be translated. Through a commensurate negotiation with otherness – a process that is manifestly different to the self-destructive forms of repetition that have been performed in the play – Lena creates a situation that permits the possibility of articulating her identity and of coming to terms with herself: it is at this point that Lena begins to remember and repeats to Outa episodes from her past.

As Derrida points out in 'Signature Event Context', the act of repeating differently structures writing and the processes of identity in general: it is a matter of iterability, which derives from the Sanskrit word for 'other' – '*iter*, once again, comes from *itara, other* in Sanskrit'. In a manner that complements and glosses the Sanskrit word for translation, *anuvad*, iterability operates according to a 'logic which links repetition to alterity'.[121] This logic pivots on the figure of the 'third', and identity, in this understanding, is not a matter of an other, but of another other. Identity comes about not through claims of origin or source, but through the possibility of repetition, and repetition assumes a third (even if there is none immediately available), as Derrida illustrates through the logic that governs the mark of identity that is the signature, which is a singular event that must be repeatable in order for it to function as an identifier of the absent signer.[122] Inherent in this process, as in the case of the act of repeating differently that is translation, is not just the other, but also a third. In the earlier sections the third was the interpreter and the parasite; here the third also assumes the role of the witness: s/he who can repeat in the place/absence of the subject.

While Boesman treats Outa as a parasite, Lena charges him to become her witness. She insists, for instance, on displaying her bruises to him and relating her past through them. Her battered body has become a form of substitute memory, a damaged way of remembering herself: '[l]ook, *Outa*. I want you to look. [*Showing him the bruises on her arms and face.*] No, not that one. That's an old one. This one. And here. Just because I dropped the sack with the empties' (*BL*, p. 214). She hopes that, as a witness, Outa might, in the future, repeat or give an account of what he has seen; he might, in other words, cite Lena and so affirm her existence. The hope embodied in the immediate presence of this third empowers Lena, and she confronts Boesman about his brutality. Telling him that she does not need his blows anymore to know that she exists, she asks him if he hits her to confirm that she is still there. This goes to the heart of Boesman's confused need for Lena, and he is entirely lost when Lena instructs him to turn his aggression towards himself and to hit his own life. At this point, his brutal dependency is unveiled, and he asks Lena to tell him where his life is: '[s]how it to me! Where is it?' (*BL*, p. 236).

> I look, and I see you. I listen, I hear you.
> LENA: And when you hit ... ?
> BOESMAN: You. You cry.
> LENA: You hear that too?
> BOESMAN: Yes. (*BL*, p. 236)

Lena's newly-won capacity is evidenced in the fact that, rather than supinely accept this, she is '*almost inarticulate with outrage*' (*BL*, p. 236) and the source of this ability to be angry, to feel that she has a say in her own life, can be gauged by her immediate recourse to Outa as a witness. She tries to wake Outa up, and orders Boesman to

> [s]ay it in the *kaffertaal*. 'You hit me for nothing.' Say it!
> BOESMAN: No.
> LENA: Then let him see it.
> [*She crawls to Boesman in an attitude of abject beggary.*]
> Hit me. Please, Boesman. For a favour. My last one, 'strue's-god. Hit me now.
> [*To the old man.*] I've shown you the bruises. Now watch. (*BL*, p. 237)[123]

Despite his inability to speak the same language and his position in his society, Outa is not just a passive witness, were such a thing conceivable: he inspires action and indeed confessions. Aided by this third, Lena has begun the process of remembering; in Freud's terms, she has literally become more 'conversant' with the resistances in her life.[124]

In this light, the role of 'witness' is not immediately connected to empirical notions of truth and falsity, but involves a process of negotiation with and through the other. This is most evident when, out of supreme frustration, Boesman beats up the already-dead Outa. Outa's last act as a witness is an effective one in that it provides a crucial interlude of self-reflection for Lena, as she is given an image of what it looks like when Boesman beats her up. It permits her to witness herself: '[s]o that's how you do it. I know what it feels like. Now I know what it looks like' (*BL*, p. 243). On an empirical level, though, as Lena notes, the bruises Boesman has left on Outa's body will falsely convict him for the murder. He, like Manus at the end of *Translations*, must flee, because the political structures to promote the honest witnessing Lena craves are lacking. In apartheid South Africa, the concept of justice was warped out of all recognisable shape and, on an official level, the difference between truthful and false witnessing was inconsequential. Pointing out empirically what is and is not the case is an essential act of defiance of a regime, but it is also not completely sufficient, as a rotten political structure can defy these facts. 'Good faith', as Derrida argued, is 'the condition of testimony', and this is what is absent in degenerate political systems. Just as history was described in the last chapter, political action can likewise be said to emerge from an interpretative witnessing of facts, and so the ethics – the structure of fidelity – under which this is conducted is fundamental.[125] Translational fidelity is examined in the following two chapters, which unfold how this concept informs Field Day's vision of ethical political action in a Northern Irish context. What has become apparent in this chapter is that Field Day's vision of a possible Irish home is one that provides an environment in which the faithful witnessing of contradictions might prosper; these plays suggest that through the processes of imagining such a site the Irish might be brought home to their histories. Similarly (and utterly differently), in stark instructions to the actors who were to perform in a 1971 London production of *Boesman and Lena*, Fugard places the ethical emphasis not on the creation of a physical home or final resting place, but on the importance of an other to act as witness: 'Boesman and Lena have seen how they are going to die ... they are going to die among the dustbins. Will there be a witness?'[126]

3
Loyal Translations: The Spirit and the Letter

Introduction

> Every culture resists translation, even if it has an essential need for it. The very aim of translation – to open up in writing a certain relation with the Other, to fertilize what is one's Own through the mediation of what is Foreign – is diametrically opposed to the ethnocentric structure of every culture, that species of narcissism by which every society wants to be a pure and unadulterated Whole. [...] The essence of translation is to be an opening, a dialogue, a cross-breeding, a decentering. Translation is 'a putting in touch with' or it is *nothing*.
>
> — Antoine Berman[1]

Within the terms of the crude sectarian binary that has controlled life in Northern Ireland, Field Day's unequivocal opposition to the politics of unionism has meant that it has been automatically placed in the Catholic nationalist camp. Consequently, it has been regarded it with deep suspicion by large sections of the Northern Protestant community. However, Field Day has had a considerably more complex relationship with this tradition than this binary allows. While all rejected the politics of unionism, three of the original six directors of Field Day were from this community, as were the contributing playwrights Stewart Parker and Derek Mahon. Furthermore, Field Day's output indicates a sustained interest in Protestant modes of thinking and especially in the drama of Presbyterianism, as is most plainly evidenced in the 1985 publication of three pamphlets devoted to these topics: Terence Brown's *The Whole Protestant Community: The Making of a Historical Myth*; Marianne Elliott's *Watchmen in Sion: The Protestant Idea of Liberty*; and R. L. McCartney's *Liberty and Authority in Ireland*. While McCartney's piece is ultimately a critique of the Republic of Ireland, the other two pamphlets unveil aspects of the historical traditions of Northern Protestantism that help elucidate the link between Protestant thinking and unionist political theology; a nexus that has set the agenda for politics in Northern Ireland, and which is at the centre of the two plays in this chapter.

Field Day's attitude to Protestantism must also be related to the fact that perhaps the most compelling historical model for the company was the republicanism of the United Irishmen, which is encapsulated in Wolfe Tone's anti-sectarian credo: '[t]o unite the whole people of Ireland, to abolish the memory of past dissensions and to substitute the common name of Irishmen in place of the denominations of Protestant, Catholic and Dissenter'.[2] As the last play staged by the company – Parker's *Northern Star* – illustrates, the United Irishmen were largely led by Northern Presbyterians who saw how officially sponsored sectarianism was devastating Ireland. But not only was this broad, rational, anti-colonial political philosophy congenial to Field Day, so too was the Protestant-inflected insistence on the priority of the individual conscience that drove the leaders of the United Irishmen. Clashes between the public and the private have animated the plays considered thus far, and at the centre of both dramas here is an *agon* between the claims, on one side, of individual will and ethical responsibility and, on the other, the demands of the law, the state and the community. In Catholic orthodoxy, this conflict would be considered more a sign of potential apostasy than of spiritual formation, as Stephen Dedalus reminds us. Protestant theology, however, opens up a space in which the individual's relationship with the Bible can challenge the authority's (the church's, the state's) interpretation. Such conscientious challenges have been largely muted in Northern Ireland since the time of the United Irishmen, as the Protestant community, to a very large degree, has calcified around a seemingly unified identity in the face of the imagined and real threat from Catholic nationalism. Both these plays call attention to this inherently conflictual tradition lodged at the core of Northern Presbyterianism in the hope (which is particularly evident in *Pentecost*) of reanimating these ethical and political debates.

It can be inferred that Field Day also wished to translate the dynamic of this conflict between individual conscience and authority into discussions of Catholic identities and political forms. But this must remain a conjecture, as Field Day never engaged in a sustained critical examination of Catholicism and its political manifestations in Ireland. There are, for instance, no pamphlets devoted to specifically Catholic modes of thinking (the one that comes nearest is Kearney's *Myth and Motherland*), nor is there a Catholic equivalent in volumes I–III of *The Field Day Anthology* to Paulin's section on 'Northern Protestant Oratory and Writing 1791–1985'. While it was suggested in 1984 that the fourth set of pamphlets should deal with 'Catholicism and Liberty', this never came to fruition.[3] This serious gap in Field Day's programme, which McCormack drew attention to as early as 1986, creates the impression of a complacent equivalence of Irish and Catholic experience (and so, without naming it, the other dramas and pamphlets might be considered to be engaging with Catholic experience).[4] It also implies that Protestantism is examined because it is the tradition that needs to engage in self-reflection and embrace change. While the power structure of Northern Ireland meant that the unionist community did indeed need to move if any change was going to be instigated, Field Day's failure to hold Catholic moral and political thinking up to critical scrutiny marred its engagement with Northern Ireland, as it left the company open to the suspicion that it was acquiescing to the sectarian demand of loyalty to one's tribe. It also compromised any desire Field Day might have had to have a critical voice in the political life of the Republic: it is hard to imagine that the issue of gender would have escaped under the *Anthology*'s radar if the company had been engaging in a sustained analysis of Irish Catholicism, the central emphasis of which in the twentieth century was on the control of sexuality.[5] In fact, this omission can be read as a symptom of the company's prevailing political disinterest in the Republic and its overriding cultural affiliation to the North, where Catholicism, although notably conservative, could possibly be viewed as a focal point of subaltern resistance. Having said that, the company's output suggests that while the directors did not accept the validity of the unionist state, they were more inclined to search for enabling political resources in the residues of the Northern Presbyterian inflected republicanism of the United Irishmen than in the Catholic, cultural nationalist mode of republicanism that shaped the Republic.[6]

Both plays here were originally due to be staged as part of double-bills, and while *The Riot Act* did tour with *High Time, Pentecost* was performed alone. It was to be performed alongside Frank McGuinness's *Carthaginians*, a reflection on the trauma of Bloody Sunday. Field Day's failure to put on this play, as well as McGuinness's *Observe the Sons of Ulster Marching Towards the Somme* and Rudkin's *The Saxon Shore*, has been taken as symptomatic of the company's limited political outlook by the two dramatists and by certain critics.[7] Yet, in all three cases, the archives show that these decisions were primarily taken on theatrical grounds. Ironically, one of the few thematic reservations that were expressed by the directors about *Carthaginians* was that its use of the idea of Carthage was too strong an echo of the end of *Translations* and so, as Deane put it, 'it may be a little too heavily sub-Frielian and self-consciously a Field Day play'.[8] More importantly, several directors pointed to problems with the text;[9] when these were put to McGuinness, he 'would not hear of rewrites', and the production collapsed.[10] Three years earlier, in 1984, McGuinness had sent Field Day *Observe the Sons of Ulster Marching Towards the Somme*, an innovative, empathic and critical engagement with the traumatic events of the Somme and how they contributed to the creation of the central unionist myths of loyalty and sacrifice. While it interested the company, the directors had no doubts 'about its impracticality' in terms of the company's resources, which placed severe limits on the sort of reworking and developing of scripts that they felt was called for in this case.[11] In 1994, Field Day finally produced a McGuinness play – his version of Chekhov's *Uncle Vanya* – which is discussed in the Epilogue.

However, the decision in 1983 not to produce Rudkin's *The Saxon Shore* has attracted the most critical attention. In September 1981, Friel wrote a letter to the other Field Day directors suggesting possible future activities; amongst these he listed commissioning a play and 'finding some means of celebrating the best of the Northern Protestant tradition'.[12] Over the next year or so, these two ideas would come together when, after being commissioned to write a play, the English playwright David Rudkin sent the company one that engaged indirectly with Northern Protestantism. In fact, *The Saxon Shore* does not mention Northern Ireland at all: it is a form of supernatural historical play, revolving around two communities living either side of Hadrian's Wall in the twilight of the Roman Empire in Britain. Celts live in the North, and, on the other side, are the Saxons, who have been planted in Britain by the Romans to hold and govern the place for the empire.[13] The allegorical parallels between this situation and that of Northern Ireland are apparent, and are reinforced throughout the play, perhaps most tellingly in Agricola's sermon in Act I, scene 3, which, in its interweaving of God, law, cultivation, work and covenant, reiterates the fundamental tenets of Northern Protestant identity. Agricola also articulates the confused and fractured identity of these colonials, who feel that they have been set adrift by the empire, which, in a neat reversal of the logic of Northern Ireland, is Rome. Through this form of irony, *The Saxon Shore* works to make the 'siege mentality' understandable and, at the same time, to challenge its assumptions. When Field Day turned it down, Rudkin, many of whose ancestors hail from the Northern Irish Protestant community, 'felt rejected politically'.[14] In the absence of a detailed reason for the rejection from the Field Day directors, he believed that the decision was informed by a fundamental hostility to the voices of unionism. Indeed, the questions raised in the last chapter about the company's decision to turn to *Boesman and Lena* as a distant yet immediate political drama are given a different, wry inflection in Rudkin's suggestion that this choice encompassed yet another form of substitution: '[a]s a white South African, Fugard [...] "is the next best thing to an Ulster Protestant"'.[15] The implication being that it was only in this translated form that Field Day was prepared to encounter Northern Irish Protestantism.

A rather different picture emerges from Richtarik's study. She concludes that, in many ways, quite contrary motives precipitated this decision, as there would appear to have been some unease in Field Day's ranks about whether Rudkin's portrayal of the Saxon/Protestant community would cause offence to unionists. The main character, Athdark, defends his community by changing at night into a werewolf and going across the border to kill Celts, and Rea told Richtarik that 'Brian [Friel] would have been [...] sensitive about them [Northern Protestants] being portrayed as ravening wolves'.[16] Grant concurs with this reading of the episode, and interprets it as a perpetuation of the dismal politics of Northern Ireland that the company set out to question:

> the potential for giving offence to Protestant audiences seems to me a quite likely concern of the Catholic directors, who might well be made to appear sectarian by letting loose against the Protestant community one of its own – a renegade member – all too willing to bite the hand that fed him. [...] The failure of nerve by Field Day to some degree abets – however unconsciously – the sectarianism that Field Day abjures and which, by a subterfuge all too familiar in the culture at large, it hopes to manage by avoidance.[17]

Since the publication of Grant's book, Field Day's archive and Friel's papers have become available and, in their light, Grant's analysis, to some extent, appears to participate in the rationale it is critiquing. What emerges from the minutes of Field Day meetings is that the decision not to produce Rudkin's play was made almost entirely on theatrical grounds. Having read the first part of the play, the consensus at the meeting (9 January 1983) was that *The Saxon Shore* 'would be [...] potentially interesting and certainly difficult'. There was, though, 'considerable anxiety [...] about the whole venture, even though no-one doubted Rudkin's ability or professionalism'. While the historical and political context of the play was mentioned – '[s]ome members were worried about the historical accuracy of embattled Saxons getting news of the Fall of Rome; others worried about the Fenian Druids (FD for short)' – the focus was on the dramatic problems it posed: '[i]t was agreed that a director and/or actor/actress of incandescent genius was needed'. At the following meeting (5 June 1983) it was decided to reject the play, as having read the second part of the play, the directors were

> unenthusiastic. Tom Paulin said that it might make a good radio play. SH [Heaney] concurred and expressed doubts about its possible success on the stage. SD [Deane] said it lacked incident and was incantatory. BF [Friel] thought that the second part did not redeem it. The metaphor of the werewolf, the wound that was healed by the priestess, the cannibalism, was strained and did not have dramatic focus.[18]

Rea continued to advocate the project, stressing Rudkin's ability to re-write, and suggesting that Field Day should be prepared to work with 'difficult, non-popular scripts'. 'Nevertheless, the difficulties with this Rudkin script were felt to be too great and it was suggested that he should be told that FD could not take it in its present form.'[19] The company obviously felt that it was under no obligation to make its reasoning clear to anyone outside its meetings, and so speculation grew around this sensitive decision.

While the suspicion with which the unionist community regards outsiders who offer representations of it is common to all groups, what is more notable is that it also has a deep-seated aversion to self-representation. Launching the third set of Field Day pamphlets, the Rev. McCaughey noted that, '[a]lmost any protestant educated in Northern Ireland in the 30's and 40's as I was, knows that our history was [...] a well-kept secret'.[20] Field Day was determined to engage with the Northern Protestant community: even though its tours made losses by playing in unionist towns, the directors 'pursue[d] a deliberate policy of visiting a Protestant venue'.[21] But finding a suitable way of breaking this silence at the heart of this community was remarkably complex; fraught, as the Rudkin episode illustrates, with both predictable and unpredictable pitfalls, particularly as any attempt to think across community borders was endlessly grinded through an interpretative machine that functioned according to the ruthless sectarian logic of Northern Ireland.

Translation and the familiar other: *The Riot Act*

> There is a sense [...] in which it is natural and economic to return to 'Antigone' each time conflicts of a historically and psychologically analogous order [...] recur.
>
> — George Steiner[22]

> The Presbyterian is happiest when he is being a radical. The austere doctrines of Calvinism, the simplicity of his worship, the democratic government of his Church, the memory of the martyred Covenanters, and the Scottish refusal to yield or dissemble – all these incline him to that difficult and cantankerous disposition which is characteristic of a certain kind of political radicalism. His natural instinct is to distrust the outward forms of civil government unless they are consonant with his religious principles. On the other hand, his situation and his history in a predominantly Catholic Ireland have bred in him attitudes which seem the opposite to these, making him defensive, intolerant and uncritically loyal to traditions and institutions.
>
> — A. T. Q. Stewart[23]

Fugard has called *Antigone* 'the greatest political play of all time',[24] and a year after it was intimated by the Northern Irish Arts Council that his version of Sophocles's play, *The Island*, might be too political, Field Day returned to *Antigone* in 1984 with Tom Paulin's *The Riot Act*, which toured as the first part of a double-bill alongside Derek Mahon's *High Time*.[25] It was one of four different versions of *Antigone* to be produced in Ireland in 1984. But, as Murray comments, it is perhaps surprising that, in light of the 'Troubles', this play should have had to wait until then before being rendered in an Irish context.[26] Condensing the clash of justice and political order down to its essentials, *Antigone* provides, as Steiner notes, an economic, yet problematising understanding of political conflict that clearly appealed to Field Day. In fact, in 1983 the directors not only thought of replacing Rudkin's play with *The Island*, but another mooted possibility was Anouilh's *Antigone*.[27] Anouilh's 1942 production in Nazi-occupied Paris is perhaps the most famous interpretation of the play in the twentieth century: while the German officers attending it viewed it as a relatively depoliticised example of Western culture, the French members of the audience took a message of defiance from it. As this illustrates, *Antigone*'s power as political drama derives from the manner in which it is open to different interpretations, and there is more than one way of understanding how *The Riot Act* relates to Northern Ireland.

It is well known that this play is, on one level, 'an extension of Paulin's quarrel with Conor Cruise O'Brien', as Deane put it.[28] O'Brien (who the Field Day directors briefly considered asking to translate Anouilh's *Antigone*)[29] turned to Sophocles's tragedy in a 1968 lecture – at the inception of the 'Troubles' – to argue for the essential justice of Antigone's stance.[30] However, six years later, in *States of Ireland*, a classic revisionist text, he reprints a significantly modified version of this in order to repudiate its conclusion. The question for O'Brien is no longer about the state's legitimacy, rather it exclusively concerns the preservation of law and order: the maintenance, in short, of Creon.[31] In his essay 'The Making of a Loyalist: Conor Cruise O'Brien', Paulin critiques this revised interpretation as a gross misreading of the play and indicative of O'Brien's movement towards his position as defender of the Union.[32] The trajectory of Paulin's political allegiances moved in precisely the opposite direction from O'Brien's. While O'Brien would end up as a staunchly unionist opponent of the 1998 Belfast Agreement, his family roots are to be found in the foundations of the Irish Free State. Paulin moved from a unionist upbringing to a position where he believed that Northern Ireland would be better served in a united Ireland. In accounting for this move, Paulin placed a central emphasis on a break with O'Brien's views, which 'began to seem sloppy and unconvincing'.[33] Fundamental to Paulin's political thinking is an eighteenth-century conception of republicanism, which he never felt was possible in the cultural nationalist form that the Republic of Ireland assumed; when he realised that it could equally not be accommodated in a unionist Northern Ireland, he was drawn to the indeterminate conceptions of identity advocated by Field Day, which he believed provided the grounds for a non-sectarian republicanism. In his own words, his

> critical position is eclectic and is founded on an idea of identity which has yet no formal or institutional existence. It assumes the existence of a non-sectarian, republican state which comprises the whole island of Ireland. It also holds to the idea of sanctuary and to the concept of 'the fifth province'.[34]

This background suggests that *The Riot Act* should be read simply in terms of Catholic protesters for civil rights opposing the dictates of the repressive Protestant state. Paulin, though, looked to encompass a wider range of interpretations, explaining to Friel as he was composing it, that 'Antigone is obviously Bernadette [Devlin], but with a Paisley tone at times [...] More and more I'd want to give Creon, not just a De Valera touch, but something episcopal, papal – a cap with Masonic symbols on it.'[35] While the Creon of the final draft of *The Riot Act* does not display such a hybrid character, Paulin's musings suggest how the play could be recast in terms of a crisis between authority and rebellion in Irish nationalism: between those such as de Valera who took the reins of power in the post-colonial state and, on the Antigone side, the more radical elements who believed that Robert Emmet's epitaph could not be written until there was a united Ireland.[36] Death and martyrdom have long been central to nationalism, as has been the specific crux of the play – how to bury one's dead. Throughout the conflict, funerals were a form of public theatre – occasions that afforded the IRA and other militant groups the space to stage a performance of their presence with minimal interference from the forces of the Northern Irish state. Even closer to the heart of the play are the cases of the so-called 'disappeared': people the IRA abducted and killed in the early 1970s because they were deemed to be traitors (which indicates how the IRA could also assume the role of Creon).[37] The whereabouts of these bodies were not revealed, and Derrida's diagnosis of Antigone's efforts to come to terms with her father's lonely death in another land speaks to the distress experienced by the relatives of those who were killed: '[w]ithout a fixed [*arrêté*] place, without a determinable *topos*, mourning is not allowed'.[38] Notably, in terms of the action of *Antigone*, the IRA was impelled to provide information on the locations of these 'graves' as part of the peace process.

Nonetheless, the most satisfying approach to Paulin's reframing of *Antigone* is in terms of a Protestant/unionist allegory: in *The Riot Act*, Creon is resolutely cast in terms of the Northern Irish state, while Antigone is of the same family as Creon and recognised the legitimacy of the state up to the point of Polynices's death. Like nationalism, unionism can be thought to be divided between state-focused and more radical elements – in this case, between the traditional governing class of unionism that claims its right to rule in the sober, legalistic terms of a social contract, and the ecstatic, evangelical aspect of Northern Presbyterianism, which has been a powerful force in working-class unionism and whose adherents believe that they have kept faith with the principles of unionism.

Joe Cleary has also read this play in terms of unionism, understanding it as a 'psychodrama in which Antigone acts as a figure for the compelling emotional loyalties that continue to bind Paulin to his ancestral community, while Creon stands for everything that makes it impossible to accede to those forces of attraction'.[39] Paulin has indeed phrased his cultural inheritance in an entirely equivocal fashion, admitting that his 'way of thinking and arguing is simplistic and confrontational, seeing everything in terms of polar opposites. [...] It is puritan but with an emphasis on communication, but only to one's own. There is a monomaniacal aspect.'[40] Although unionism appals him on a political level, Paulin is ultimately fascinated by the Protestant imagination and, in particular, its tradition of radical, republican thinking. Almost all his works testify to this, in particular his book on William Hazlitt and *Minotaur* – a collection of essays that casts his relationship with O'Brien in an almost Oedipal light, as in it he sought to 'consider literary texts in relation to Protestant or puritan ideas', in a way that might 'complement Conor Cruise O'Brien's remarkable study of the Catholic imagination, *Maria Cross*'.[41] Cleary develops his reading by positing that Creon represents 'the self-righteous intransigence' of unionism that Paulin finds repellent and, as such, is 'a rather heavy-handed Paisleyite caricature'.[42] While Creon does exhibit these qualities, Antigone more fully expresses Paisley's attributes, as Paulin intimated in his note to Friel cited above. In his essay 'Paisley's Progress', Paulin describes Paisley as 'a natural overreacher who has no regard for the ideas of balance, decorum and limitation which are strong features of English culture'.[43] Paisley gives voice to the precisely deeply embedded aspects of the religious worldview of his culture: a 'Protestantism which is pushing deep into the territory of mystery and mythology [...] a celebration of chthonic forces and a rejection of secular and utilitarian values'.[44] In place of rational argument, this Presbyterianism privileges 'an apparently unstructured, intensely emotional experience. "Justification," he [Paisley] argues, "is heart work as opposed to head work." This assertion of emotion over intellect is both authoritarian and romantic.'[45] These unaccommodating, earthy and emotional characteristics that produce an unwavering sense of justification are Antigone's traits. She insists on burial and is, in turn, buried alive, and personifies what Akenson has described as the 'autochthonous system of local law that had nothing to do with the law as set out by the state' that prevailed through most of Protestant Ulster in the nineteenth century.[46]

Thinking of Paisley as Antigone is an inherently ironic gesture, as he has been unequivocal about the position of women: 'the Bible states clearly that the man is the head of the woman and Christ is the head of the man'.[47]

For the most part women have played ancillary roles in unionism, and have been expected to know their place.[48] However, Janice Holmes has pointed out that the theatre of nineteenth-century Presbyterian revivalism – the bedrock of Paisley's religious practice – afforded one of the few stages outside the domestic where Northern Protestant women were granted a certain freedom of expression.[49] While the permitted hysteria of these occasions – the speaking in tongues and so forth – conformed to gender expectations, this movement (like the other Irish tradition of the keening of the dead, which also speaks to the action of *The Riot Act/Antigone*) afforded moments of liberation in which women spoke in a non-sanctioned fashion for the community.[50]

Creon's political control, on the other hand, makes his identification with what was until 2003 the state-face of unionism – the Official/Ulster Unionist Party and/or British Secretaries of State for Northern Ireland – relatively straightforward. As Roche notes, in Creon's first speech, 'we hear a verbal medley of the two reigning powers in Northern Ireland, Westminster and Unionism'.[51] It is pure politician-speak, and lays heavy emphasis on the necessity of loyalty to the state, which is the rationale behind his decision that '"Polynices shall be deemed guilty of treason and refused all burial"' (*RA*, p. 17). Explicitly reading from a legal decree, Creon impresses on everyone that his loyalty is inseparable from the letter of the law: his present state has come about, '[f]or purely technical and legal reasons' (*RA*, p. 16). Loyalty to the state, which he commands, comes before any other consideration: 'if ever any man here should find himself faced with a choice between betraying his country and betraying his friend, then he must swiftly place that friend in the hands of the authorities' (*RA*, p. 16). His reliance on the letter of the law as the basis of the social contract through which he claims authority is representative of 'liberal unionism'. In the words of Aughey, this views itself as holding to an 'imperial notion of "*civus Britannicus sum*"', which, he believes, 'has transformed itself into the democratic ideal of different nations, different religions and different colours, all equal citizens under the one government'.[52] But, as Norman Porter has pointed out, the apparent hospitality within this ideal is built upon an inherent hostility to those outside the boundaries of allegiance to the Union:

> [n]ationalists [...] are to be persuaded to accept the integrity of Northern Ireland either by having their fundamental objections ignored or by being (intellectually or politically) bludgeoned into submission. [...] liberal unionists, with their gaze fixed unwaveringly on the goal of unsullied political Britishness, seem oddly oblivious to the political responsibilities of living in a deeply divided society.[53]

Whatever the inability of nationalists to recognise a Platonic notion of Britishness, this concept is placed under more threat when sections of the unionist community bring it into question by unveiling fissures in its façade of universal allegiance to the same conception of a just government. For this reason, an Antigone-like intervention must be quashed.

These ruptures are exposed through an act of translation. Indeed, *Antigone* is nothing other than a drama of translation: translation signifies moving a dead body from one state to another – the central act of this tragedy and the source of the conflict between Antigone and Creon. Moreover, *Antigone* holds a crucial position in the history of translation, as Hölderlin's translation, in particular, has come to represent an apogee: for Berman, it constitutes nothing less than '*the space of modern Western translation*'.[54] Standing as, in Steiner's words, 'a "tragedy of translation" as well as supreme tragedy "in translation"', it comes closest to articulating the ideal act of restitution that completes his hermeneutic cycle.[55] It achieves this by embodying the significance of the action of the play – its radical ruptures – in the contours of the translation; in it, 'meaning', as Benjamin perceived, 'plunges from abyss to abyss until it threatens to become lost in the bottomless depths of language'.[56] Composed as Hölderlin teetered on the brink of schizophrenia, his translations of Sophocles, as Berman put it, 'submit themselves to the "violent motion of the foreign language"', and in this way perform 'the relation of languages, relations of *coupling* and *differentiation*, of *confrontation* and *hybridization* – to be more precise: The relation of the mother tongue to other languages as it is enacted in translation, and *as it determines the relation of the mother tongue to itself*.'[57] Hölderlin's translations, in other words, incorporate the crisis of fidelity engendered by the act of translating Polynices. Through the translational encounter with the other, a rupture within the self is revealed, and this estranges the concept of a self (to which fidelity is due; which might offer fidelity) from itself. Because fidelity in translation cannot be defined, fidelity is what is at issue in translation: the ethics of translation, as Berman points out, 'consists of defining what "fidelity" is'.[58] It is in terms of this process of construction and fracturing of fidelity that the unionist allegory in *The Riot Act* must be understood.

'As the constant object of dialogue and *agon*', the corpse of Polynices is, as McCall has suggested, the 'real hero' of the tragedy: 'the decomposing source of the play's conflicts and catastrophes'.[59] The disintegrating remains of the traitor Polynices render explicit the impossibility of locating it as a verifiable source of authority, or as a whole, unified, pure, original to which a translation can faithfully return. It incarnates Benjamin's proposal that 'a translation issues from the original – not so much from its life as from its afterlife', and 'marks [its] stage of continued life'.[60] The original, in this understanding, is that which is always coming into being through its translational afterlife: in *The Riot Act*, Polynices's significance is being created by Antigone and Creon, and their two attitudes to translation articulate aspects of the unionist conception of fidelity.

In its own fashion, *The Riot Act* foregrounds the inherent, ghostly, anachrony of translation: while the names of the characters and the action of the play are those of Sophocles, the audience is also, at the same time, brought to somewhere in contemporary Northern Irish Protestantism, as is signalled by the initial stage directions which call for the set to include: '[*t*]*riangles, masonic symbols, neo-classical architrave*' (*RA*, p. 9).[61] Furthermore, this is also explicitly a drama of language – more precisely, of forms of Hiberno-English. While Creon's loyalty to the letter of the law is articulated in a formal and standard English, Antigone's and Ismene's speeches are littered with what might be termed 'ulsterisms': 'some screggy, smelly crowd'; 'they'll sleg us then' (*RA*, p. 12); 'the scrake of dawn' (*RA*, p. 20) and so forth. Although Robert Johnson has claimed that this is Paulin's own 'sort of Northern Irish dialect',[62] it nonetheless reflects Paulin's assertion in his Field Day pamphlet that 'dialect is notable for its intimacy and for the bonds which it creates among speakers [...] while [...] standard alternatives can be regarded as coldly public and extra-familiar'.[63] *The Riot Act* is, as Cleary points out, a domestic and family tragedy, and dialect stresses the family links between the two sisters.[64] Creon, though, is also part of that extended family, and so his locutionary choices create a 'clash' that speaks of, and emerges from, what Paulin calls, 'complex loyalties'.[65]

Dialect words and phrases are 'literally homeless' according to Paulin, and the drama of translation in *The Riot Act* is also phrased in terms of hospitality and hostility.[66] For Creon, Polynices's treason has legally debarred him from the hospitality afforded by a burial. Antigone, on the other hand, strives to provide accommodation for Polynices through an act of chthonic fidelity that she believes supersedes all manmade laws:

> Down in the dark earth
> there's no law says,
> 'Break with your own kin,
> go lick the state.'
> We're bound to the dead:
> we must be loyal to them. (*RA*, p. 27)

Later, as she is being taken away to her live burial, Antigone calls for a form of righteous translation by requesting her ghostly family to render her appropriate hospitality. Hers is a summons to familiar others (both propinquity and the meaning of 'familiar' that raises a spirit that one can call upon should be heard here), to her past as possible hosts of her future in the 'otherworld'.

> Laius, Oedipus, Jocasta,
> Etecoles and Polynices,
> I call on you, my kin—

may you bid me welcome
where I'm going now (*RA*, p. 47)

Being brought to death because of her attempts to render hospitality to the afterlife of Polynices, and having previously worked to ease her father's exile by accompanying him right up to his final 'home' in *Oedipus at Colonus*, Antigone is now, at the threshold of her death, demanding her rights under the spirit of the law of hospitality. While the spirit of hospitality would see the act in absolute terms – as the opening of one's house to the anonymous other without any notion of a reciprocal pact – Antigone calls attention to the letter of an economy that is inscribed within this duty. Derrida maintained that 'absolute hospitality' – which speaks to the spirit of the act – and 'hospitality by right' – which follows the letter – are 'as strangely heterogeneous as justice is heterogeneous to the law to which it is yet so close, from which in truth it is indissociable'.[67] An indissociable opposition characterises the relationship of the spirit and the letter of hospitality: just such a dynamic defines the relationship of Antigone and Creon and can be seen to structure unionist identities.[68]

Cleary, as has been noted, identified Creon with Paisley, and indeed Creon is first heralded in the play as 'the big man', which is Paisley's *nom de guerre* (*RA*, p. 15).[69] This was politically prescient, as since 2003 his Democratic Unionist Party has been the major unionist party, and from 2007 to 2008 Paisley was First Minister in the devolved Northern Irish assembly, which was created by the 1998 Belfast Agreement. Crucially, however, this Creon-like role came about because of Paisley's Antigone-like opposition to the Agreement, for which he was initially ostracised by the majority of unionists. But, as is the case with the troublemaker Antigone, unpopularity is 'one sign of authenticity' in this community, as Akenson argues, and Paisley has been sustained by a long history of disparagement from his own community.[70]

Moreover, like Antigone, Creon is prepared to break the law. When his son Haemon openly challenges his position, his insistence on the unambiguous letter of the law is revealed as an act of arbitrary interpretation; the law, in other words, is his interpretation.

I made that law—
the law is clear;
she broke it,
on her head be it.
She betrayed the state
And she betrayed me (*RA*, p. 37)

As Haemon tries to bring his father around through the force of public opinion, the idea that Creon's right to rule is rooted in a contract with the people is shown to be a sham.

HAEMON
The people of Thebes
would contradict you.
CREON
You'd have them tell me
what laws I'll make? (*RA*, p. 39)

This is yet another Field Day scene of 'problematic filiation', in which Haemon's 'mad anger / at his father' culminates in his suicide (*RA*, p. 57). On a political level, the erasure of the positions represented by Haemon and Antigone has been the way in which the indissociable opposition in Northern unionism has been resolved: since the founding of Northern Ireland, the Antigone disposition has most often deferred to, or been silenced by, the needs of the Creon class, because it was felt that it would destabilise the image of the Protestant community's right to rule. However, it is clear that Antigone's position is ultimately the basis upon which Creon's state is built, precisely because, as Tiresias warns him, her actions are sanctioned by the gods. Similarly, Paisley's political longevity was ultimately due to the fact that he was felt to be the voice of a deeper Protestant, unionist conscience. McClelland has contended that social contract theory, the basis of constitutional unionism, 'can be used to argue anything',[71] and so, as Bruce argues, 'loyalism depends on its religious base. [...] beyond evangelical Protestantism, no secure identity is available'.[72] From this perspective, the clash between the two dispositions in the play concerns the impossibility of constructing an enabling narrative that can accommodate unionism's religious and legal sources.

As McCall notes, in *Antigone*, 'the attempts to bury (translate, render) [Polynices] all end in failure, and at the end the corpse gets torn to pieces, rent rather than properly rendered to Hades'.[73] In terms of the Northern Irish allegory, neither attitude produces a faithful translation because both, for apparently contrary reasons, evince a distrust of language. In submissions to the broad-ranging Opsahl Inquiry, various Protestant churchmen gave blunt accounts of the exaggerated literalism that characterises Presbyterianism: 'Presbyterians [...] have little time for ambiguity. [...] We don't want to read between the lines'; 'the Protestant use of language, is very straight up and down. [...] There is a literalism, sometimes an over-literalism, about Protestantism and Protestant reaction to things political.' Alongside this is a distrust of what is seen as Catholic casuistry: 'Northern Protestants [...] believe that Catholics do not say what they mean, that they are profligate with words, past masters of the art of the fine point, the innuendo and the half-truth.'[74] Such an attitude to language expresses the indissociable opposition of contract theory and evangelicalism, which is characteristic of a covenantal culture, as Akenson illustrates: he describes Northern Ireland as 'a semisacralized entity [...] founded morally on the terms of the Ulster

Covenant, and pragmatically on the promise that it would be a Protestant state for a Protestant people, a Chosen People'.[75] The 1912 Ulster Covenant was a renewing translation of the Solemn League and Covenant between England and Scotland of 1643, the history of which Elliott records in her Field Day pamphlet. Both covenants display a legal passion: an insistence 'that the world may bear witness with our consciences of our loyalty'.[76] In effect, the literal words of the contract assume a biblical quality because they are subjected to a mode of religious reading that renders the word spiritual. Such a spiritual literalism denies temporality, the possibility of change across time, by reducing language to its 'base metaphor, the covenant', as Akenson claims.[77] Apparently rigid, such a view of language nonetheless allows for forms of politically opportune moves, as it is at the basis of what is known as unionism's 'conditional loyalty', which is succinctly detailed by George Graham in an Antigone-like fashion: '[m]y loyalty is to the British Throne being Protestant. I have no loyalty to any Westminster government.'[78] The British crown is granted a metaphysical status, while the British government is merely of the world and hence a possible source of betrayal.

Evangelicalism, though on one level apparently imbued in text and the past, offers a completely a-historical perspective. Just as Antigone summons her dead ancestors to host her in the future, the evangelical eye focuses on the world to come.[79] This gaze shapes a discourse that loosens itself from the bounds of context, and in which interpretation takes the form of divinely ordained truth or, in other terms, assumes the status of a performative. For Paulin, this view of language emerges from the ever-present drama of witnessing that takes place in the Presbyterian church.

> The puritan aesthetic that shapes vernacular discourse is shaped on an anti-aesthetic of 'truth' or immediate divine intervention. On the surface there is usually an appearance of directness and informality, but it is a carefully calculated illusion which can make the preacher resemble a method actor in his urgent authenticity and professions of personal sincerity. The church becomes a theatre where we witness an intense drama of consciousness and where a powerful sense of the absolute importance of the present moment – the urgent *now* of utterance – springs from the fusion of audience attention with the preacher's voice and gestures.[80]

Analogous to Antigone's practice of translation, this resolves the *agon* between the call of the letter and that of the spirit precisely within the performance of the divinely inspired interpreter. The very staging of this role – the act itself – is calculated to dismiss the claims of the historical sense of words: the literal words become no more than necessary, temporary vehicles of the spirit. Paulin locates the inspiration for this performance in a written text that expresses a profound distrust of written texts. He holds that it is

> a version of the argument in the third chapter of the second epistle to the Corinthians where Paul rejects actual inscribed letters through an imaginative assertion, which abolishes the distinctions between readers of written texts [...] and the writers of such texts. He asserts that 'ye' – the Corinthians he is addressing – are 'our epistle written in our hearts', and then draws a distinction between letters written in ink or 'in tables of stone' and those letters written 'with the Spirit of the living God ... in the fleshy tables of the heart.' This is an antinomian, born-again attitude which dedicates itself to 'the new testament; not of the letter, but of the spirit: for the letter killeth, but the spirit giveth life.' Paul contrasts this new spiritual text with the 'ministration of death, written and engraven in stones' (i.e. the Mosaic Law). This inscribed law is a veil that remains 'untaken away in the reading of the old testament'. Paul abolishes it and offers instead a new libertarian speech, 'great plainness of speech' as he terms it.[81]

Rather than being based on the letter, Protestant literalism expresses a discourse that has gone beyond the letter, and is a direct articulation of the spirit. It is an ideal that abolishes any conception of the past or of the word as other, as it presupposes that the word has already been lodged in the elect and requires no further interpretation.

If any family warns against the temptation of assuming correct interpretation it is Antigone's, as nothing other than a literal understanding of origins brought about its tragedy. Antigone's father, Oedipus, went to the Oracle because he wanted to determine his origins. In doing so, he set in train the original story of 'problematic filiation' that hollows out the notion of origins, precisely because he assumed the role of a correct interpreter both with the Oracle and, later, with the riddle of the Sphinx. The allure of this position, and of the perfectly coterminous notion of a definable origin, is reproduced in the most influential rendition of this story: Freud's version of *Oedipus Rex* in *The Interpretation of Dreams*.[82] Freud retells the story in order to illustrate that it was not the will of the gods that produced the tragedy, but sexual repression. That is, to employ Freud's own terms, the manifest content of his reading, but the latent content is precisely a crisis of interpretation, as his paraphrasing underlines how the tragic actions spring directly from moments of apparent correct interpretation or, more precisely, because of the possibility of multiple interpretations of origins and lineage. Instead of interpretation leading to origins, power, authority and light – the position that Freud wants to claim for himself – Oedipus blinds himself when he finally moves beyond misunderstanding. Uncannily, the source of this crisis of interpretation – the belief in an unambiguous, definable origin – almost at once manifests itself in terms of the problem of translation in Freud's text, as he begins the next chapter by invoking the idea of translation in order to describe the operation of what he terms the 'dream-work'.

> The dream-thoughts and the dream-content are presented to us like two versions of the same subject-matter in two different languages. Or, more properly, the dream-content seems like a transcript of the dream-thoughts into another mode of expression, whose character and syntactic laws it is our business to discover by comparing the original and the translation.[83]

The problem is immediately obvious: how can the original be compared with the translation, when it is precisely the original that must be arrived at by the analyst or interpreter? What this structure posits, despite Freud's assumed intentions, is that the original can be nothing other than a possible interpretation of the translation; it is a case, as Adam Phillips put it, 'of translation without there being anything like an original text there to translate'.[84]

In like fashion, both Antigone and Creon are determined to translate Polynices into an inviolable original in ways that would allow them to lay claim on him as a body of meaning within the terms of the narratives of their identities as sister and as law-maker.[85] In doing so, they reveal the problematic fidelity at the heart of a covenantal apprehension of the world, in which language is subjected to forms of spiritual literalism, or literal spiritualism, that hermetically entomb the 'original' in ways that deny it the possibility of an afterlife, of re-translation. Though apparently rooted in the past, these interwoven attitudes abolish any sense of temporal history, and so ignore their own otherness. In his Field Day pamphlet, Terence Brown noted the reluctance of unionism to engage faithfully with its complicated, entangled historical origins and called for a recognition on the part of that culture of the diverse nature of its roots. He suggests that a more faithful story of this past would include an account of the experience of homelessness; in recounting this, a more open concept of hospitality that might accommodate the decomposing original at the heart of this play could be fostered.

> A people who have known resistance as well as dissent, rebellion, dispute, religious enthusiasm in the midst of rural and urban deprivation, have an interesting story to tell themselves – one of essential homelessness, dependency, anxiety, obdurate fanaticizing, sacrifices in the name of liberty, villainous political opportunism, moments of idealistic aspiration. And in the telling of it they may come to realize at last where they are most at home and with whom they share that home.[86]

Reimagining the framework: *Pentecost*

> The crisis consists precisely in the fact that the old is dying and the new cannot be born; in this interregnum a great variety of morbid symptoms appear.
>
> — Antonio Gramsci[87]

Kirkland argues that Gramsci's concept of the interregnum provides a useful approach to the historical narrative of Northern Ireland in light of the 'Troubles'. Figuring the interregnum as a 'struggle between two identities' in which links with what has gone before are fractured but not entirely severed, Kirkland believes it produces a state of 'instability and flux' that unveils the workings of subjugation, and so allows for 'a possibility close to Benjamin's "Messianic cessation of happening, or put differently, a revolutionary chance in the fight for the oppressed past"'.[88] This describes the ambition of Parker's *Pentecost*: set in the midst of a crisis that lays bare the oppressive nature of the Northern Irish state, the play seeks to articulate a way of living through its 'morbid symptoms' (which are manifested in forms of living death, impotence and dead babies), by unearthing in these very symptoms the seeds of a possible future. Like Benjamin, Parker calls upon a messianic conception of history as a way of rethinking the oppressed past and present; specifically, he summons up and rethinks the resources latent within evangelical Presbyterianism. While *The Riot Act* was read as articulating the failure of the a-historical vision that drives this tradition to provide for a liberating translational afterlife, *Pentecost* seeks to revalue its immediate now-ness as a process of continual rebirth that can frame an inclusive ethical approach to living in the present.

Although Parker was not a Field Day director, he shared the company's view that the theatre could provide the site of an alternative politics. For Parker, the role of the dramatist in the 'Troubles' was to provide a vision of 'a working model of wholeness' for the society; a wholeness that does not occlude the contradictions of the society, but encompasses and thrives on them.

> Drama can [...] contain the conflicts and contradictions, the cruelty and the killings, the implacable ghosts, the unending rancour, pettiness and meanness of spirit, the poverty of the imagination and evasion of truth which unites our two communities in their compact of mutual impotence and sterility – all in a single image.[89]

Parker's conception of the cultural role of drama is particularly distinguished by an emphasis on 'play', which, as he put it, 'is how we experiment, imagine, invent, and move forward'.[90] Play reaches beyond the bounds of historical events to forms of resolution and becomes, in *Pentecost*, a prism through which the inherent theatrical immediacy and the crucial illogic of evangelical Presbyterianism is reviewed.

Growing up in a Belfast unionist family, in the shadow of the Harland and Wolff shipyard, Parker experienced the disjunction between the promise of future possibilities inscribed in industrial modernisation and the chthonic ghosts brought forth by the evangelical imagination.[91] This background placed him in an angular relationship with what he saw as the 'nationalist rationale' of Irish theatrical tradition, and left him with no choice but to experiment: 'to invent the theatre all over again'.[92] He did this by remaining faithful to the contradictions of East Belfast: his dramas bring the enclosed world of always-besieged Presbyterianism into play with the world that was being opened up by the ships and planes that were being produced in Belfast. These plays are both local and far-reaching; in Andrews's terms, they have 'a general appeal and relevance radiating outwards, spoke-like, from these intimate origins'.[93] However, the movement does not operate in one direction only: in *Pentecost* there is a shuttling between politics and various forms of myth, with the result that the myths are rendered local and the politics of Northern Ireland takes on a global shading.

Pentecost can be read as a distorted sequel to *The Riot Act*. As Steiner notes, Antigone's final invocation of the gods releases them into the city and, while the potential of such a visitation is necessary, its manifestation is eminently hostile to its mortal inhabitants:

> in *Antigone*, Sophocles confronts the limitations of the city of man, of the state as the genius of man [...], with the homecoming of the god, a homecoming compulsively inherent in religious ritual and in the extremity of human supplication. Such epiphany is the ecstatic expectation, the desideratum of the human spirit when this spirit aspires to its own fulfilment, when it strives to return to its own pre-civic sources of being [...]. But how, except in a destructive fire-storm, is Dionysus to inhabit Thebes? Can there be any coexistence, other than suicidal, between transcendence and *civitas*?[94]

These are the questions that Parker looks to answer in *Pentecost*, which is set during the Ulster Workers' Council strike of May 1974: a Protestant fundamentalist sabotaging of its own, Protestant, civic structures. Not that there was much in the way of functional civic life in Northern Ireland in 1974: by this stage in the 'Troubles' over 1,000 people had been killed and 'somewhere between 30,000 and 60,000 people in the Greater Belfast area were driven to leaving their homes, at that date possibly the largest enforced movement of population in Europe since the Second World War'.[95] The strike was a deliberate, and successful, attempt on behalf of the unionist community to destroy a possible solution to the crisis. It was called to oppose the British government's creation of a power-sharing Executive for the Northern Ireland Assembly and the Sunningdale Agreement of December 1973, which made provision for a limited cross-border Council of Ireland. Choosing the known comforts of the siege mentality over moves towards forms of civic partnership, unionists fell back into their citadel where the only relationships permitted were with each other and with the gods. Within a week, the strike had brought Northern Ireland to a virtual standstill, with the strikers becoming 'the *de facto* government in many areas'.[96] In stark contrast to their reactions to nationalist protests, the police and army remained inactive while barricades were erected, allowing the unionist mob to take over the increasingly rubbish-filled streets. Although they were laying the siege, the strikers insisted that they were under siege and, as ever, this construction brought the different unionist classes together; as Lee notes, 'there was little political hostility to the strikers among middle-class Protestants. When the crunch came, they all belonged to the same community.'[97] Having no opposition from civilian authorities, the strike achieved its goal, and on 28 May the Executive collapsed, with the result that, in Bruce's succinct terms, '[p]ower-sharing was at an end'.[98] The principles of power-sharing and limited cross-border cooperation were finally acceded to by this community in the 1998 Belfast Agreement, which Seamus Mallon called 'Sunningdale for slow learners'.[99] The twenty-four years of this education cost another 2,000 deaths.

These mythological and political frames are interwoven into the most explicit frame in the drama: the house within which it is set, which is the last occupied house in a Belfast terrace, located in a no-man's land between the warring communities. A drama of hospitality is played out in this site that is enduring a process of literal deconstruction, both externally and internally. This unstable, malfunctioning, anachronistic, fragmenting, unwelcoming site, located in a violently contested interstitial zone, recalls the situation of Northern Ireland itself, which was, in Dawe's depiction of Belfast during the strike, 'a ghost town; [...] a bizarre disconnected statelet'.[100] For Brown, the set 'suggests the frozen stasis of a society',[101] but it also functions, in Roche's words, 'as a kind of stay or refuge, an asylum or temporary holding-ground, one step removed from the (war) zone of historical circumstance'.[102] In effect, Parker wants to initiate a process that shifts the significance of the house from the first to the second of these descriptions; to imagine it as some form of unlikely 'fifth province'. For this to happen, the dilapidated framework of the house must become the place where reading takes place; rather than signifying a definitive halting of meaning, the frame becomes a site of interaction between that which is within and that which is outside its bounds. It becomes, in Derrida's terms, the *parergon*: that which 'is against, beside and above and beyond the *ergon*, the work accomplished, the accomplishment of the work. But it is not incidental; it is connected to and cooperates in its operation from the outside.'[103] It is a boundary that, in its act of defining, confuses what might be thought of as antinomies, such as that which obtains between spirit/letter and play/history, where '/' designates a site of complex, conflictual, translational interaction. In the play, the identities of the characters within the house are determined through the ways they relate to the history and politics without via the house: as Andrews points out, each of the four characters is 'a victim, a reflection, a symptom. The sickness of a society is enacted in this enclosed form.'[104] This describes a *mise en scène* that takes the form of a *mise en abîme*, with the characters reflecting the pressures of the context. Having set this up, Parker seeks to reverse this direction and to review the political/historical from the perspective of the personal/body.

Two of the characters are Catholics: Lenny, a shiftless musician who has inherited the house on the death of its tenant, and his estranged wife, Marian, who is in the process of purchasing it from him. The other two are Protestant: Ruth, a friend of Marian's who has escaped from her violent policeman husband, and Peter, who has returned from Birmingham to his home town for a visit. As such, the house has different significances: for Lenny, it is an inheritance and a source of much-needed cash; Marian sees it as a relic of the past; Ruth flees to it as a shelter from her husband; and, for Peter, it is a cheap lodging. These calls for diverse forms of hospitality raise the insistent quandary of who can play the role of the host and offer hospitality, and this dilemma is played out in terms of the spirit and letter of the law. Lenny holds the keys under the letter of the law – 'I've inherited the house, by law it has come to me' (*P*, p. 174) – however, he is in the process of selling it to Marian. She is a host in need of hospitality, who is attacked by the women on the barricades as a 'Fenian hoor of a squatter' (*P*, p. 222). As putative host, Marian feels that she has been forced to offer hospitality to the others, whom she calls 'refugees' (*P*, p. 208). She lets them stay under express terms: Ruth is informed that if she wants to stay, she must abide by 'ground rules' (*P*, p. 191), while Peter is asked to provide a report on the structural stability of the house. Peter's status as visitor-native renders him suitable for the role as inspector of the framework: he has returned from Birmingham and casts the most critical eye on the state of his home, a place he repeatedly refers to as 'Lilliput'.[105] He is unsure why he has returned, dismissively claiming at one point that he was moved by 'the spirit' (*P*, p. 205).[106] However, Peter's is not the most interesting spirit to test the framework of this house.

Marian has begun to archive and catalogue the life of Lily Matthews, the recently deceased tenant, with a view to turning the house over to the National Trust as a testimony to a 'community of experience' (*P*, p. 208). She believes that she understands and nurtures the spirit of the house, telling Ruth that Lily is 'our hostess [...] It's her house' (*P*, p. 186). Indeed, the most apparent onstage claim on the house is registered by Lily's ghost. Ghosts populate what Parker called his 'Theatre of the Ludicrous': he held that '[a]ll plays are ghost plays' and, in a manner that recalls Frank Hardy in Friel's *Faith Healer*, claimed that the dramatist is a medium who conducts the antagonistic spectral voices that 'issue forth and take their shape from a conflict within the writer himself'.[107] Through ghosts, the past intrudes on to the stage in ways that accommodate both logical and illogical modes of understanding and defy teleological conceptions of history. Ghosts fail to respect boundaries, and in traversing these they carry meaning from one context to another and bring it into conflict with other interpretations. Their motion puts under question the role of the frame itself, or, more accurately, it highlights its central function in determining meaning. *Pentecost* is the last in Parker's '[t]riptych' of history plays that explore aspects of Irish Protestant traditions.[108] As noted, *Northern Star*, the first of these plays, was the last Field Day production; the other drama, *Heavenly Bodies*, riffs on the life of the nineteenth-century dramatist Dion Boucicault. In all three plays, history disturbs 'prophesy[ing] and bicker[ing]' ancestral ghosts into being; these link the dramas, not in a rational fashion, but through a shared fragmentary coding: '[g]hosts are uncompleted souls: witness the Phantom Bride in *Northern Star*, handing on to the Phantom Fiddler in *Heavenly Bodies*, and he in turn to *Pentecost*'s Lily Matthews'.[109] Through Lily, Parker offers what he hopes is an enabling refiguration of the spectre of Irish Protestantism; a rethinking through which 'the cycle of retribution' that these ghosts embody is 'laid to rest, in the only way [Parker] can foresee as having any possible meaning'.[110]

Lily first appears as a 74-year-old woman, and her opening line to Marian is: 'I don't want you in my house' (*P*, p. 181). She is the ghostly embodiment of hard-line Northern Irish Protestantism and is horrified at the idea of a Catholic in her house. Her history of the house frames it as an expression of Northern Protestant experience. Its foundations are in the trenches of the First World War, as her husband Alfie set it up on his return from Passchendaele as a *revenant* – '[b]ack from the dead' (*P*, p. 182). Unionists viewed the sacrifice of the First World War as a testament to their loyalty to the British crown and, under the terms of their covenant with Britain, it was to secure them their own home after the war. However, while this was granted, the British continued to assume the role of absentee landlords, and the 1974 strike was one of the many fraught moments in their relationship with their tenants. For the British, the strike defied the terms of the tenancy, as is recalled by the British Prime Minister, Harold Wilson, in a speech that is relayed as a direct political/historical intervention into the house and the play through the ghostly medium of radio. Casting the British as indulgent landlords, Wilson depicts the strikers as parasites whose existence is granted by the British democratic systems they are attacking. By characterising them as 'spongers' (*P*, p. 214), this speech flew in the face of unionists' perception of their pact with Britain, and has generally been regarded as an unhelpful contribution to the crisis – Lee has described it as a 'futile and self-indulgent sneer' on Wilson's part.[111]

Similar strains marked the conflicting understandings of the 1643 Solemn League and Covenant, as Elliott's Field Day pamphlet illustrates. The Scottish Presbyterians, who brought the Covenant to Northern Ireland in 1644, maintained a 'strict adherence to the letter of the law', and 'assumed the banner of pure protestantism against the episcopalians'.[112] On the other hand, the English regarded the Covenant in a more temporal fashion and 'its terms were not honoured'.[113] The execution of Charles I provoked an early crisis, as the literal expression of this accord was embodied in visible loyalty to the 'person and authority' of the king.[114] Accusing the English of breaking the Covenant, the Irish Presbyterians appealed above their heads to the eternal verities in a document entitled '*A necessary Representation of the present evils and eminent dangers to Religion, Laws and Liberties* (15 Feb. 1649)'. Milton responded like Wilson, describing the Irish Presbyterians as 'upstarts' and 'a generation of highland thieves' who were 'neighbourly admitted [...] by the courtesy of England to hold possessions in our Province', but who 'have [...] proved ungrateful and treacherous guests'.[115] What this history illustrates is that while unionists see themselves as preservers of the fundamental principles of the British state, the English disregard such 'originary' claims and instead frame their relationship in the more mundane terms of a form of a tenancy agreement. They have little affection for Northern Ireland, and often find unionists awkward and costly lessees, ironically because of the way in which they insist on following so closely the letter of the contract: unionists, as was noted in the reading of *The Riot Act*, have construed their legal covenants as expressions of eternal spiritual truths. But they have also persevered with this frequently unhappy relationship out of a fear that the spectre of Catholicism might, like Lily, return and reclaim its lands; as Elliot noted: 'protestant survival depended on catholic suppression. It did so because protestant power was novel and based on a confiscation of property.'[116]

To protect themselves from the ever-present threat of this spectre, unionists styled Northern Ireland as a relic of seventeenth-century belligerent Protestantism. In a variation on the idea of moving dead bodies, relics describe a form of translation which disavows difference, as the spirit or the essence of the saint is present or preserved in even the smallest fragment of her/his body. Despite *sparagmos*, the fragment represents the whole: it is, as Norfolk puts it, '[s]ynecdoche with a vengeance'.[117] Marian's first intention is to freeze the house as a relic, believing that this would embody a hospitality towards the spirit of the past. However, she comes to reject this idea when she discovers that this spirit, in the form of Lily, is not an unchanging, stable entity. Initially anything but a welcoming (g)host, with each return she makes to the stage – each time as a younger version of herself – more of her life is uncovered, and she becomes a more sympathetic character. Her afterlife is, in Benjamin's terms, the setting in which her history is accreted: it is revealed that she had a private life that no one knew anything about while she lived; a life that ran contrary to her achieved appearance.

The need for privacy and the need for it to be disrupted are debated in various ways throughout the drama. Privacy is a refuge, a condition necessary for communicating with oneself and with one's intimate others. It runs contrary to hospitality, yet is also a mark of it, as it could be said that the more privacy the host gives the guest, the more hospitality s/he is also offering. The threats to privacy in the play come not only from the streets without or from the friends pitching up at the house, but also from the processes of historical research. Marian's archival work is prompted by her desire to communicate with Lily, to understand her beyond her role as a representative of hard-line unionism, and as Marian notes, it was also called for: by keeping a diary, Lily was looking for it to be read, and for her past to be re-evaluated. Lily, though, is appalled that Marian should have gained access to her diary, which, like the house, is '*old and disintegrating*' (*P*, p. 229). As Marian negotiates the ethical crux of how to read and how to accommodate faithfully the spirit of the past, the house becomes a site of reading and somewhat more and other than an expression of a unionist version of Northern Ireland. The turning point in Lily's history is an instance of hospitality: an English airman, Alan Ferris, was taken in as a lodger during the Depression. Unbeknown to anyone, Lily had an affair with Ferris and a child, which she delivered herself and left at a Baptist church door. With this affair, the 'letter' of the contract of hospitality – explicit in Alan Ferris's role as a lodger – is subsumed by the spirit. Indeed, Lily expressly conceives of him in biblical terms, as 'the fair-skinned archangel' who turned out to be 'a dark angel. Angel of death. Agent of Satan. He swept me up, high up, took me up into the sky ... and then he dropped me. [...] Left me falling' (*P*, p. 232). Having agreed to let him come and stay in her house as a paying lodger, Lily then welcomed him into herself as a form of spirit. She portrays this as an excessive act: like Dionysus's entry into the city, it is an unbalancing, impossible surplus that cannot be contained. Yet, it did result in a child, in a possible future, and the ambiguous nature of this story sets the scene for the act of recuperation that is attempted at the end of the play.

First, though, to move towards this, a new way of understanding this ghost is required: a new reader; a new witness. Up to this point, Lily has relied on the spirit to fulfil this role, as this would preserve her privacy, but rather than serving as a balanced witness, the purely spiritual is unbearable:

> nobody ever knew only the Lord our God knew and His eye was on me all right, burning into the very soul of me, He alone was witness to the torment that I've suffered every living hour in this house where the very walls and doors cry out against me [...] nobody to help me, only me here in this house, gnawing and tearing away at my own heart and lights, [...] until I was all consumed by my own wickedness, on the inside, nothing

> left but the shell of me, for appearance sake [...]. At least I never let myself down [...]. Never surrendered. Not one inch. I went to my grave a respectable woman, Mrs Alfred George Matthews, I never betrayed him. That was the way I atoned [...]. I done him proud. He never knew any reason to be ashamed of me, or doubt my loyalty. (*P*, p. 231)

This speech is littered with unionist nostrums – 'no surrender', 'not an inch' and 'loyalty' – which the story leaches of spirit: while she went to her grave a respectable relict, Lily was a rhetorical husk. An understanding of the role of the spirit such as this deprives the living person of spirit and, as Lily shows, this deficit engenders a spectral afterlife. The problem is that the living do not understand these ghosts as symptoms of a process of devitalisation, but rather as evidence of the inviolable spirit animating unionist rhetoric. It is in this manner that what Parker called the spectral 'cycle of retribution' circulates through his three history plays, and it is 'laid to rest' through Marian's simple response to Lily's story, which betokens a necessary step towards communication and possible reconciliation with herself and with others: '**Marian** *takes* **Lily**'*s hand and holds it against her own heart.* **Marian:** Forgive me, Lily' (*P*, p. 232). Reading is initially invasive and acquisitive, as the processes of translation show; for it to be an ethical act, it must incorporate a restitutive form of forgiveness. Both the asking for forgiveness and the granting of it – to herself and to others – have been missing in Lily's spirituality. As Lily does not reappear after this and is presumably at rest, the play reaches something of a conclusion with this act. However, the drama moves on, venturing deeper into an exploration of how this ethical process might uncover some form of hope within the conception of spiritualism that animates the mob on the barricades.

Parker has written that he saw 'no point in writing a "plea" for unity between prods and taigs. What use has piety been?' Instead, his ambition was to find a way of 'actually embodying that unity, practicing that inclusiveness, in an artistic image; creating it as an act of the imagination, postulating it before an audience'.[118] The final scene of *Pentecost* is perhaps the most vivid example of this praxis, but critical reaction to it has been quite negative. Andrews, for instance, reads it as a good-natured attempt to give a climactic, ringing affirmation of the power of reconciliation that is ultimately 'somewhat clumsily forced and impatient. The tensions concentrate toward a final, overwhelming stasis, a consummate declaration – rather than a consummate act – beyond which the audience must not think.'[119] It is set in the early hours of Pentecost Sunday; the Northern Ireland Executive has collapsed and the strike has come to an end. Light is now returning as the workers return to the electricity and gas installations. The Pentecostal setting is underscored in the stage directions: the '*blinding searchlight of* [a] *hovering helicopter*' and the '*glimmering of distant bonfires*'

(*P*, p. 232), which have been lit in celebration, should be seen. There is an aura of incongruous celebration – the celebration of destruction – that is captured by Peter in an image that sets out the dominant theme of this scene: 'you'd think they'd given birth, actually created something for once, instead of battering it to death, yet again, the only kind of victory they ever credit' (*P*, p. 234).

Harris has noted that 'Parker's writing is filled with sacrificial images', and *Pentecost* strives to offer a radical rethinking of the way in which this act has been interpreted in Northern Ireland.[120] Both *The Riot Act* and *Pentecost* unambiguously link religion and the sacred with violence; this nexus (with intertwined etymological roots) has been examined by Girard in *Violence and the Sacred*.[121] For Girard, religion and law function through the ritualised redirection of violence, and nowhere can this be better evidenced than in the central rite of sacrifice, the purpose of which is 'to protect the entire community from *its own* violence [...] to restore harmony to the community, to reinforce the social fabric'.[122] To control vengeance, the victim is taken from the margins of society – 'prisoners of war, slaves, small children, unmarried adolescents, and the handicapped'[123] – emphasising once again the central role of the frame: society's future is seen to depend upon those who exist in its *parergonal* space. While each year the effigy of Lundy (the 'traitor' of the siege of Derry) is ritually cast into the flames by unionists, maintaining Catholics as scapegoats has been key to keeping the unionist community together; as Girard noted: '[o]rder, peace, and fecundity depend on cultural distinctions; it is not these distinctions but the loss of them that gives birth to fierce rivalries'.[124] In this light, the creation of the Northern Ireland Executive and the Sunningdale Agreement precipitated a form of, in Girard's words, 'sacrificial crisis' within unionism, as the gap between it and its chosen victims had grown too small.[125]

The enaction of this 'sacrificial crisis' is most vibrantly portrayed in the 'Dionysiac outbreak' in Euripides's *The Bacchae*.[126] Dionysus, the god loosed upon the city by Antigone, embodies, in many ways, the two dispositions of unionism: in Girard's terms he is 'a dual figure. On the one hand, there is the Dionysus of the Maenads, the defender of divine and human laws, the jealous guardian of legality; on the other, there is the subversive *agent-provocateur* of the tragedy.'[127] Parker presents the UWC strike in terms of the ritualised mob violence of the bacchantes: Marian is specifically set upon by a group of women who attack her with their bare hands – '[i]t was all women—shrieking and squealing and scrabbing at me' (*P*, p. 223). Euripides's tragedy is resolved through a sacrifice: the women rip apart Pentheus, 'an outsider' who bears notable similarities with Dionysus.[128] This ritual *sparagmos*, which 'reenacts with scrupulous exactitude the mob violence', brings the violence to an end.[129] For this reason, Dionysus is termed by Girard '*the god of decisive mob action*' as, '[h]e claims legitimacy not from his ability to disturb the peace but from his ability to restore the peace he himself disturbed – thereby justifying, a posteriori, having disturbed it in the first place'.[130] The 1974 strike was a moment in which unionism revealed that its legitimacy to rule rested on a commensurate claim to such a Janus-faced authority.

Surprisingly, *Pentecost* does not simply jettison the sacrificial model that it has used to frame the strike, but in the final scene works to translate it into another, internal form. The four characters present narratives that, in different ways, involve the necessity of radically altering one's perceptions in order to create the possibility of reconciliation. The men's tales of failed escapism are rejected by Marian and Ruth. They insist, like Antigone, on a deeper understanding of religion; as noted, women have played a *parergonal* role in Presbyterianism, by performing, amongst other things, the Pentecostal glossolalia. Marian's story is an account of her interactions with Lily that, in its telling, indicates how the crucial ethical move she has made – the unbidden asking for forgiveness – has allowed her to work towards an act of achieved reimagination and a faithful translation. Having faced up to the task of coming to terms with Lily through reading the letter of her diary and engaging with her spirit, she can offer a version of her experience through a form of translational substitution that is anything but metonymy. Instead, what she brings into play is a form of self-sacrifice on the part of the translator: Marian's voice takes the place of Lily's. She imagines Lily alone in the house, experiencing the bombing during the Second World War as the sexual ecstasy and spiritual death that was brought to her by the Dionysian figure of Ferris. She faithfully employs the evangelical and legal language of unionism, but in doing so she also provides another version of Lily – one that allows Lily to be more of, and other than, herself. Richard Sieburth has read Hölderlin's translations as performing an impossible process of reciprocal sacrifice: 'the original is sacrificed [...] in the translation, just as the translation is in turn sacrificed back to the original so that the latter, now renewed and transformed, may continue on its course.'[131] A similar interaction takes place here: Marian's life has been drained of spirit, but she has learnt from the process of translating Lily ('her Other and double' as Michael Parker calls her) that her responsibility for this woman's afterlife calls on her to live her life as fully as possible.[132] She has thus decided not to give the house over to the National Trust. If Northern Ireland has been constructed and maintained as a relic that provides refuge from Catholicism, what Lily has taught Marian is that a relic provides a very limited form of refuge: it can take the form of a prison, not only for the Catholic population, but also for those who have constructed it. The '[s]trange logic' that Derrida sees as underwriting 'the laws of hospitality' is invoked here: the host has been 'awaiting his guest as a liberator', but this liberation takes the form of a series of 'substitutions [that] make everyone into everyone else's hostage'.[133] Parker's next move is to rethink the meaning of evangelism through this model of translational self-sacrifice.

Ruth's narrative is a word-perfect recitation of the biblical account of Pentecost. The centrality of this text to Northern Irish Protestantism is underlined by the fact that Peter, the non-religious Protestant, quickly chimes in and the two of them tell the tale in a call-and-response fashion. Its account of wind rocking a house and ecstatic divine fire bears an unmistakable resemblance to Marian's version of Lily's experience. However, glossing this secular form of translation with the literal rereading of the biblical scene of the Pentecost to illuminate its wider significance precisely to secular life creates a problem of substitution, which Peter touches on when he sarcastically notes that a replacement for 'the Prince of Peace Himself' is required (*P*, p. 243).

Marian proposes a solution to the casting of what Grant has termed this 'secular Pentecost'.[134] She offers a 'living' example of the 'kind of christ in every one of us' (*P*, p. 244), by turning to the shared experience of childlessness that has pervaded the drama (a motif that is also to be found in other Field Day plays, as is discussed in the following chapter). These absent children are presented not simply as aspects of individual, internal experience, but are explicitly linked with the political situation. Northern Ireland has not provided conditions fit for their birth or survival: 'what' as Marian asks 'could it possibly mean—a future? In a place like this?' (*P*, p. 212). Lily forsook the child she had with Alan Ferris, and subsequently had no other children, as Alfie had returned impotent from the hell of the First World War. Ruth has had three miscarriages, for the first of which her violent policeman husband was to blame, and is at a point of desperation: 'if I can't have a child I won't live!' (*P*, p. 212). Marian and Lenny had a child together and, while it was alive, their marriage worked; when it died at five months, the marriage too began to die. Yet, this shared experience of loss and hopelessness provides the empathy that enriches Marian's encounter with Lily, and Marian's translation of Lily's experience has provided her with a way of hosting and talking for her dead son, for the future. She brings up her dead son and he haunts the end of the play, bearing a heavy political and religious symbolic load. Not only was he was named Christopher, in which his 'Christ-ness' is obviously inscribed, but he was also born during the most intense period of the start of the conflict, in August 1969: the British government decided to send the British army to Northern Ireland on 14–15 August 1969, in circumstances, described by Lee, 'of apparently incipient pogroms' against the Catholic community.[135] In this light, the manner in which Marian now reconceptualises her relationship with her dead son charts an ethical process to which a whole society might turn.

> I denied him. The christ in him. Which he had entrusted to my care, the ghost of him that I do still carry [...] The christ in him absorbed into the christ in me. [...] I want to live now. I want this house to live. [...] We don't just owe it to ourselves, we owe it to our dead too ... our innocent

dead. They're not our masters, they're only our creditors, for the life they never knew. (*P*, pp. 244–5)

The spirits of the dead children take the place of the tongues (of fire) which entered those at the original Pentecost and, as the Holy Ghost provided visions of what was to come, the hosting of these spirits provides a way of reaching towards the future. This is a scene of translation as an act of extreme hospitality, which re-reads the positive spirituality and, in particular, the message of rebirth contained within the Pentecost in terms of a renewed afterlife: as the site for a new understanding of history.

However, in order to be the 'guest[s] (*hôte*) [who] become[ ] the host (*hôte*) of the host (*hôte*)', these dead children are first hosts in the sense of sacrifice victims, from the Latin *hostia*.[136] They are the already-sacrificed scapegoats of the 'Troubles', and the effort that is called for by Marian – which the play emphatically endorses as it closes with this speech – is to recognise this sacrifice as a sacrifice and to honour it by maintaining a hope in the future. In this way, the drama wants to get beyond the implacable cycle of violence inherent in the religious, sacrificial model of society, without discarding this model. While this is a brave engagement with the 'morbid symptoms' of the 'interregnum', the contradiction of looking to break a cycle of violence through the very model that generates it, disables this as a vision for Northern Ireland. Moreover, placing Marian's translational encounter with Lily within a religious framework that remains faithful to notions such as Paisley's evangelical claim that 'true faith is a martyred faith', drains their interaction of much of its political salience.[137] The violence that the religious model calls for is displaced chronologically (the sacrifices have already taken place) and becomes a matter of individual, internal interaction with one's most intimate other within, albeit supplemented with an awareness of the fact that others are also hosting these spirits. What was crucial in Marian's translation of Lily was precisely her taking the place of the dead woman, a process that required her to cross the political frames of Northern Ireland. Pursuing this communication into a biblical setting loses the immediacy of this political act, and places it in a symbolic order that cannot fully account for it, with the result that, rather than experience a Dionysian double-birth, the dead children are, in some fashion, resacrificed and reinterred under the weight of the responsibility of acting as passive symbols of the interaction of the political and religious frames of Northern Ireland.

4
Reciprocation and Resolution

Introduction

> The enactment of reciprocity in order to restore balance is the crux of the *métier* and morals of translation.
>
> — George Steiner[1]

Steiner's four-part hermeneutic cycle of translation has haunted the readings in this study and, as has been noted, the ethical crux of this model lies in the ways in which the final motion – restitution – is interpreted.

> The translator, the exegetist, the reader is *faithful to* his text, makes his response responsible, only when he endeavours to restore the balance of forces, of integral presence, which his appropriative comprehension has disrupted. Fidelity is ethical, but also, in the full sense, economic. By virtue of tact, and tact intensified is moral vision, the translator-interpreter creates a condition of significant exchange. The arrows of meaning, of cultural, psychological benefaction, move both ways. There is, ideally, exchange without loss.[2]

While *Pentecost* offered a final vision of a resolution, or a hopeful wish for a resolution in a future time beyond that of the play, the two translations considered here provide resolutions on their stages. These resolutions can be understood as acts that rebalance the economy of fidelity that Steiner describes. With the notable exception of the Choral intervention at the conclusion of *The Cure at Troy*, these plays seem far removed from the 'Troubles', yet the crises that demand resolution arise because fidelity has been equated with resentment, the state and unquestioned authority – forms of fidelity that look to elide the other, and which have structured life in Northern Ireland. As the forgiveness Marian asked for from Lily in *Pentecost* illustrated, a more ethical concept of fidelity requires a type of reciprocation that goes beyond the terms of this established, disabling economy. Heaney's play further pursues this idea and suggests that the economy of forgiveness supplements a fidelity to the word with a call to the other to trust your word. While fidelities have proliferated in Northern Ireland, trust has been missing, and both plays explore the difficulties of creating conditions of trust. Ultimately, the ethical restitutions in these plays concern expressions of trust in the other, which is the first motion in Steiner's hermeneutic cycle. This generates a circular motion, and despite the appearances of final, achieved, resolutions, the settlements these dramas propose are, in some manner, incomplete, and calls for future reperformances of the reciprocation of trust – for re-translations – can be heard.

Between hope and history: *The Cure at Troy*

> History says, *Don't hope*
> *On this side of the grave.*
> But then, once in a lifetime
> The longed-for tidal wave
> Of justice can rise up,
> And hope and history rhyme. (*CT*, p. 77)

Field Day celebrated its tenth anniversary in October 1990 with Seamus Heaney's *The Cure at Troy*, and this verse, which constitutes the play's key moment of resolution, is the most frequently cited passage from the company's entire output. Its position in contemporary political rhetoric is such that 'few dramatic texts', Denard has argued, 'can claim to have acquired such prominence in the political affairs of modern times'.[3] This process began in December 1990, shortly after the play's first performances, when it was employed by Mary Robinson, the first woman to be elected President of Ireland, in her inauguration speech (to which Richard Kearney contributed ideas). Subsequently, the idea of rhyming hope and history became a motto for the Northern Irish peace process, which gathered momentum with the announcement of the first IRA ceasefire in twenty-five years in August 1994. Politicians such as Bill Clinton, Jacques Santer and Gerry Adams, alongside innumerable journalists, turned to this concept to express the expectations and ideals that informed the process.[4] However, its use has not been confined to an Irish context and it has been availed of to encapsulate a more general mood of millennial possibility and longing in titles such as Nadine Gordimer's *Living in Hope and History: Notes on Our Century* (1999) and Bill Clinton's *Between Hope and History: Meeting America's Challenges for the 21st Century* (1996). These travels might be thought to express a form of translational reciprocation, as the impetus to cast the play's resolution in these terms came from beyond Irish shores: from 'the extraordinary events of late 1989' that swept across Eastern and Central Europe, according to Heaney.[5]

In many respects, this passage embodies Field Day's indeterminate conceptions of the interaction of art and society, which were discussed in the Introduction. As the varied political responses provoked by this rendering of a symbolic political act suggest, it has helped to organise and articulate sensations that were immanent in society: in and through poetic articulation a new, alternative way of thinking is opened that is also a form of recognition. Heaney has glossed this verse with his thoughts on the political role of art in *The Government of the Tongue*; these speak to how the indirect consequences of art in the world emerge from the ways in which art communicates directly with the self.[6]

> Here is the great paradox of poetry and of the imaginative arts in general. Faced with the brutality of the historical onslaught, they are practically useless. Yet they verify out singularity, they strike and stake out the ore of self which lies at the base of every individual life. In one sense the efficacy of poetry is nil – no lyric has ever stopped a tank. In another sense, it is unlimited. [...]
>
> [...] poetry [is] a break with the usual life but not an absconding from it. Poetry [...] is arbitrary and marks time in every possible sense of that phrase. It does not say [...] 'Now a solution will take place', it does not propose to be instrumental or effective. Instead, in the rift between what is going to happen and whatever we wish to happen, poetry holds attention for a space, functions not as distraction but as pure concentration, a focus where our power to concentrate is concentrated back on itself.[7]

The crisis in *The Cure at Troy* takes place in the 'rift' between the interpersonal, hopeful, 'wish [for something] to happen' and the implacable demands of history – of 'what is going to happen' – and the resolution depends on figuring this gap as the location of poetry. If, as McCormack has claimed, 'solutions are necessarily determined by the range and limitation of the language in which the crisis is explored', the drama invokes the impossible, poetic idea of rhyming hope and history precisely to push the limits of language so that alternative solutions might be imagined.[8]

This utopian endeavour takes place in and through an act of translation: *The Cure at Troy* is a translation of Sophocles's *Philoctetes* (Deane dubbed Sophocles 'Field Day's Greek dramatist').[9] The idea of rhyming hope and history is nothing less than a solution to a problem of translation. The Choral ode is Heaney's largest diversion from Sophocles: in Sophocles's play, the resolution is explicitly brought about by the intervention of Hercules *ex machina*. Such a conclusion is out of place in the late twentieth century, and Heaney looked to articulate a dénouement that negotiated between the transcendental and the human, or political, forms of resolution; between the tensions, in short, that animated the plays considered in the last chapter. In his own words, he 'attempted to present the conclusion as the inevitable culmination of an honestly endured spiritual and psychological crisis rather than as the result of a supernatural intervention'.[10] Up to this point, however, the translation follows Sophocles's version closely, and so the culmination must be understood in terms of the processes that lead to it, which are particular to the Greek world of the drama.

The action of *The Cure at Troy* and *Philoctetes* concerns the attempts of Neoptolemus and Odysseus to get Philoctetes's bow, without which, it has been foretold, Troy will not be captured. Philoctetes had been abandoned on the island of Lemnos by Odysseus and his Greek companions because his wounded foot was creating an unendurable stench on their ship. He has spent a decade in this enforced exile with his painful, festering foot and with an even more powerful festering hatred of the Greeks, in general, and of Odysseus, in particular: a destructive cycle of hatred, which he has only been able, like Boesman, to turn inwards, on himself. Despite this, he is desperate to return home to Greece, and this is, as Odysseus knows, the vulnerability that must be exploited if he is to be moved from Lemnos to Troy. Aware that Philoctetes would never accede to anything coming from him, Odysseus's plan is that Neoptolemus, Achilles's son, should meet and trick Philoctetes. Upon encountering him, he should act as if he were on his way back from Troy in a rage against the Greeks, in general, and Odysseus, in particular, for having duped him into joining the war, and because Odysseus had taken Achilles's armour for himself. By repeating Philoctetes's resentment, an empathy based on a shared hatred will be propagated between Neoptolemus and Philoctetes. It is this self-destructive form of repetition that is condemned by the Chorus in the first lines of the play:

> Philoctetes.
> Hercules.
> Odysseus.
> Heros. Victims. Gods and human beings.
> All throwing shapes, every one of them
> Convinced he's in the right, all of them glad
> To repeat themselves and their every last mistake,
> No matter what.
>
> People so deep into
> Their own self-pity, self-pity buoys them up.
> People so staunch and true, they're fixated,
> Shining with self-regard like polished stones.
> And their whole life spent admiring themselves
> For their own long-suffering.
> Licking their wounds
> And flashing them about like decorations. (*CT*, pp. 1–2)

As the performer of Odysseus's script, Neoptolemus becomes the third in this scenario: the interpreter, the communicator and, finally, the one who interrupts this simple mirroring of hatred. He does this by engaging in and then reframing this act of duplicity.

The spectral presence of Odysseus as the great manipulator of language has been noted in *Translations* and *The Communication Cord*. In both plays, he marks the limits of the counterfactual resources of language; he represents, in other words, a moment of ethical decision. This is rendered explicit here, as his linguistic tactics can be considered both a liberating stratagem and a hegemonic ruse. Neoptolemus must face up to this ethical crux in his role as the translator, 'fit to read between the lines / For the message' (*CT*, p. 11). At the outset, he is reluctant to assume this task, because lying, he believes, 'undermines your life' (*CT*, p. 10); he prefers the more 'honourable' course of using force to gain the bow. Odysseus, however, advocates deception as a considerably more economical and efficient method, knowingly informing him that,

> [...] the very people
> That go mad at the slightest show of force
> Will be eating from your hand if you take them right
> And tell the story so as to just suit them. (*CT*, p. 9)

This argument carries very little weight for Neoptolemus, who resolves his initial misgivings through his loyalty to the state: he only agrees to take this approach when Odysseus convinces him that to do so is in the interest of Greece (thus implicitly linking deception with a fidelity to the state). However, this facile resolution is unsustainable: when Neoptolemus encounters Philoctetes, he also confronts the fidelity due to the other and to language (the two are inextricably linked), and this sets in train an ethical conflict that, as Weissberg points out, 'turns the drama of Philoctetes into Neoptolemus' tragedy'.[11] It becomes, in other words, a translator's tragedy.

It is through a ritualised scene of hospitality that Neoptolemus confronts the dilemma of finding a resolution between the competing claims of the state, the other and language. As Belfiore has argued, 'the formal *xenia* relationship between host and guest has an essential role in the interactions of Neoptolemus and Philoctetes'. The *xenia* was

> a formal relationship entered into between two members of different social groups, and manifested by the exchange of goods and services. [...] *Xenoi* were obligated to do a wide range of other services for one another. [...] To harm a *xenos* was a particularly heinous crime, an offense against the gods as well as against human decency.[12]

In the fluid economy of the *xenia* relationship, the roles of host and guest are not mutually exclusive. On encountering Neoptolemus, Philoctetes, like all hosts, wants to know his name and from where he has come. These questions are laced with hope and expectation, and are properly speaking requests; immediately, as it were, the 'guest becomes the host's host'.[13] Philoctetes wants to exchange words, to communicate with a human he can understand – with a Greek – and so he opens up an economy of hospitality on this level. Neoptolemus enters into this on the level suggested to him by Odysseus: pretending to be himself and yet someone else, and acting as if he knew nothing of Philoctetes. This ignorance amazes Philoctetes and his anger at his plight increases, as he now believes that his very name has been obliterated. Having thus unbalanced Philoctetes, Neoptolemus evinces a rather sure touch in his role as a duplicitous in-between, and pushes home his advantage by voicing his anger at Odysseus, serving up to Philoctetes the story of his supposed betrayal. In offering him news from the war and an opportunity to voice his hatred of Odysseus, Neoptolemus increases his putative host's debt. So when Philoctetes reveals his most important request, he assumes the pose of the supplicant *xenos* – 'I'm on my knees to you' (*CT*, p. 27) – and appeals beyond this economy, or rather to its core – to the empathy that is its basis:

> [...] when I ask
> What I am asking of you now, remember
> Your own father and mother.
> You know how your heart lifts when you think of home?
> Well, think what it's like to be me here,
> Always homesick, abandoned every time.
> Take me with you. As a passenger. (*CT*, p. 26)

After feigning hesitation, Neoptolemus agrees, and Philoctetes phrases his joy as a debt that exceeds economics and exchange: 'how will I ever manage / To pay you back?' (*CT*, pp. 28–9). This is immediately thrown into relief by the mundane economy of the Merchant – 'a trader [...] / In whatever's going. Like information' (*CT*, p. 32). A partner in the subterfuge, the Merchant warns Neoptolemus against taking Philoctetes with him. Neoptolemus dismisses him, confirming that his relationship with Philoctetes is now established beyond the bounds of such an ordinary economy: '[t]his man and I are two fast friends' (*CT*, p. 32). As Gould noted, the *xenia* process changed the status of the participants 'from *xenos* in the sense of "outsider" to *xenos* in the sense of "guest", one who in the future may be addressed as *philos*'.[14] As this move is not immediately registered in a change of name, it is another ambiguous process that pushes the economy of translation to its limits and, in so doing, highlights the ethical stakes that are involved in the act.[15]

It is only once this status has been established that the question of the bow can be broached. Philoctetes recounts its history, telling Neoptolemus how the dying Hercules gave it to him in gratitude for setting his funeral pyre aflame – a provenance that invests the bow with a godlike quality that would appear to exclude it from common exchange.[16] According to Belfiore, in the *xenia*, 'gift was promptly followed by counter-gift of commensurate worth', and within its terms, Neoptolemus asks for, and is given, permission to hold the bow.[17]

> You of all men have the right to hold
> Philoctetes' bow. What's mine is yours.
> You gave to me, I give to you ...
> (*The bow is proffered, elevated and held significantly between them.*)
> [...]
> [...] and the reason why
> Is the reason I got it from Hercules
> In the first place: generous behaviour. (*CT*, p. 37)

However, the creation of what Neoptolemus calls an 'economy of kindness' (*CT*, p. 37), sanctioned and authorised through the afterlife of a god, describes a scene of Sophoclean irony. The supplicant, homeless (Philoctetes describes Lemnos as 'a home where I never was at home' (*CT*, p. 29)) host is to be granted his request from the false guest, because what he wants can be gained through the processes of the *xenia*, not because of its moral force.

At the same time, Neoptolemus is conscious of this irony: his awareness that his 'gift' to Philoctetes is falsely motivated is sharpened precisely by Philoctetes's genuine reciprocation, which moreover takes the form of the bow he inherited from Hercules for having fulfilled the dues of friendship.[18] In direct contrast, Neoptolemus's false story concerned Odysseus's supposed usurpation of Neoptolemus's rights of inheritance: his incorrect taking of the dead Achilles's armour. While this mirroring tale lubricated the cogs of his deceptive machinations, the *xenia* exchange has placed him in a situation in which he cannot easily evade moral responsibility. In this respect, the *xenia* resembles the 'potlatch' system that Mauss analysed in Maori culture, which also exceeds the usual definitions of an economy:

> to accept something from somebody is to accept some part of his spiritual essence, of his soul. To retain that thing would be dangerous and mortal, not only because it would be against law and morality, but also because that thing coming from the person not only morally, but physically and spiritually [...] exert[s] a magical or religious hold over you.[19]

Neoptolemus's engagement in this ritual exchange ultimately defines his own good faith. His predicament is further heightened by the fact that he also fully recognises the imperatives of the state that has urged him into this deception. In this, he differs from Antigone and Creon, who each represent one form of authority alone. His dilemma reaches a crisis point when

Philoctetes suffers a fit induced by the pain in his foot, which leaves him unconscious. Philoctetes entrusts Neoptolemus with the keeping of the bow, and Neoptolemus realises that to take it in this manner would be 'easy and meaningless' (*CT*, p. 46): to defy the ethical pact he has entered into can only, he determines, hollow out the duty he owes to his state.[20]

Fundamentally, the *xenia* depends on being faithful to one's word; to appropriate the bow in this fashion would render language itself meaningless. In fact, a tragedy of language is enunciated with Philoctetes's cries of pain. As Weissberg notes, Herder turned to these cries to conceptualise an extreme, 'originary' moment of language, as a cry of pain – a 'sound that does not have a home and country' – 'moves the feeling out of the body, pierces the distinction between the inside and the outside, the suffering individual and the possibility to being heard by the other'.[21] Existing at the limits of linguistic expression, which are the limits of translatability, these cries perform the insoluble difficulties of communication that, as Weissberg argues, constitute the essence of language.

> With Philoctetes' cries, Sophocles transforms the 'friendliest of tongues' into a Greek that is apt, by its sound alone, to provoke the terror and pity of tragedy. This tragedy of language, telling the story of pain and its inexpressibility, is in turn hardly translatable. [...] But the difficulty of translation points precisely to the transformation Sophocles has achieved. Philoctetes' pain is no longer a matter of human limitations alone. It touches the question of the possibilities of expression. In Sophocles' drama, language moves into the foreground; linguistic articulation itself is thematized.[22]

This is graphically evident in Sophocles's text where Philoctetes's cries brutally interrupt his discourse in such a manner that the violence is evident even to those who cannot read classical Greek. Philoctetes's experience of pain and trauma exceeds linguistic expression – '[t]here are', he says, 'no words for it' (*CT*, p. 41) – and the breakdown of his language brings home to Neoptolemus the stakes involved in the translation he is attempting. He is recalled to the duty due to language at the moment in which he believed he had successfully set in train the translation of Philoctetes from Lemnos to Troy. While his duplicitous translation strategy might have had an economical efficiency to it, he now has no doubt that its inherent lack of faith and trust render it insufficient.

What is at issue here is the ethical relationship between words and actions, which structures, and is structured by, interpersonal relationships. In many respects, the *xenia* is an instance of what Austin called performative utterances: speech acts such as promises and apologies that are, in themselves, deeds or events.[23] Performatives cannot be said to be true or false by any objective criteria; rather, their meanings or enactments depend on an environment of trust, which may have a formal or informal structure. It is only in the context of trust that performatives can make things happen; but trust itself is created through language, and words, as Philoctetes knows, are not always kept. When he recovers (into) language – a linguistic renascence of sorts, in Scarry's terms[24] – his first thought is of betrayal; despite the vows they had exchanged, he is surprised to find Neoptolemus where he said he would be: 'I never thought you'd stand your ground with me' (*CT*, p. 47). The tragedy of language in this drama concerns negotiating the paradox that while trust provides the fragile conditions necessary for the possibility of meaningful communication, by being created through language, trust is always susceptible to being rendered void by language.

'Meaning? What meaning? How could there be meaning?' (*CT*, p. 50). This is Philoctetes's response when Neoptolemus reveals that their interactions have been based on a deception. Having quickly learnt anew that language, which he craved to exchange with Neoptolemus, is also the mode of the betrayer, Philoctetes wishes to be rid of language, declaring that his communication with the animals on the island is more meaningful because it does not depend on trust: '[a]ny kind of trust is a mistake!' (*CT*, p. 52). To restore trust, Neoptolemus must renounce the thing that he most wanted – the bow. He makes to return the bow to Philoctetes in an action that is to be seen as a direct counterpart of the initial exchange. However, this act of reciprocation is interrupted by Odysseus who comes *'between them'* (*CT*, p. 53) and halts the return of the bow in the name of the pragmatic concerns of the state.

Before seeing him, Philoctetes recognises Odysseus by his voice, and the curse he hurls at the Greek who knew best how to manipulate language is aimed directly at this power: 'I'd give the whole agony of my life just to see you cut down in the end, and your tongue ripped out of you like a bleeding ox-tongue' (*CT*, p. 57). With this intervention, the political and the private, interpersonal relations come into direct conflict.[25] These are bound in conflict by language itself: the claims of Greek – the mother-tongue and the language of the state – both links and puts these characters at odds with each other. While Philoctetes holds that trust in the other is an ethical prerequisite for meaningful communication, Odysseus insists that loyalty to one's own people or state guarantees a system within which language assumes meaning; for him, the possibility of alternative meanings represents the potential for profitable duplicity, rather than a call for the enactment of trust.

It falls to Neoptolemus to fashion a route out of this deadlock. His task as the translator who understands the claims of both discourses is to make a choice between their competing claims: it is impossible to accommodate both as they are formulated at this juncture. While his first instinct is to acquiesce with the dictates of the state, he quickly revises that notion, realising that without trust there can be no resolution, and so he determines to return the bow and complete the *xenia* ritual: 'I am going to redress the balance. / The scales will even out when the bow's restored' (*CT*, p. 65). When Odysseus challenges this by invoking his authority as a representative of the Greek people, Neoptolemus counters with Antigone's argument: '[t]he jurisdiction I am under here / Is justice herself. She isn't only Greek' (*CT*, p. 67). Nonetheless, he never disregards the needs of his state; once he has seen off Odysseus, he asks Philoctetes if he will join them at Troy. When he refuses again, Neoptolemus offers him the bow, swearing that the offer is real and unconditional. Philoctetes cannot trust his words alone: '[s]wear, oh, you'll swear! It's only words to you' (*CT*, p. 70). However, the transaction is completed, and having thus restored himself in Philoctetes's good faith, Neoptolemus is now in a position to tell Philoctetes that his obstinacy has done him no good. By refusing to trust others, despite and in the face of the treachery that is always possible, he has metamorphosed into one of the wild animals with which he has communed for ten years.

> Human beings have to bear up and face
> Whatever's meant to be. There's a courage
> And dignity in ordinary people
> That can be breathtaking. But you're the opposite.
> Your courage has gone wild, you're like a brute
> That can only foam at the mouth. (*CT*, p. 72)

The courage he speaks of involves confronting the aporia that trust, which is a prerequisite for meaningful communication, must be created through communication, which always contains the possibility of treachery. To bring this home to Philoctetes, Neoptolemus goes beyond the exchange dictated by the *xenia* (which was fulfilled by restoring to Philoctetes his bow) and declares that he will also keep his promise to return him to Greece.

Despite this, Philoctetes does not budge from his position of refusing to aid the Greek army at Troy. He only does so after the Chorus intervenes in the place of Hercules *ex machina*, and proposes a solution that, in effect, reiterates and reframes the central role of trust in human affairs. To recall, this speech is Heaney's solution to a problem of translation: it is intended to be a more psychologically plausible resolution for a late twentieth-century audience, that nonetheless remains faithful to the dynamic that has been built up throughout the drama. However, the most disconcerting aspect of this solution is the form in which the contemporary enters: the *ex machina* irruption of references to the 'Troubles' into the Greek warp and weft of the play.

> The innocent in gaols
> Beat on their bars together.
> A hunger-striker's father

Stands in the graveyard dumb.
The police widow in veils
Faints at the funeral home. (*CT*, p. 77)

This is jarring and suggests a misleading Northern Irish allegory that Heaney has had to refute by pointing out that 'while there are parallels, and wonderful suggestive ones, between the psychology and predicaments of certain characters in the play and certain parties and conditions in Northern Ireland, the play does not exist to exploit them'.[26] More broadly, these references, in conjunction with those to 'miracles / And cures and healing wells' (*CT*, p. 77), might suggest that Heaney has translated Hercules's divine intervention into Christian terms. However, as is intimated by the vague allusion to 'a god' (*CT*, p. 77) in its penultimate stanza, rather than offering a resolution in purely Christian terms, Christianity is called upon as a symbol of an ethics – a law that exceeds that of the state – to convey what is fundamentally a utopian site, one that can only be conceived of in language, through the impossible rhyme. Heaney was conscientious in doing his 'Field Day "homework"' (his description of the play), as this is the fullest single exposition of the 'fifth province' in Field Day's output.[27]

In the Introduction, the problem of accounting for the transcendental aspect of the 'fifth province' was discussed, and in Heaney's solution to a problem of translation, the transcendental, in a manner similar to Kearney's conception of the 'fifth province', is rephrased as that which makes translation impossible and necessary. More specifically, considering that the writ of a *deus ex machina* is, in effect, a pure performative, the idea of rhyming hope and history calls for the economy of translation to be exceeded through a commensurate performative act if a faithful translation is to be essayed. As in *Pentecost*, forgiveness – which is to be found '[o]n the far side of revenge' (*CT*, p. 77) – is the performative that is called for by this verse. Echoing Neoptolemus's move beyond the bounds of the *xenia*, just such an unbidden gesture is required to generate conditions in which the trust necessary for dialogue and communication to take place in Northern Ireland might be fostered. The dimensions of such a translational environment describe perhaps the most politically meaningful rendition of the 'fifth province'.

As Derrida reads Portia's speech in *The Merchant of Venice*, forgiveness is beyond all economy, beyond the law, beyond even the implicit exchange invoked in asking for it: it is '*more than* just, more than justice or the law. It rises above law or above what in justice is only law; it is beyond human law, the very thing that invokes prayer.'[28] It betokens divinity, being only 'possible in doing the impossible', and indeed 'must announce itself as impossibility itself'.[29] Derrida cleaves to this radically pure concept of forgiveness because to do otherwise would be to submit to the economy of the 'power-to-forgive'.[30] It is precisely just such a pure forgiveness that is called for by the Chorus. As Caputo argues in specifically Christian terms, pure forgiveness operates in 'a temporality of trust', which essentially cancels history.[31]

> Forgiveness is aimed at dismissing the past. Forget the past. Dismiss our past just insofar as we have dismissed the past of others. [...] If our redemption from the future is to live without anxiety, the only redemption from the past is forgiveness. [...] Vengeance and resentment chain us to the past, forcing us to go over it again and again, [...] whereas forgiveness releases and sets free.[32]

Phrased in this manner, it becomes possible to conceive of hope and history rhyming. Through the act of forgiveness, history is replaced with the temporality of hope, and so hope comes to rhyme with hope. It would be fair to say that, in general, the logic of substituting history with hope informs the use of the phrase in contemporary political rhetoric. However, the fact that this is a poor rhyme signals that there is something facile about this construction. To an extent, it is susceptible to Lloyd's critique of the sense of resolution that is found in Heaney's *oeuvre*, which he sees as being achieved by 'reduc[ing] history to myth'.[33] However, while this play reaches towards a form of a-historical continuity, it cannot, in the final analysis, fully articulate this. Instead, the drama is marked by a tension between what might be called the pure and the economic dimensions of forgiveness.

Hope and history will only rhyme if the act of forgiveness is thought of in the pure fashion that Derrida insists is worthy of the name of forgiveness. Kearney's critique of Derrida's unlimited view of hospitality, which insists on a commensurate, impossible purity, can be employed here. Kearney argues that there is a legal and ethical requirement to distinguish between different types of others: a 'need to differentiate not just legally but *ethically* between good and evil aliens'.[34] Similarly, there is a need to call for and to grant forgiveness within legal limitations, and within the constraints imposed by the past, rather than dismissing them. Just as Neoptolemus never loses sight of Troy and is fully cognisant of the problems he is creating by defying his obligations to his state, the economy-beyond-economy of the performative act also needs to relate to, rather than simply dismiss, the economy it is transgressing. Such a dynamic is registered in the exchange of gifts in the play. As Caputo points out '[a] gift is a gift for Derrida when it is removed from the circle or circulation of giving and paying back (remuneration, retribution), of action and proportionate reaction, when we let the gift be.'[35] Speech and the bow, the exchanged gifts of the *xenia*, would not qualify as gifts in this definition, as neither exists in the innocent, a-contextual dimensions it charts. Speech, indeed, was valued precisely as a mode of exchange, and the double-edged nature of this gift has been noted. While the bow belongs to an 'economy of kindness' that would render it a gift in the pure sense of the word, its meaning is also tied to a 'teleological causality to the future', from which such a gift should be removed.[36] After all, this is the bow that will end the Trojan War, or, to put it another way, will hasten the fall and the sack of Troy. As such, the 'cure' that it brings about – the eventual end of the war – is only accomplished by the destruction of one side; a point that places an even larger question mark over the references to Northern Ireland in the Choral intervention.[37]

In this light, the bow is a *pharmakon*, something that can act as a cure and/or as a poison. Like *xenos*, this is yet another term that describes an indissociable opposition between two seemingly contrary processes or states, and presents commensurate problems for a translator; as Derrida put it, 'the choice of only one of these renditions by the translator has as its first effect the neutralization of the citational play'.[38] A way of overcoming this crux of translation is suggested by the manner in which the *pharmakon* figures in Plato both as a form of supplement or limit to memory – without which memory would not be possible – and as that which 'dulls the memory'.[39] Similarly, in order for this term to be translated, a possible meaning has to be somewhat forgotten, or put in parentheses. Unlike the forgetfulness insisted upon by pure forgiveness, this would be a forgetfulness that does not entirely forget and is not a-historical, but which forgets enough, in a certain moment, in order to move forward. The ethical responsibility falls on the translator, whose choice is an enactment of her/his trust. Just as a performative requires an appropriate context for it to function as a performative and so complete the context, the gift and forgiveness, thought of in an analogous construction, are required precisely in order to further communication: although they gesture beyond it, they are ultimately a function of human exchange. This engenders an irony, which, as Heaney has noted, informs this play: 'Philoctetes is "cured" but cured into the very loyalty and solidarity which Neoptolemus had to flout in order to bring the cure about.'[40] Hence, while *The Cure at Troy* strives to hold 'in a single thought reality and justice', it ultimately suggests that this might occur only in fleeting moments.[41] Rather than simply rhyming, it is only if hope and history are understood as containing the possibility of rhyming that the utopian drive of this idea (and of the 'fifth province') has meaning and stays relevant to political discourse.

Gender, trust and irony: *High Time*

> All oppositional politics [...] move under the sign of irony, knowing themselves ineluctably parasitic on their antagonists. Our grudge against the ruling order is not only that it has oppressed us in our social, sexual and racial identities, but that it has forced us to lavish an extraordinary amount

> of attention on these things, which are not in the long run all that important.
>
> — Terry Eagleton[42]

Derek Mahon's *High Time* replays in a comic mode the clash of authority and justice that was performed as tragedy in *The Cure at Troy* and *The Riot Act*, with which it toured in 1984. The comedy is generated in this translation of Molière's farce *L'Ecole des Maris* through the use of irony, which also gives it a political edge. As O'Brien has noted, the failure of colonial authority in Ireland rendered irony the most 'congenial' mode of political expression:

> [h]ypocrisy, the permanent and universal element in the ideologies of the ruling classes, seeks to mask the gap between profession and action, to cover the realities of social and political struggle with an illusion of harmony. Irony uses the language of hypocrisy [...] with a calculated excess, so that, as the realities show through, the pretences come to seem ghastly.[43]

As has been most apparent in the two translations of Sophocles, the language of political authority betrays a lack of trust (in the other; in its own discourse); in *High Time*, justice is brought about by employing this deficit against the forces of authority. At the same time, this play ironically suggests the limitations of a liberation achieved in this manner: it is too closely tied to the logic of the oppressor to provide for an entirely enabling freedom.

Authority in *High Time* is represented by two middle-aged brothers, Tom and Archie, who have in their care two late-teenage wards, the sisters Isabel and Helen. On the death of their father, they were left, as children, to the brothers:

> [t]hey're without parents, and our friend their father
> committed them to us in his last hour;
> we were meant to marry them or, if not *us*,
> marry them off to others in due course:
> he meant to make us, from their childhood on,
> fathers and husbands to his orphaned children
> with all the authority that that implies.
> You took on that one and I took on this. (*HT*, p. 15)

Such a structure speaks to the *xenia* relationship that was enacted in *The Cure at Troy*; as Belfiore noted, '*xenia* relationships were inherited. *Xenoi* protected one another's children, and often acted as foster parents.'[44] What is described here is a 'scene of ethical allegory', in the sense that, as Avital Ronell notes, '[t]he fundamental structure of our being-for-the-other is that of adoptive parent'.[45] This can be understood as a form of translational substitution: a translation is both a form of substitute and an original text, and the translator must assume, for a period, the position and responsibilities of the original author. In the play, the father's deathbed wish explicitly charges the brothers with the task of moving from the status of substitute to that of another type of original: from the position of *in loco parentis* to husbands. In this light, the ethical dilemma for the brothers concerns the negotiation of the paradox of standing in the place of the other while assuming the role of an original. The boundary between these two states is determined by trust and fidelity, and a conflict based on definitions of these terms lies at the heart of this comedy, as the brothers have two differing conceptions of how to answer faithfully the call of the dead father and his trust in vouchsafing his daughters to them.

Just such a 'scene of ethical allegory' was employed for the act of translation by the Earl of Roscommon in his seventeenth-century *An Essay on Translated Verse*. In it, the translator takes the place of his friend the author, and assumes responsibility for protecting and nurturing the moral well-being of his offspring – the texts – while they undergo translation.

> [...] chuse an *Author* as you chuse a *Friend*.
> United by this *Sympathetick Bond*,
> You grow *Familiar*, *Intimate* and *Fond*;
> Your *thoughts*, your *Words*, your *Stiles*, your *Souls* agree,
> No longer his *Interpreter*, but *He*.
>
> With how much ease is a *young Muse Betray'd*,
> How *nice* the *Reputation* of the *Maid*!
> Your *early*, *kind*, *paternal* care appears,
> By *chaste Instruction* of her *Tender Years*.
> The *first Impression* in her *Infant* Breast
> Will be the *deepest*, and should be the best
> Let no Austerity breed servile *Fear*,
> No *wanton* Sound offend her *Virgin-Ear*.
> Secure from *foolish* Pride's *affected state*,
> And *specious Flattery's more pernicious Bait*,
> *Habitual Innocence* adorns her *Thoughts*
> But your neglect must answer for her *Faults*[46]

Roscommon, as Vance noted, embodied 'some of the perplexities and aspirations of the Irish English writer down the ages'. In a manner that bears notable parallels with the Hugh O'Neill of *Making History* and the protagonists of plays considered in the following chapter – Oscar Wilde, Brendan Bracken and William Joyce – he straddled the Anglo-Irish divide. According to Vance, he was '[c]osmopolitan, English and Irish, prescriptive and insecure, devoted to a classical idea of art, civility and transhistorical permanence as a mode of refuge from Irish realities'.[47] In his prefatory eulogy to the second edition of the *Essay*, Dryden phrased Roscommon's Anglo-Irish background in terms of translation.

> How much in him may rising *Ireland* boast,
> How much in gaining him has *Britain* lost!
> Their island in revenge has ours reclaim'd,
> The more instructed we, the more we still are sham'd.
> 'Tis well for us his generous bloud did flow
> Deriv'd from *British* Channels long ago,
> That here his conquering Ancestors were nurst;
> And *Ireland* but translated *England* first:
> By this Reprisal we regain our right,
> Else must the two contending Nations fight,
> A nobler quarrel for his Native earth,
> Than what divided *Greece* for *Homer*'s birth.[48]

Ireland is viewed here as a foster-parent of sorts for Roscommon; his original Englishness becomes evident through his work on translation which, like Dryden's, worked to construe Britain's new empire as an improving and refining evolution of the classical world. But, in contrast to Roscommon's own Irish translation, in this manifestation of the 'transhistorical permanence' of empire, the translation is superior to the original: as Roscommon put it, English translations improved on 'what was writ Before' by blending 'Judgment' into the headier Mediterranean ingredients of the classics.[49] Lori Chamberlain has noted that the traditional hierarchy between production and reproduction in translation is marked by rendering one element 'original and "masculine," the other [...] derivative and "feminine"'.[50] To claim that the English translation is superior to the original subverts this hierarchy and, indeed, Dryden employed a form of gendered oxymoron to account for it: claiming that Roscommon improved on classical translations by infusing in them a 'Manly sweetness'.[51] While, as Roscommon's *Essay* illustrates, authority is invariably figured as being male, the gendered tropes that have been employed to represent the act of translation also betoken, Chamberlain argues, a 'struggle for authority and the politics of originality informing this struggle'.[52] This not only describes the action of *High Time*, but different versions of this tension can be traced in Field Day's output more generally.

Although the issue of gender and Field Day is most readily associated with the controversy that surrounded the publication of *The Field Day Anthology of Irish Writing*, the enterprise's other productions also betray distinctive gendered traits. All of the playwrights considered here are men, and all the translations are of plays written by men. In addition, fourteen of the fifteen pamphlets are by men. Therefore, on the level of authority there was an abyssal gender gap in Field Day's output in the 1980s and 1990s. On a few occasions this issue cropped up in Field Day discussions, most often at the instigation of Tom Paulin. In a memo to the other directors (17 August 1984), he noted that '[p]amphleteers need not be pure specimens of Irish manhood' and suggested inviting Edna Longley to write a pamphlet.[53] He returned to this theme on 23 January 1987, as he was concerned that Field Day was perceived as a male preserve and he urged the other directors to consider 'address[ing] certain issues – gender roles, patriarchy, family law, etc.'.[54] At a meeting later in the same year (30 August 1987), he proposed a pamphlet series on 'Women and Irish Society'; however, '[t]here was little enthusiasm from the other directors', who preferred a series on nationalism or colonialism.[55] The possibility of commissioning a play from a woman playwright, such as Anne Devlin or Christina Reid, was briefly considered and quickly dismissed at the 28 May 1988 meeting.[56] It was only in the wake of the *Anthology* controversy that the company took the issue seriously, with Deane admitting to the other directors that the blindness to gender in the *Anthology* was 'no doubt [...] a serious weakness'.[57] In order to rectify this there were suggestions at the 20 January 1991 meeting to invite women to join the board of directors. The names suggested included Mary Holland and Rosaleen Linehan, but no decision was reached as, at this juncture, the company was treading water.[58]

At the same time, on a thematic level, gender plays an explicit structuring role in the dramas in a manner that plays on and subverts the patriarchal expectations inscribed in the model of restitution that Steiner turns to in *After Babel*: '[t]he general model here is that of Lévi-Strauss's *Anthropologie structurale* which regards social structures as attempts at dynamic equilibrium achieved through an exchange of words, women and material goods'.[59] As Chamberlain contends, in Steiner, 'when communication is at issue, that which can be exchanged is depicted at least partially in male terms [...] while when "restitution" is at issue, that which can be exchanged is depicted in female terms'.[60] The Field Day plays tend to hollow out this stereotype as they perform versions of its drive. In *The Cure at Troy* two men perform a restitutive exchange only after they have been prompted to do so by a feminine Chorus, rather than Sophocles's ultra-masculine Hercules; as Heaney put it, he intended that the Chorus would articulate, amongst other things, 'a gender-politics aspect, insofar as the militaristic, male-bonding world of the Greek army is challenged by the anima [...] impulse in Neoptolemus'.[61] In like fashion, the other Field Day plays tend to put a gendered model such as Steiner's under the sign of irony, as many of the male characters are precisely impotent and incapable of exchange. Furthermore, even when they are not the main characters, women have generally been the dominant personalities in the plays considered to this point: in contrast to the ineffectual men, they are the ones who do. In *Translations*, Maire's attempts to find an accommodation with her changing circumstances are brought into relief by the drunk, weak and idealistic males around her. Likewise, in *The Communication Cord*, the female characters are less ridiculous and more grounded than the men. *Three Sisters* is, of course, an almost entirely female play in which the male characters exist on the margins. As Fugard noted, in *Boesman and Lena*, Lena's feminine ability for 'survival and defiance' stands in direct contrast with the 'passive, most times impotent male'.[62] The importance of Antigone as a symbol of woman's resistance to male, official power has already been underlined. Both motions of the resolution offered in *Pentecost* are emphatically feminine – Marian's communication with Lily and her secular rereading of Ruth's biblical story – and are directly contrasted with the inept male interventions. While this is undoubtedly connected to the prevailing emphasis on home and the domestic, it does not betoken a passive disposition; as is most radically illustrated by Antigone, in maintaining the more domestic, earth-bound duties and laws, these women come into conflict with the patriarchal, machine-like state.

However, two of the three plays considered in the next chapter feature men in dominant roles. Roughly speaking, this represents a shift in emphasis between earlier and later Field Day productions, as the majority of the plays considered so far have come from the first half of the 1980s. Like *Making History*, many of the dramas produced in the second half of that decade are history plays and, in these, men are historical protagonists. In this light, the women in Field Day plays tend to symbolise a hope that might defy the processes of history, and that is always in threat of being extinguished. One of the striking features of these dramas has been the number of dead children in them. As was noted, this theme is central to *Pentecost*, where all the female characters have suffered losses of children; in *Translations*, the baby at whose christening Hugh gets drunk in the first act, is reported dead before the play ends; the most intimate moment in *Boesman and Lena* is when Lena recounts to Outa the death of her child; while in *Making History*, Mabel dies giving birth, and her last reported words are: 'Hugh would never betray his people' (*MH*, p. 317). She was, as O'Donnell's epithet had it, 'a very loyal wee girl' (*MH*, p. 309), and the hope that women represent in these dramas is tied to the ways in which, despite and in the face of the persistence of disabling political structures, they recall others to the duty owed to forms of trust and fidelity that go beyond those imposed by patriarchal social systems. As Field Day distrusted the structures of political authority in both parts of Ireland, this precarious, feminine hope is central to its postcolonial intervention, which privileges the processes of the always contradictory work towards fidelity in a translation over the rule of the authoritative original. Through yet another Field Day scene of 'problematic filiation', this dynamic is played out in *High Time*, where, once again, an unequivocal resolution proves elusive.

Tiffany has pointed out that '"treason" derives from the same Latin verb that gives us the word "tradition": *tradere*',[63] and, in *High Time*, Tom's repression of his ward, Isabel, is motivated both by a fear of being betrayed and by the desire to conform to a traditional understanding of society.

> She'll listen to no trash from some young clown
> or go out traipsing up and down the town.
> The flesh is weak; I know how people talk.
> No horns for me, thanks. Since it's her good luck
> to marry me next month, I must be strict
> and leave no doubt in *that* respect.
> ISABEL
> You don't suspect me [...]?
> TOM
> Shut your gob, my dear;
> I'll let you know when to go out, d'you hear? (*HT*, p. 16)

Archie, on the other hand, has read in a more liberal manner the trust the father placed in him. He grants his ward Helen considerably more freedom, allowing her to go to pubs and discos and letting her know that he will welcome her boyfriends into his house.

> Her father wanted us to marry, but
> I don't intend to make an issue of it.
> I'm well aware she's thirty years my junior
> and I'm content to leave the choice to her.
> [...]
> I'd rather see her settled happily
> with someone else than miserable with *me*. (*HT*, p. 18)

On an allegorical level, what is enacted here is the balance the translator must maintain between the trust placed in her/him by the author with her/his trust in the text itself. Tom literally interprets the call of the dead father and practises a word-for-word style of translation, whereas Archie favours a freer interpretation. These contrasting choices are further reflected in their respective ways of dressing. Tom rebukes Archie for his style:

> [t]here's nothing more ridiculous, I fear,
> than a mature man striving to appear
> youthful. He'll always get it wrong, of course,
> and end up looking like a pantomime horse (*HT*, p. 12)

He, on the contrary, dresses according to the dictates of tradition: '*I* choose to dress with taste and dignity, / not in a daft shirt but a suit and tie / and stout brogues like our father used to wear' (*HT*, p. 13). Rather than trust Isabel, Tom attends to the call of a tradition laid down by dead patriarchs.

Drawing on the model of his own dead father, he reads the trust placed in him by Isabel's dead father as a call to act as the dead father alive; in effect, to deny this death, to suspend time. Archie's defence of his sartorial choices rejects this fictional temporality:

> I think it only reasonable to conform
> to current taste, and not defy the norm.
> You can go too far either way, of course:
> one shouldn't ever be conspicuous
> either in clothes or speech but, as I say,
> adopt the changes custom throws our way. (*HT*, p. 12)

It comes as little surprise that Tom is betrayed by Isabel with the result that, in his eyes, he betrays his friend's trust, whereas Helen stays 'faithful' to Archie and agrees to marry him at the play's conclusion. By seeming to ignore the demands of the original author and placing trust instead in the text to be translated, the non-literal translation is, in this light, the more faithful of the two translations.

In many respects Archie's style of translation resembles the one employed by Mahon: his 'free translation' is an act of trust in Molière's ability to live and resonate in the present, and to adapt to contemporary tastes and mores.[64] While the transposition of the basic scenario – the bequest of two young daughters to two middle-aged bachelors – to late twentieth-century Ireland strikes an anachronistic note, Mahon believed that 'Molière is at home everywhere and in every age.'[65] He and the directors of the play 'never considered doing *High Time* in period: a stylized costume drama seemed like a sterile exercise'. After considering different contexts, most especially 'May 1968', they decided on '*now* [...] for immediacy instead of nostalgia'.[66] As Mahon wrote to Friel (22 March 1984), while the 'oddness of the ward-guardian set-up' called 'for a willing suspension of disbelief' on the part of audiences, it was nonetheless 'justified by my version's self-guying tone'.[67] By drawing attention to its inherent anachronism, it accentuates its status as translation, and the ironic humour this produces opens up a way in which it can function in the present. There is an inevitable sense of disjunction when characters such as Val and Ernie, who are described as '*a frightening punk apparition*' (*HT*, p. 22), speak in rhyming couplets, and Mahon utilises this as a source of ironic, self-reflexive comedy, through a clever and continual shift of rhetorical registers. Val, for example, moves from contemporary street talk – '[h]ey, Fart-Face!' (*HT*, p. 22) – to a somewhat over-elaborate form of request: '[p]lease / forgive me if I interrupt your reveries' (*HT*, p. 23). On other occasions, the exigencies of the rhyme compel a contorted phrase which is ironically commented upon by the actor: 'VAL: Certainly, sir; what can I do for you? TOM: Nothing; the foot is on the other shoe. (*registers puzzlement at this line*)' (*HT*, p. 28). According to O'Toole, this 'artificial style' is 'a kind of modern equivalent of the stylisation that Molière himself would have employed'.[68] By trusting that the text can live in the present, it is being faithful to Molière. In fact, *L'Ecole des Maris* was itself a free translation of Terence's Roman comedy *The Brothers*, which was, in turn, a translation of Menander's Greek comedy *Adelphoe*, and so the call for translation within Molière's text is radically different from the summons that was imposed on the brothers: it comes from a tradition of thriving in the present. Indeed, the 'now-ness' of Mahon's text could even be interpreted as a call for a re-translation, as it now comes across as being very much of the 1980s.

The tradition from which this play derives is also the tradition that has provided our concepts of irony; as de Man has noted, '[m]ost discourses about irony are set up' in terms of 'the traditional opposition between *eiron* and *alazon*, as they appear in Greek or Hellenic comedy, the smart guy and the dumb guy'. Within this structure, the audience must 'keep in mind that the smart guy [...] always turns out to be the dumb guy, and that he's always being set up by the person he thinks of as being the dumb guy, the *alazon*'.[69] This dynamic is played out in *High Time* through the triangle of Tom, Isabel and Val. Despite the fact that they have only had visual contact, Isabel and Val find themselves in love with each other. However, they cannot be sure of the other's feelings because Tom has forbidden Isabel to communicate with anyone without his supervision. He is so smugly confident of his control over Isabel that he cannot imagine her having the opportunity to disobey him. Tom's trust in his own lack of trust places him squarely in the role of the *eiron* or 'smart guy', and precipitates his undoing, as Isabel devises a stratagem that takes advantage of it. She informs Tom that Val has been trying to speak to her, and asks Tom to tell him to stop. He is only too delighted to do so, as Isabel's story confirms the efficacy of his way of bringing her up. Hence, he informs Val that Isabel,

> [...] like the good girl she is, and one
> devoted to me from childhood on,
> she told me what you're up to; and moreover
> (*scoldingly*)
> that, since you've set yourself up as a lover,
> she understands full well what you desire.
> She sees it in your eyes, she tells me; you're
> quite wrong if you imagine you can make
> your feelings plainer. You're an open book.
> [...]
> [...] She says to cease
> looking and smiling; take some other course. (*HT*, pp. 29–30)

Employing Tom's distrust to undermine him, Isabel thus informs Val that she appreciates his attention, and encourages him to press his suit in another manner. She trusts that Tom's literalism will render him blind to the message he is carrying from one side of the stage to the other and that Val will be able to read between the lines. This literally metaphorical process continues and develops: from carrying over verbal and written messages, Tom even brings the lovers together, and so effects his own 'betrayal'. In this manner, *High Time* performs the broad ethical message that one must trust the other's trust or good faith, and be wary of blindly following tradition, as its dictates do not necessarily provide a blueprint for life in the present.

Grant argues that a similar impulse is to be found throughout Mahon's *oeuvre*, which is distinguished by 'a cool and ironic play of intelligence measured against the pull of old allegiances'.[70] Thought of in this fashion, the process set in train by Isabel has an obvious resonance in Northern Ireland. But irony is also, as de Man had it, a form of parabasis: a type of disruption and disillusionment within a narrative.[71] Therefore, in Hayden White's words, irony

> represents a stage of consciousness in which the problematic nature of language itself has been recognized. It points to the potential foolishness of all linguistic characterizations of reality as much as to the absurdity of the beliefs it parodies. [...] The trope of Irony [...] provides the linguistic paradigm of a mode of thought which is radically self-critical with respect not only to a given characterization of the world of experience but also to the very effort to capture adequately the truth of things in language.[72]

Irony, in short, is the trope of the knowledge of the impossibility of full knowledge, and, as such, it inevitably engenders radical doubt. So while the audience of this play watches, with the comedic pleasure of foreknowledge, an ironic process in which a man who believed he knew everything is shown to be a fool, this process cannot but induce a commensurate doubt into an understanding of the play, as it is impossible to gauge where this disruptive irony might stop. It might, for instance, be argued that one of the ironies of this play is that the two brothers are essentially quite similar, as both wanted the same thing.[73] Moreover, Isabel's way of escaping from Tom's authority is to marry another man and so reinforce patriarchal structures of authority, which is also the upshot of Helen's 'faithful' translation. Viewed from this perspective, the relationship of the allegory to Northern Ireland is not straightforward: it does not offer an uncomplicated vision of liberation.

Paulin has claimed that Mahon's work is infused with a utopian impulse, a 'dream of a perfect elsewhere'.[74] Similar to the manner in which the 'fifth province' is a refuge from, and an intervention into, the 'Troubles', this describes an escape from, and an engagement with, Northern Ireland. Mahon's 'ironic detachment' from his home-place (to use Grant's terms) is deeply rooted in his interest in translation.[75] This is not just a matter of engaging with other languages and literatures, but an incorporation of the disruptive motions of translation into his poetic practice in such a way that his work expresses the uncertainties of Northern Ireland. As Goodby argues, the 'Troubles' find disruptive expression in Mahon's practice of self-translation – his frequent revisions of his own poetry:

> the more the poet attempts to finalise meaning the more he destabilises it. [...] The irony is that continual revision makes textual authority weaker, not stronger. More importantly, Mahon's procedures reveal a violent instability inherent in the text which mimics the violence from which, as aesthetic object, the poem claimed detachment even as it recognized the Troubles.[76]

Such a practice articulates the volatility of the 'Troubles' through the tensions that inhere in irony between knowledge (typified by Sophoclean irony) and doubt (which irony cannot but induce). The only way to deal with the radical and seemingly self-engendering doubt that irony and the 'Troubles' create is through trust, which is what has been absent in Northern Ireland. Trust is at the core of any meaningful political initiative: '[m]ost institutions', as Ricoeur points out, 'rely fundamentally on the trust they place in the word of the other'.[77] Irony, on the other hand, as White puts it 'tends to dissolve all belief in the possibility of positive political actions'.[78] Considered in this light, these two plays neither dismiss the possibility of politics nor advocate a blind trust; rather, they ultimately keep irony and trust in play. In different ways, Philoctetes and Tom learn that trust is not simply an oppositional counter to irony, but is itself fundamentally ironic, as within trust there has to be an element of non-understanding: it is an act made in the knowledge of the lack of full knowledge of the other.

5
Masks: Men and History

Introduction

> [O]ne of [Field Day's] abiding ambitions [is] to make available the possibility of a sense of freedom which is not dependent for its sense of belonging to a tradition that is or was or is about to be betrayed. The breaking of the bond between notions of freedom and notions of treachery, between those who are civilians only because there are others who are barbarians, would be, in Field Day's view, a salutary and maturing achievement.
>
> — Seamus Deane[1]

Negotiating the need for treason has been a key theme in the dramas considered so far, and was crucial to the resolutions offered in the two plays considered in the last chapter. Treason involves playing a role, and the dramas in this chapter suggest that, because identity is always performative, any construction of the self incorporates an element of treason. 'Person', after all, derives from the Latin *persona* – 'a player's mask' – and what is at issue in these plays is the difficulty of fashioning a self that is a faithful traitor in the context of the seemingly implacable machinery of history. *Double Cross* and *Saint Oscar* offer versions of, in Rea's words, 'the lives of extraordinary Irishmen struggling with confusion and contradiction'.[2] These men want to reimagine their pasts; acts that involve, on some level, forms of betrayal. However, within the dynamics of the Anglo-Irish relationship, the possible forms that this treason might assume are seen to be limited; it is remarkably difficult to break out of the hold of the binary colonial logic that devalues one culture while vaunting the imperial one. Just as the black man sought to don a white mask in Fanon's analysis, these protagonists, in different, even conflicting, ways, aspire towards English ideals.[3] Wilde, in Eagleton's *Saint Oscar*, is particularly attracted to the theatricality of the upper-class English. Benefiting from a less interrupted cultural experience, this identity was better rehearsed than the more disjointed expressions of Irishness, and was preferable to Wilde on these grounds – as an achieved performance, not as a more authentic culture. Bracken and Joyce in *Double Cross* also thrive on this sort of theatricality, which granted them the freedom to assume different personae. However, unlike Wilde, they needed Britishness to express an inviolable authenticity, despite the fact that they continually expose its performativity. Neither therefore escapes the bonds of the colonial stereotype; such a liberation requires, as the example of Wilde shows, at least a doubled treason: to remain a faithful traitor, the *traduttore traditore* does not just cross one border once, but is constantly in movement between different states. While the characters in the third play, *The Madame MacAdam Travelling Theatre*, are entirely fictional, the play is loosely based on Anew McMaster's travelling theatre company in Ireland in the mid-twentieth century. Reversing the journeys taken in the other dramas, here an English touring company is a guest in the strangely liminal Irish Free State during the Second World War; a place that exhibits, through its attitudes to the troupe and to the theatre, aspects of what Jaaware called the 'double, often contradictory thing that is called the post-colonial consciousness'.[4] As in *The Communication Cord*, what is portrayed here is a post-colonial state that has spray-painted colonial stereotypes green rather than face up to the task of exploring alternative modes of identity.

Two of these plays are by Thomas Kilroy, the last director to join the Field Day board and the only one from the Republic. From its inception, though, he was highly supportive of, and interested in, the Field Day project: in 1980, he hailed the opening night of *Translations* as 'an event of some political importance to the whole country'.[5] His work also manifests a considerable thematic overlap with the Field Day plays, as it is distinguished by a sustained interest in doubled, split and conflicted characters; indeed, he has produced dramas about Wilde and Hugh O'Neill, and a version of Chekhov.[6] On joining the enterprise, he infused a new energy into the theatre company, which had begun to flag somewhat as Field Day's attentions were being increasingly diverted towards its publishing wing. To a considerable extent, his plays set the theatrical agenda for Field Day in the second half of the 1980s: while his work complements that of his good friend Friel, his 'intensely self-reflexive, strikingly modernist drama', as Pelletier notes, 'expand[ed] [Field Day's] theatrical vocabulary', and gave a renewed edge to its imaginative politics of contradiction.[7]

Traduttore traditore: *Double Cross*

> For, as Rosenzweig says, to translate is 'to serve two masters': the foreign work and the foreign language; one's own public and one's own language. A double fidelity is needed,

> then, incessantly threatened by the specter of a double treason.
>
> — Antoine Berman[8]

Irish nationalism creates situations of 'problematic filiation' for Brendan Bracken and William Joyce in *Double Cross*; it estranges them from their families and, as a result, both strive to create filiative connections to a British home. Caught between Ireland and Britain, they are also, like O'Neill in *Making History*, explicitly caught between their lives and their representations. But, in contrast to the naturalism of Friel's play, both are dead and alive (or undead) onstage: they are their dead selves reviewing episodes from their lives, which they perform as if alive. As such, *Double Cross* is a ghost play that examines the generation of different types of ghosts, rather than a straightforward account of both characters' lives. For this reason, it requires a basic knowledge of their backgrounds, and the brief biographical sketches that were provided in the programme for the first performances in 1986 are included in subsequent published versions of the play; as Kilroy noted, 'one of the problems with' the Sophoclean irony that operates in this drama, 'is that it relies upon a kind of complicity with your audience or with your reader'.[9]

A prosperous man, Brendan Bracken's father, J. K. Bracken, was one of the seven founders of the Gaelic Athletic Association in 1884. He was also a Fenian for a time, which led to him being denounced from the pulpit as being 'worse than a Protestant or even an atheist' by his local parish priest in Templemore, County Tipperary.[10] Because of the remarkable divergence between their politics, Bracken's relationship with his father haunts *Double Cross*. However, as his father died in 1904, when Brendan was three, he never really had an immediate filiative relationship with him. Instead, in the play, his father represents an affiliation to Irish nationalism – Bracken describes him as having 'the face of a condemned people' (*DC*, p. 45). This link would have been fostered by his mother, Hannah Agnes Ryan, whose political outlook replaced that of the 'law of the father' in Bracken's upbringing: she took a particularly dim, and snobbish, view of J. K.'s involvement in nationalist movements and discouraged him from participating in them.[11] While Brendan imbibed her anti-nationalism, he also provided her with many problems, as he was unruly, given to drama, enterprise and fighting – as a teenage gang leader, he almost killed someone in a fight – and, in exasperation, she sent him to Australia, to be supervised by a priest who was a relative.[12] On his return, in 1919, he was more or less turned away from the door of his mother's new house, which she had set up on remarrying in his absence. Nonetheless, as Lysaght noted, Bracken continued to 'love[ ] and admire[ ]' his mother 'with a rare intensity', even if 'she did not respond' and 'remained "distant as the evening star"'.[13] In *Double Cross*, the lack of balance in this relationship is deeply entwined in Bracken's disabling attitude to his past: it is presented as a symbol, and as a possible generator, of the restrictive limits to his performance of the self.

These limits are by no means obvious, as after leaving Ireland for England, Bracken embarked upon a truly remarkable self-translation: from inveigling himself into a minor public school, he went on to become a newspaper baron, Conservative MP, Minister of Information and Viscount Bracken of Christchurch. He achieved this through his capacity for audacious lying and a commensurate, boundless enthusiasm and energy for performing; he was a 'self-made' man, who insinuated himself into the upper echelons of English society. A major step in this process is revealed in a letter to his mother: 'I've decided to hitch on to a political star, [...] Winston Churchill.'[14] Bracken energetically pursued Churchill's acquaintance after their first meeting in 1923, bursting into his family life, much to Churchill's amusement and his wife's consternation: while his youthful zest and loyalty appealed to Churchill, he 'was all that the English upper-class lady who was Winston Churchill's wife found distasteful: brash, pushing, over-familiar, disrespectful of persons and conventions'.[15] By the time Churchill became Prime Minister, Bracken was one of his closest confidants and a member of his 'kitchen cabinet'; and, in 1941, Churchill asked him to take on the role of Minister of Information, the position he holds in *Double Cross*.

Admittance into Britain's upper classes is normally through birth, and members of this class could only make sense of Bracken's incongruous position as an outsider so close to the centre of power by positing a filiative link: the rumour spread that he was Churchill's illegitimate son, the result of one of Churchill's colonial sojourns in South Africa or Ireland.[16] But, as Castlerosse puts it in the play: '[t]he thing about the English is that they're too damned polite to ask who you are or where you're from' (*DC*, p. 37), and Bracken instinctively knew how to exploit the hypocritical basis of this social group. Anxious to find a substitute father, he propagated this rumour, even stealing photographs of Churchill and displaying them 'like family snapshots in his flat', as Churchill's wife reported.[17] Churchill did little to deflect the rumours and, according to his eldest child Diana, his children 'grew up actually believing that Brendan was their father's son'.[18]

In conjecturing that Churchill was 'the living embodiment of that eighteenth-century England of Bracken's day-dreams, bestriding the modern age', Boyle points toward a possible motive informing Bracken's choice of Churchill as a substitute father.[19] At the heart of Bracken's eighteenth-century reveries was Edmund Burke, who, like Bracken, left Ireland to make his name in England. Churchill, in this light, embodies a patrimony stretching back to Burke that Bracken wants to inherit. In order to create this 'natural' succession, he had to elide the constructed nature of this project and, indeed, the warnings about the limits of this approach evident in Burke's thinking and career: while he reinvented himself through a process of 'grandiose self-gentrification', as Lock puts it, Burke 'never [...] achieved complete acceptance or integration into English society, remaining something of an Irish outsider'.[20]

'"It's a family portrait," [Bracken] was wont to tell guests whose attention had been directed to Romney's portrait of Edmund Burke, which was his most prized possession.'[21] A reproduction of this portrait hovers over *Double Cross*, and Bracken's Oedipal struggle in the drama – his wish to replace his dead father with a substitute father who might be worthy of his idealised mother – brought him into conflict with the laws of this chosen father. As Deane notes, for Burke, 'the real world and the represented world are clearly different from one another and the real world is the senior of the two. [...] This is the heart of the Burkean aesthetic. It is profoundly anti-romantic. According to it, no representation can or should compete with the actual.'[22] Such an aesthetic found political expression in the priority that Burke ascribed to inherited authority, which compounds Bracken's dilemma. This is alluded to in an oblique manner in *Double Cross*. Having learnt that his friend Beaverbrook has sent a reporter to Dublin who has unearthed files in Dublin Castle on his father, Bracken takes momentary refuge in a Burkean trance: '"[t]o be bred in a place of estimation, to see nothing low or sordid from one's infancy—to be habituated in the pursuit of honour and duty—"' (*DC*, p. 44). Bracken only gets this far, but that which is not heard, which completes this quotation from *An Appeal from the New to the Old Whigs* (1791), speaks directly to Bracken's problematic filiation: 'these are the circumstances of men that form what I should call a *natural* aristocracy, without which there is no nation'.[23] He does not finish the quotation because Beaverbrook interrupts him by recalling that Gandhi had 'once quoted Edmund Burke at me' (*DC*, p. 44), an intervention that threatens to shatter Bracken's iconographic construction of Burke as a father. While he claimed that the quotation concerned 'the British inheritance' (*DC*, p. 44), the reference to Gandhi would have reminded Bracken that Burke also denounced the colonial activities of Warren Hastings and the East India Company in precisely these terms: they were ruining the natural aristocracy of India.[24] Furthermore, Burke employed these same arguments against the administration of Ireland under the Penal Laws, describing the 'unnatural' authority of the Protestant Ascendancy 'junto' as, 'nothing like an influence obtained by virtue, by love, or even by artifice and seduction'.[25] Burke's heritage is therefore ambivalent: although by no means anti-imperialist, the father of British conservative thinking was also, precisely through his privileging of forms of 'natural' authority, 'the father of Irish national sentiment', as Deane notes.[26]

Burke's imperial outlook eschewed a simple 'civilians and barbarians' binary, as, according to Gibbons, he believed that '[c]ivilization did not proceed along a one-way street, bringing the Enlightenment from the centre to the periphery, but travelled in both directions, with each side standing to gain from an encounter with "the other".'[27] Bracken's version of Burke, though, had to be unambiguous, and an unambiguous rejection of that which his father represented, as is evident in his violent reaction when Beaverbrook asks if he knew Gandhi.

> BRACKEN: (*Sudden turn so that it shocks* BEAVERBROOK *and drains* BRACKEN) I don't need to know him. I know his type. That sanctimonious, ascetic face with its watery eyes, running with pious self-denial. Peasants in the fields. What an image for the world to imitate! I despise him, begging bowl in one hand, dagger concealed in the other. Determined to prove the superiority of the primitive. I would happily trample him into the ground. What do these people know of law? Of grace? Cultivated living? They would overrun us, mark you. With their foul smells. Their obscene rituals. Animalism. (*DC*, pp. 44–5)

At the same time, Bracken's political attitudes owed something to the tradition he abhorred. Stanley Baldwin called Bracken one of Churchill's 'faithful chelas', and 'chela', according to Lysaght, is a 'Hindustani word for a person occupying a position between that of a slave and a disciple'.[28] It expresses the loyalty of one possessed, who is property and, for Lysaght, there is 'no contemporary English analogy [...] capable of doing justice' to this concept; rather, such 'unconditional loyalty' to a passionate defender of the British empire can only be understood in terms of pre-/anti-modern, anti-colonial, political formations: '[i]t belonged to another age and another civilization. It had about it the fidelity of the Gaelic clansman. The men in Ireland who stood by De Valera, or Parnell before him, would have understood.'[29] While the mode of Bracken's loyalty is inherited from his father, its expression is always a betrayal of his father. Hence, his acts of treason invoke the ghost of his father, who, of course, denies Bracken the place he is trying to fashion for himself in a Burkean natural aristocracy.

Moreover, Bracken's performance sabotages his objective: his triumph in bluffing and bantering his way into the English upper-classes unveils the entirely fabricated character of the concept of a natural aristocracy. Yet, he cannot maintain an ironic distance from this *milieu*; he wants the myth to be real and to play an integral role in it. He therefore rejects the status of a vital colonial catalyst that Beaverbrook ascribes to him: '[t]he Englishman always respects the man who comes in from the colonies. Always has. Always found a place for him. New blood. New energy. Men tempered by the fire at the front. Men who make the choice' (*DC*, p. 43). Such a *parergonal* role would represent, in terms of Burke's famous binary, a version of the sublime, which is, as Eagleton has argued,

> a negation of settled order without which any order would grow inert and wither. The sublime is the anti-social condition of all sociality [...] Its social connotations are interestingly contradictory: in one sense the

> memory trace of a historically suppressed barbarism, it also has something of the challenge of mercantile enterprise to a too-clubbable aristocratic indolence. Within the figure of the sublime, warring barons and busy speculators merge to prod society out of its specular smugness.[30]

Rather than the dynamism of the sublime, Bracken hankered after the mimetic indolence of the 'naturally' beautiful, and he laboured constantly to claim his place in the 'settled order'. In particular, Bracken employs the 'unnatural' tools of language to attain a 'natural' position.

Read as a strategy of interpretation, the Burkean notion of a natural aristocracy would tend to insist on a confluence of the signified and the signifier: as the meaning of the man is predetermined by the circumstances of his family, there should be no room for an alternative interpretation. Gandhi would have reminded Bracken that even within this concept that apparently extols inevitability, there is the possibility of reinterpretation. This exegetical point reopens the gap over which Bracken's identity floats. Through a counterfactual use of language, he has reinvented himself in order to become worthy of an ideal, substitute father: deception and artifice have been employed to demonstrate his fidelity to such a 'natural' state. Furthermore, his project of erasing all marks of his own translation, specifically all traces of his father, draws constant attention to a most obvious translation effect – the split between the signifier and signified, which he has excessively and explicitly exploited to create himself – with the result that he does not betray his father, but himself.

While Beaverbrook assures Bracken that he has consigned his history to his 'Deadman's Chest' (*DC*, p. 45),[31] Bracken has already interred his past in a form of inner crypt, as is attested to by the occasional reappearance of its chief inhabitant, his father. The crypt, for Derrida, is a form of parasite, 'a foreign body included through incorporation in the Self', which 'always marks an effect of impossible or refused mourning'.[32] Mourning is impossible because the occupant of the crypt is kept there as undead: '[t]he inhabitant of a crypt is always a living dead, a dead entity we are perfectly willing to keep alive, but *as* dead, one we are willing to keep, as long as we keep it, within us, intact in any way save as living.'[33] Incapable of mourning him, Bracken keeps his dead father alive as symbol of his dead self. While all his actions are dictated by the need to bury, to place out of sight, his past, these also grant it a life of sorts, which manifests itself in the 'phoney' quality of his discourse – it always sounds somewhat wrong. The occasion of his father's *Hamlet*-like appearance in the play, on the roof of the Admiralty Building during the Blitz, drives this phoniness to its most heightened pitch. On the roof, he enters into a conversation with a Warden, and even such a casual acquaintance receives the benefit of Bracken's fabrications, which, perhaps inevitably, concern betrayal. Having told the Warden that his brother had '[d]ied splendidly' (*DC*, p. 50) in the RAF, he changes tack completely when the Warden remarks that his son is also in the RAF; his brother immediately becomes a traitor who had, in the Warden's words, '[g]one over to the other side' (*DC*, p. 50). His brother, Peter, plays a shadowy, threatening role in *Double Cross*, as another manifestation of the spectre of his father and, after an explosion knocks out the lights, these spectres emerge from 'the other side'.[34]

> BRACKEN: (*Deep fear*) Daddy! Please—no! Mammy, Mammy! (*At first, faltering*) My father, my father who is, my father was—(*Low, strong Tipperary accent, boyish tone*) Me father was wan of the lads, so he was, wan of the hillside men. He took the oath. He was out in the tenants war of eighty-nine. (*Shift. Heavy adult voice*) Bejasus I was. I knew the treason prisoners of sixty-five. They were all great friends of yer father, so they were, the men that rotted away in Pentonville and Portland but bechrist their time will come again and when it does—
> [...]
> (*Irish accent, as before*) Come here, Peter, me son. Come here a minnit, me little man. You'll stand up for Ireland, won't you, boy, when you grow up, not like that brother of yours in the corner, Brendan-Brendy, the little scut, Mammy's pet, always whinging and bawling, four-eyes—(*English accent*) Leave me alone! Leave me alone! (*DC*, p. 51)

In this Babel of spectral voices, it is difficult to determine who is speaking through Bracken. It would appear that there is a young Bracken and his brother – one afraid of his father; the other proud of him. There is also his father upbraiding him and what is presumably an old IRB companion of his father: a spectre who promises the return of other spectres. As Derrida suggested, '[a]t bottom, the spectre is the future, it is always to come, it presents itself only as that which could come or come back'.[35] In this crepuscular light, Bracken's fear of a re-emergence of his past is expressed through a constant need to re-kill his father: an impossible process that keeps his father alive as dead.

Barbara Johnson has suggested that the doubled fidelity called for in the act of translation renders the translator a 'faithful bigamist, with loyalties split between a native language and a foreign tongue'. A bigamist is also 'necessarily doubly unfaithful', and so a translator 'must push to its utmost limit the very capacity for faithfulness'.[36] As is apparent in the letter to his mother with which the 'Bracken Play' draws to a conclusion, Bracken is incapable of a doubled fidelity/treason; instead, he recreates his mother into a symbol of perfect fidelity – a representation of a past to which he can manifest unequivocal loyalty. His version of his mother's fidelity is clearly an invention and an erasure of her otherness – just as the Burke who would serve as his substitute father was an attenuated Burke. Unable to come to terms with the fiction of identity, Bracken's self-translation cannot faithfully betray either his 'origins' – his problematic 'mother tongue' – or the English upper classes. Instead of a form of faithful bigamy, Bracken's excessive fidelity assumes an incestuous, Oedipal shading in the play. The disabling mode of translation that this produces is evident in his interactions with his girlfriend, Popsie, which describe, to use Johnson's terms, a 'scene of linguistic castration'.[37]

Rather than an intense encounter with a real other (and hence with himself), sex only interests Bracken as an image of sex. Sitting on his knee in a Boy Scout's uniform, Popsie bluntly enquires if he thinks he's 'making love to a boy?' He dismisses this as 'boringly literal':

> BRACKEN: Who would want anything else with the pleasure of such a—such a—composition?
> POPSIE: Speak for yourself.
> BRACKEN: It's the image, my dear. What might otherwise be beyond our reach. I wonder if that's what it's all about? The contrivance of what is really inaccessible?
> POPSIE: What? Sex?
> BRACKEN: No. Art. Well, both as a matter of fact. (*DC*, pp. 32–3)

Sex, for him, is a manifestation of the mimetic qualities of Burke's notion of the beautiful: the composition is everything. Hence, the would-be sexual encounter comes to a jarring halt when Popsie critiques his use of language. Bracken's fastidious creation of himself as emphatically English, which involves never leaving his use of language 'to chance' (*DC*, p. 35), does nothing more, as Popsie notes, than reveal his 'native' roots. Such a critique is unbearable for him:

> POPSIE: [...] it's rather as if one were speaking to someone who was discovering the words as he went along. It's aboriginal, extremely refreshing, of course—
> BRACKEN: (*He means it*) That is the most appalling thing anyone has ever said to me.
> [...]
> It [language] is what makes me what I am! Without it, I am nothing! (*DC*, p. 36)

At the same moment, he betrays his limiting use of language: wishing to escape, he tells Popsie that he loves her; as she knows, this has nothing to do with love, but rather, 'is the signal for us not to see one another for a while' (*DC*, p. 35). Ultimately, the subservient loyalty he displays to his images of Churchill, Burke and his mother makes fewer demands on, and is less threatening to, his sense of self. These fidelities render him something of a void; as Popsie explains: 'how could one possibly give oneself wholly to someone who was never there [...]?' (*DC*, p. 30). He is a newspaper baron who created himself through his words, but they are not intended to be a mode of communication, in the sense that he is uninterested in dialogue, in engaging with the other. Rather, his interlocutors become, as Popsie notes, undifferentiated member of 'an audience' (*DC*, p. 47).

The 'Joyce Play' is an eerie, distorted, echo of Bracken's, and the counterpart to this scene is the one in which Joyce's wife, Margaret, reveals to him that she has had sex with Erich, her language student. Joyce cannot come to terms with 'the betrayal, the betrayal' (*DC*, p. 73), to which Margaret responds by reminding him of what she understands to be their shared ideals: '[n]onsense—How could I have betrayed you if we were both free? [...] my sexuality [is not] the property of you or anyone else. [...] How could one possibly betray something which simply doesn't control one?' (*DC*, p. 73). Such a concept of freedom is beyond Joyce's comprehension; like Bracken, his moves to emancipate himself from Ireland involved allowing himself to be possessed by an allegiance (to what he construed as a pure form of fidelity), with the result, as Margaret's equation suggests, that his fidelities find expression through betrayal. In common with Bracken, he employs language to conceal a void, which his use of language does nothing other than reveal. Margaret recounts that Joyce 'talked non-stop for twenty-four hours' after she had revealed the affair; then, '[f]inally, it stopped. That sickening spill of words. I thought I should never hear that voice stop. But it did. It was like an engine running down. There was always some gap between what he said and what he really felt. When that gap widened all that was left to him was speech' (*DC*, p. 75). This excessive logorrhoea is an attempt to vomit a parasitical crypt buried deep within the self and which structures a chasm within his language. The presence of this crypt becomes apparent through his interactions with his ghostly other and double, Bracken.

Again echoing Bracken, Joyce's use of language is not about communication with the other, and so it is not surprising that their exchanges do not constitute a dialogue, but take place, in particular, over the radio. Both halves of *Double Cross* emerge out of a Babel of radio static, the French term for which is *parasite*. Static is not so much an outside agency interfering with the medium, but rather it is the medium itself; it is, it could be said, radio revealing itself as a parasite. The links between the parasite and translation were explored in the reading of *Three Sisters* in Chapter 2, and the Nazis sought to harness radio's translational, parasitical properties. Eugen Hadamovsy, the Chief Programme Director of the *Reichsrundfunkgesellschaft* (the propaganda service for which Joyce worked), saw his operation as conducting a form of germ warfare, with radio as an unstoppable virus that defied boundaries:

> we are possessed in it [radio] of a miraculous power – the strongest weapon ever given to the spirit – that opens hearts and does not stop

> at the borders of cities and does not turn back before closed doors; that jumps rivers, mountains, and seas; that is able to force peoples under the spell of one powerful spirit.[38]

Radio was an ideal tool for propaganda because of 'its capacity for speaking simultaneously to untold millions, each of whom felt addressed as an individual', as Hobsbawm noted. He added, via McLuhan, that 'the medium, not the message' is key to this effect; in other words, what makes radio a potent force is its status as a parasitical mode of translation, not what it conveys.[39] This rendered it the medium *par excellence* of fascism, which depended upon forms of emotional resonance, rather than on any coherent message: as Herf put it, fascism involved the 'reconciliation of technology and unreason'.[40] McLuhan, indeed, has argued that Hitler's political existence 'is directly owing to radio and public-address systems', which he sees as the technology of primordial instincts: as an 'extension of the central nervous system', radio knits the inner selves of those who hear it into a mass, unified, force.[41]

> Radio affects most people intimately, person-to-person, offering a world of unspoken communication between writer-speaker and the listener. That is the immediate aspect of radio. A private experience. The subliminal depths of radio are charged with the resonating echoes of tribal horns and antique drums. This is inherent in the very nature of this medium, with its power to turn the psyche and society into a single echo chamber.[42]

This curious passage suggests that radio renders the private public by transporting the listener to a past that, rather than being unique, takes the form of a type of Jungian collective unconscious: while listeners believe that they are being addressed individually, this virus induces a form of infinite doubling that merges into a single echo. The stage of *Double Cross* becomes an 'echo chamber' of this sort, as both Joyce and Bracken not only echo each other, but their discourses also echo the occupants of their parasitical crypts.

An echo is a passive and distorted form of repetition, and the objective of Nazi propaganda, as Hitler put it in 1933, was to reduce people to such a state: '[t]he enemy must be demoralized and driven to passivity [...] Our strategy is to destroy the enemy from within, to conquer him through himself. Mental confusion, contradictions of feeling, indecision, panic – these are our weapons.'[43] As has been documented, the most insidious aspect of Lord Haw-Haw was not the content of his transmissions, but the fact that he managed to insinuate himself into the British body-politic from Germany; this translation, in itself, fostered disorientation.[44] Indeed, Joyce, as Cole put it, 'was a man who is remembered for what he did not say'; particularly during periods of heightened anxiety, such as the Blitz, many British people, including sections of the government, believed they had heard Joyce say something he had not uttered.[45]

> ACTRESS: Each morning each person seemed to have a different story of what Lord Haw Haw had said the night before.
>
> ACTOR: Two people in the same room before the same wireless would report two different versions of what had been heard— (*DC*, p. 61)

He was usually thought to have said something that required inside, local knowledge, which led some to believe that his information came from 'fifth columnists', Nazi operatives working like parasites in Britain – precisely the sort of disconcerting, distrustful, reverberations his work sought to induce. In particular, Joyce was, in Cole's words, 'supposed to be uncannily well informed about British clocks'.[46] As he puts it in the play, 'people become agitated when their clocks are interfered with' (*DC*, p. 61), and it is suggested in *Double Cross* that the disorientating anachrony that this parasite induces has correspondences with the description of the function of radio offered by the Futurists Marinetti and Masnata in 1933: '[t]he reception amplification and transfiguration of vibrations emitted by living beings living or dead spirits dramas of wordless noise-states'.[47] As a medium that can bring back the dead past, radio, like the appearance of the ghost of Hamlet's father, knocks time 'out of joint'. Just as Joyce's radio persona – Lord Haw-Haw – originally referred to 'a fairly large and mixed collection of individuals' rather than to him alone, all the manifestations of the spectral in *Double Cross* – the dead-as-alive protagonists; the revisiting of Bracken's father through Bracken; Bracken's letter to his dead mother – are marked by a contradictory doubling and splitting of time analogous to that experienced by those who heard Joyce's broadcasts.[48] In this manner, the protagonists express the ways in which their parasitical crypts doubled and split their identities (a process the arbitrating double act, Actor – Actress, terms 'the Double Cross Effect' (*DC*, p. 62)). Through his reading of the ghosts in *Hamlet* and *Ulysses*, Kearney contends that '[d]eficit of mourning triggers fits of doubling'.[49] Bracken and Joyce manifest this symptom because, rather than mourn, they want to (re-)kill aspects of their past with their every utterance, with the result that they are rendered passive, distorted echoes of the occupants of their internalised crypts; as Derrida noted, 'the cryptic safe can only maintain in a state of repetition the mortal conflict it is impotent to resolve.'[50]

While Kilroy has posited that both protagonists had 'the compulsion to eliminate, or, if you will, to betray their ancestors', this is not quite so straightforward in Joyce's case, as his relationship with his father was of a different order to that of Bracken's.[51] As Cole notes, Joyce's father, Michael, 'was staunchly pro-British'; whereas the Royal Irish Constabulary considered the father of the future British Minister of Information to be too dangerous to the British crown to hold a gun licence, the Galway County Inspector of Police affirmed that Michael Joyce was, 'one of the most respectable, law-abiding and loyal men in this locality'.[52] This was the father of the man who, as Rebecca West (who features in *Double Cross* in the guise of 'A Lady Journalist') put it, came to define the meaning of the term traitor for the English.[53] Joyce described his upbringing in 1940 in this manner: 'I was brought up by my parents in a creed of fanatical patriotism which the English people found very hard to understand. From my earliest days, I was taught to love England and her Empire. Patriotism was the highest virtue that I knew.'[54] While Bracken constructed his life against his father's ghost, Joyce wanted his life to be a literal translation of his father's convictions. But just as an excessively faithful translation will be a betrayal, his extravagant fidelity to an idealised conception of Britishness produces a destructive treason.[55]

Between 1916 and 1921 Ireland's colonial relationship with Britain underwent a seismic shift and, in these changed circumstances, Joyce found that his form of loyalty could only find expression through informing. His father's properties were burnt out and when the Black and Tans were introduced in 1920, the teenage Joyce offered his services to them; a scene he drunkenly recalls in *Double Cross*: 'I want to serve, sir. [...] And what can you do, boy? I know all the Sinn Féiners of the town, sir' (*DC*, p. 64). Derrida has argued the 'the crypt itself is *built* by violence', and this initiation into a life of treasonous loyalty to his father's allegiances is coloured by savagery: inextricably linked in his memory of this submission is the fact that the soldiers' truck was flowing with '[s]o much blood that I thought that something was still bleeding in there' (*DC*, p. 65).[56] While biographical accounts would indicate that Ireland played a relatively unimportant part in Joyce's adult life, *Double Cross* suggests that the violent colonial fracture he experienced there warped his subsequent expressions of loyalty.[57]

With the signing of the Anglo-Irish Treaty in 1921, it can be imagined that this 'loyal' family felt betrayed by the power to which they had given their allegiance. Left exposed, Joyce headed for England two days after the signing of the Treaty. In London, he embarked upon his violent involvement with various fascist parties, leaving for Germany just before the outbreak of the Second World War, with the intention, once again, of serving Britain. He never thought of his broadcasts for Nazi Germany as a betrayal of England, which, as Goebbels noted, was key to his success as a propagandist: '"[t]hat boy's all right!" he declared, "[a]nd besides totally incorruptible. He genuinely wants to serve England."'[58] This culminates in the mad linking of Germany and England in his final radio speech, when the defeat of Germany was imminent.

> This is not defeat.
> [...]
> I pray to Almighty God that England will be great again. May the standard of the Hakenkreutz be raised from the dust! I call on the dead sons of

> England who have died without knowing why. I call on their seed that future generations of England may ensure the nobility of the race! *Seig Heil! Seig Heil! Seig Heil!* (*DC*, pp. 25–6)[59]

Recasting British deaths in the war as sacrifices through which a Germanic England may be reborn is the ultimate expression of the wild disjunction of the signified and the signifier that structured his life. Over this scene of imagined sacrifice, the 'crooked cross' (the *Hakenkreutz*) casts a spectral shadow, forming the sign, it might be said, of the double cross under which fascism operated (which is precisely how Chaplin rendered the Swastika in *The Great Dictator* – XX). Doubling and splitting characterise Joyce's double crosses: despite believing that his father was betrayed by England in Ireland, he cannot mourn the passing of an England that had summoned such an extreme loyalty, as to do so would betray his father. Instead, he split the meaning of England and created an idealised England to which his loyalty was summoned, and another England in which this ideal had been betrayed (typically, in his fascist ramblings, by Jews). The war was confirmation for him that this second England had become dominant, and his duty was to help destroy this and restore an England that might live up to his father's loyalty. Joyce's would-be solution to his dilemma structures and maintains, precisely in and through repetition, the crypt in which resides his father's loyalty to Britain. Even at the end of the war, he cannot mourn the death of this idealised England; instead, he rends sense from language by reframing the war's conclusion as the site of the future resurrection of such an England. Finally, in a manner somewhat reminiscent of Ezra Pound's attitude in *The Pisan Cantos*, in the face of 'the enormous tragedy of the dream', Joyce translates his loyalty into a transcendent realm.[60] Awaiting execution, he tells Beaverbrook that he is reading 'Thomas à Kempis' (*DC*, p. 88), and in looking to imitate Christ, Joyce is also renouncing language as a mode of communication: having always distrusted language, this propagandist finally attempts a move to a domain that calls for an inexpressible loyalty.

Just as excessive fidelity produces treason, 'profound fidelities' can be 'produced by treason' (*DC*, p. 84), as Beaverbrook notes. By hanging Joyce, the British had to translate him into a British subject, despite the fact that he was never legally British.[61] Like Bracken, his loyalty to a fixed concept of Britain could only find expression through a fabricated identity and, in effect, a fictional character – Lord Haw-Haw – rather than Joyce was found guilty of treason and executed. As Wilde illustrates in *Saint Oscar*, all identities are inventions; it is the failure of these two extravagant remodellers of the self to acknowledge this performativity that warps their characters: they insist that their performances express an authentic identity. In *Double Cross*, radio acts as a witnessing parasite that picks up the undead sources of these disabling features: the characters betray themselves through radio, and the presence of that which has been kept secret – the crypt – is uncovered.

Having failed to deal faithfully with this other within, their encounters with others without are obstructed, as is evident in the strategies of evasion that mould their discourse. They are both very much at home in the ether where, like ghosts, they engage in one-way conversations.

Prosopopoeia: *Saint Oscar*

> For in art there is no such thing as a universal truth. A Truth in art is that whose contradictory is also true. [...] The truths of metaphysics are the truths of masks.
>
> — Oscar Wilde[62]

While Bracken and Joyce in *Double Cross* worked to occlude their Irish pasts and erase the signs of the contradictions that moulded their identities, the Wilde of *Saint Oscar* is more at home with his past, precisely because he rejects the 'normal', instinctual significance of the ideas of roots. Identity is neither fixed nor singular for him; rather it emerges from a continual performance of a series of different roles that by no means need to be consistent with each other. 'He was', as Eagleton lists, 'English and Irish, upper class and underdog, socialite and sodomite, bohemian and *bon viveur*, aesthete and republican, a respectable paterfamilias who sported with rent boys in cheap hotels.'[63] *Saint Oscar* presents its audiences with a Wilde who not only encompasses these aspects, but through whom other literary and critical figures such as Joyce, Beckett, Barthes and Baudrillard might also be heard. In doing this, the play offers an alternative rendition of 'the Double Cross Effect' (*DC*, p. 62), as these writers who were influenced by Wilde are presented as elements of his identity. Wilde does not play at being these other characters, in the sense that he does not try to imitate them. Rather, they inform the manner in which he performs versions of himself. This Wilde embodies the trope of prosopopoeia, which is, in Riffaterre's description, 'no more than a mask put on something that may not even have a face. The mask's simplicity gives free play to the viewer's imagination. All the mask does is put the reader in a specific interpretative mood.'[64] Character in this anti-mimetic structure takes on a spectral shading: prosopopoeia, according to Fontanier, 'consists in staging, as it were, absent, dead supernatural or even inanimate beings'.[65] Invoking not only the Latin link between person and mask, this also gathers in the French *personne* – 'no one'. Congruent with Butler's view, discussed in the Introduction, that identity is performative, in this construction, the identity of the mask is not determined by an authoritative wearer but by the viewer or the interpreter. Identity, in other words, depends upon the other, as Manoussakis underlines in his reading of the Greek term *prosopon*, which 'was translated by Latin fathers [...] as *persona*':[66]

> [t]o be *pros-opon* means to be on your way towards the other being. This also situates the being into a perpetual ek-sistence, a stepping-out-of-yourself and a being-towards-the-other. [...] Prosopon strongly implies reciprocity of gaze through which the self is interpellated by the other, and ultimately, 'othered'.[67]

This movement distinguishes the Wilde of *Saint Oscar* from the protagonists of *Double Cross*, who worked to avoid the other. Wilde is always potential other: his identity is a manifestation of what Eagleton saw as '[t]he imaginary', in which, 'one passes spontaneously into another, or its image, with apparently no mediation'; a movement that delineates 'a precious ethical moment'.[68] And, as Kiberd has shown in *Anglo-Irish Attitudes: Inventing Ireland* and *Irish Classics*, Wilde's protean qualities are at the heart of his emancipatory politics; indeed, it is apparent that Kiberd's work has heavily informed Eagleton's portrayal of Wilde.

While he professed to be a socialist, Wilde, according to Kiberd, was more properly speaking an anarchist, as at the heart of his political thinking was the conviction that 'all change must begin with the individual'.[69] 'Over the portal of the new world, "Be thyself" shall be written. And the message of Christ to man was simply "Be thyself." That is the secret of Christ': this is how Wilde put it in 'The Soul of Man Under Socialism'.[70] If the creation of the self is one's ultimate task, it can never be definitively accomplished, as it is worked towards through, as Wilde says in *Saint Oscar*, 'play-acting' (*SO*, p. 34), and so assumes different forms. It is only through this process that humans can move towards the idea of self-fulfilment that lies at the core of Wilde's professed utopian socialism: 'I'm a socialist *because* I'm an individualist. [...] I'm prefiguring the New Jerusalem, in which everybody will be able to be purely themselves. [...] Socialism isn't an end in itself [...]: it's just a device for turning us all into Michelangelos' (*SO*, p. 31). Socialism sets the stage upon which people can become artists of themselves. And, in the process of developing their artistic abilities, people will recognise and reject the poorly written and badly acted hegemonic, oppressive norms that allow for just one version of the self.

Wilde's individualism is, therefore, an expression of 'his ethics of "not-being"', and this is linked in the play to his martyrdom.[71] The martyr expresses her/his individualism by giving up her/his life – 'the most valuable thing he or she possesses'[72] – and it is precisely through his martyrdom that the figure of Christ represented 'the supreme Individualist' for Wilde, as Kiberd points out.[73] *Saint Oscar* further projects Wilde's posthumous assumption of the mask of the gay martyr through at least two other masks: Sartre's *St. Genet* and 'Wilde's favourite painting', Guido Reni's *St. Sebastian*, a reproduction of which was in the programme for the first performances.[74] Sebastian became a martyr because he stood up to an oppressive dictator, and Wilde believes that this highest demonstration of individualism disarms the tyrant: '[t]he martyr will always worst the conqueror. Power the conqueror can understand; it's sheer helplessness which leaves him disarmed' (*SO*, p. 49). The power of the ruler 'primarily derives', as Baudrillard notes, 'from the suspension of death. Power is therefore never, contrary to what we might imagine, the power of putting to death, but exactly the opposite, that of allowing to live'.[75] In claiming her/his own life, the martyr defies this logic and removes her/himself from the limiting meaning that was forced on her/him by the master-narrative.

Wilde's radical views on identity and the relationship of art and reality, according to Eagleton, 'prefigured the insights of contemporary cultural theory'.[76] Eagleton, who has been dubbed the 'high priest of lit crit',[77] turned to *Saint Oscar* because he 'had reached a sort of theoretical impasse which could only be negotiated through creative work'.[78] With such a provenance, it is not surprising that, as this reading illustrates, this is a play of citation (notably, in terms of the structure of identity that the play proposes, with one exception, this does not involve directly quoting Wilde). Heaney's critique of another Eagleton play – *The White, The Gold and The Gangerene,* which Field Day considered in 1991 – can be reapplied to this drama: Heaney found it 'a bit too set in its rhetorical and political ways. Subversive and paradoxical and wonderfully gifted speech by speech, but overall, a feeling of predictability – the correct subversion, the well-placed aggravation.'[79] Nonetheless, while *Saint Oscar* is 'not inherently dramatic', as Billington noted, it is equally 'not a documentary', but rather an 'entertaining piece of parodic polemic';[80] indeed, it was Field Day's last box-office success.[81]

Eagleton particularly imagines Wilde as an 'Irish Roland Barthes', whose work is distinguished by an inherent theatricality.[82] As Sontag noted, Barthes's insights are characterised by a process of splitting and doubling: he had an 'aphorist's ability to conjure up a vivacious duality: anything could be split either into itself and its opposite, or into two versions of itself; and one term then fielded against the other to yield an unexpected relation'.[83] Such a protean mode of thinking stands opposed to normative concepts, and this camp politics linked him with Wilde in Eagleton's view:

> Wilde despises Nature [...] because, as for Roland Barthes and Michel Foucault, it suggests an oppressive normativity. Nature is the family, heterosexuality, stock notions, social convention; and Wilde had only to be presented with a convention to feel the irresistible urge to violate it.[84]

Nature, understood in this manner, informed the *bourgeois* myth that Barthes cast an uncanny eye on in his essay 'Myth Today'. His reading of the manipulation of language involved in 'myth-making' speaks directly to nineteenth-century colonial views of Ireland (and their subsequent aping by the post-colonial nationalist state). The myth's 'rhetorical forms' involve, amongst other manoeuvres, 'depriv[ing] the object of which it speaks of all History', with the result that the other can only be imagined through a form of 'exoticism[:] The Other becomes a pure object, a spectacle, a clown.' Such a view is sustained by a form of '[*t*]*autology*' – 'a double murder: one kills rationality because it resists one; one kills language because it betrays one'. In doing this, '[t]autology creates a dead, a motionless world', such as the one Bracken and Joyce sought to fabricate in *Double Cross*.[85] The myth finds culminating expression in the foreclosing, positivist, '*statement of fact*', which

> bur[ies] the traces of [...] production under a self-evident appearance of eternity: it is a counter-explanation, the decorous equivalent of a tautology [...] The foundation of the bourgeois statement of fact is *common sense*, that is, truth when it stops on the arbitrary order of him who speaks it.[86]

'Myths are fictions that have forgotten that they're such' (*SO*, p. 24), as Wilde put it in *Saint Oscar*; and his project is to unearth the precisely aesthetic means through which the positivist common sense that underpins colonial and nationalist discourses is produced.

As has been argued, the construction of a self-evident history is crucial to hegemonic politics: 'the very principle of myth', as Barthes noted, is that 'it transforms history into nature'.[87] Just as Wilde detests nature, he finds much to dislike in such a history: its plot, he complains, is 'appalling thin'; it has '[n]o narrative thrust: just a lot of sub-plots carelessly abandoned, themes left hanging in mid-air, a mishmash of sensational occurrences. Wars, famines, massacres, revolutions: I've never read anything more improbable in my life' (*SO*, pp. 18–19). His nationalist mother presses the claims of realism upon him, arguing that his view would mean that, 'Cromwell was an invention; the penal laws were a fantasy; the famine was a bad dream' (*SO*, p. 25). While such an argument has some counter-hegemonic potential, it is not a sufficient strategy, according to Wilde, because it ultimately involves substituting one myth with another; like Barthes, he felt that 'the best weapon against myth is perhaps to mythify it in its turn, and to produce an *artificial myth*'.[88] Rather than succumb to the myths of Irish nationalism, Wilde points to the artificiality of the construction: 'what else are the Irish but imitation Irish? The whole nation's an imposture. I'm not false, mother: I'm true to the sham of my country' (*SO*, p. 25). By casting the country in such a performative manner, attention is drawn to the means of the production of history: '[t]he Irish have no history. It belongs entirely to the British' (*SO*, p. 25). Wilde's strategy does not avoid the political; it exposes the power structures that gird the myth. It is brought into direct conflict with these structures in his trial scene, as Wilde employs this performative strategy as the basis of his defence: he rejects the British, empirical version of truth, claiming that there is a 'fundamental incompatibility between the English and Irish notions of truth. Truth for you means whatever is the case; I am afraid that we Irish have very little idea any longer of what is the case, if indeed we ever knew' (*SO*, p. 46). He continues by invoking the dead tautological dualities that Barthes catalogued:

> no Irishman can receive a fair hearing in an English court, because the Irish are figments of the English imagination. I am not really here; I am just one of your racial fantasies. You cannot manacle a fantasy. I do not believe in your morality and I do not believe in your truth. [...] You hold that a man is a man and a woman is a woman. I hold that nothing is ever purely itself, and that the point where it becomes so is known as death. (*SO*, pp. 46–7)[89]

As can be heard, at the heart of this myth is a normative vision of gender and sexuality, which, as Wilde found out to his cost, is stringently policed. The myth does away with the body by rendering it inert within a rigid logic of 'either/or' – man/woman; father/mother – which, in common with other colonial and nationalist dichotomies, has inscribed within it a strict patriarchal hierarchy: as Wilde reminds his mother, because of – not despite – the nationalist idealisation of a feminised Ireland, '[t]he Queen Maeves are all in the cowsheds, labouring for their menfolk' (*SO*, p. 23). Men and women are not to be reduced to the myths of their natures, nor is the body an unproblematic expression of nature; rather, it is the site of the others within, as the prosopopoeia that is enacted in this drama makes clear.

Eagleton links the multiple and contradictory identities of Wilde to the fact that he grew up in the colonial context of 'the city which his compatriot James Joyce spelt as "Doublin", and everything about him was doubled, ambiguous, unstable'.[90] As Nolan has noted, in the wake of Joyce, 'the Oedipal struggle with native traditions' of 'every rebellious son of Mother Ireland' has involved a 'struggle with the ghost of Stephen Dedalus', and the doubled ghost of Joyce and Dedalus is invoked in the '*tenuously realist*' (*SO*, p. 36) trial scene when Wilde quips about his relationship with his native country.[91]

> Where I come from, we have rather more respect for our artists than the English do: we drive them out.
> CARSON: You are a foreigner in this country?
> WILDE: That's so. But I speak the language fairly fluently; you were kind enough to teach it to me. Like Caliban, your reward is that I know how to curse. (*SO*, p. 42)

While Wilde left Ireland uncomplainingly to go to Oxford in 1874, for audiences of *Saint Oscar* the allusion to Irish writers driven from their country inevitably brings Joyce to mind. Moreover, embedded in this is a reference to a scene of failed mimesis in the 'Telemachus' episode of *Ulysses*: Mulligan compares Stephen's inability to see himself in the cracked mirror to Caliban's anger at his similar failure. Moreover, this is not simply a direct reference to Shakespeare, but is, as Mulligan intimates, an allusion to Wilde.[92] What is evidenced here is prosopopoeia with a vengeance; as Wilde puts it in *Saint Oscar*: '[i]t's only theatre after all. Strip off one mask and you find another' (*SO*, p. 54). The removal of each mask multiplies rather than limits possible meanings, just as the cracked looking-glass doubles and splits that which it reflects.

Wilde's assumption of the mask of Caliban gestures to a process of citation – to a distorting hall of cracked mirrors in which Caliban, a character from a play, looks for his reflection – rather than to an appropriation of a stable identity. Even Buck Mulligan's reference is split: on the one hand, it points to Wilde's argument about the limits of mimesis in the Preface to *The Picture of Dorian Gray*: '[t]he nineteenth century dislike of Realism is the rage of Caliban seeing his face in a glass. The nineteenth century dislike of Romanticism is the rage of Caliban not seeing his own face in a glass.'[93] The 'cracked looking-glass', on the other hand, is to be found in a passage from Wilde's 'The Decay of Lying', in which normative conceptions of mimesis are inverted; where it is debated if 'Life in fact is the mirror, and Art the reality'.[94] The ghost of Wilde informs the mask of Wilde in *Saint Oscar* in this fragmentary, shifting manner; just as Wilde argued in 'The Decay of Lying' that nature comes into being through art, the identity of Eagleton's Wilde is always emerging through indirect citation. As Wilde put it in 'The Decay of Lying': '[n]ature is no great mother who has borne us. She is our creation. It is in our brain that she quickens to life. Things are because we see them, and what we see, and how we see it, depends on the Arts that have influenced us.'[95] Citation grants nature its meaning, not the other way around. This queer generative model takes the place of Mother Nature, and structures the alternative performance of 'problematic filiation' that is staged in the play. The only filial inheritance Wilde acknowledges from his mother is performance itself: 'I learned all my drama from you, mother. [...] You're playing an Irishwoman, I'm playing an Englishman; what's the difference?' (*SO*, p. 23). Moreover, the allusions to Shakespeare and *Ulysses* also hint at the spectral nature of paternity. Through a doubled Shakespeare – Shakespeare the man and Shakespeare the composer of *Hamlet* (and performer in the play as the ghost of Hamlet's father) – Dedalus constructs a tenuous argument in the 'Scylla and Charybdis' episode of *Ulysses* about the 'legal fiction' of paternity.[96] Again, in the place where origins might be expected, a process of fragmenting citation is encountered and, just as Dedalus saw the cracked mirror as a symbol of Ireland, amongst these scattered reflections, glimpses of a splintered Ireland can be perceived. But, rather than take on the role of a mother- or father-land, this Ireland accretes meanings through a process of dispersive citation; it is by bringing this to light that the Wilde of *Saint Oscar* articulates a disruptive, postcolonial intervention.

Arnold and Renan haunt the references to Caliban in *Ulysses*, and the sort of hierarchical antitheses (Hebraic/Hellene; Saxon/Celt) that sustain colonial stereotypes such as theirs can find no purchase in the ever-unfolding 'both/and' logic that fuels the dynamic of prosopopoeia in *Saint Oscar*.[97] Wilde's direct reference in his reply to Carson is a reworking of Caliban's reply to Miranda in Act I, scene 2 of *The Tempest*: 'You taught me language; and my profit on't / Is, I know how to curse'. This is the classic statement of the unbalanced exchange that takes place within the scene of colonial translation: even Caliban's 'profit' marks him out as a inauthentic copy because, as Cheyfitz points out, 'he does not know how to curse in the eloquent, or lethal, manner of his master'.[98] Wilde, though, is the Caliban who can out-Prospero Prospero, who is, in this case, being played by another 'Doubliner': the unionist leader, Carson. Wilde's Caliban is as an overly-faithful translation of English norms; Richard Wallace's critique of this performance completely misreads it, as he assumes that Wilde aspired to be an authentic part of English upper-class *milieu*: '[y]ou don't fool anybody, Oscar [...] You're much too perfect to be the real thing' (*SO*, p. 31).[99] Like Bracken and Joyce in *Double Cross*, Wilde commits an act of treason through a form of excessive fidelity; unlike them, however, his betrayal is not an unintended consequence of his self-translation, but is an elegant unveiling of the essentially performative basis of his hosts' identity: 'that's just to rub their noses in it: to show them I can handle their preposterous conventions even better than they can. That's what infuriates them so much; they can't tell whether it's praise or parody. I subvert their forms by obeying them so faithfully' (*SO*, p. 31). While Wilde, according to Yeats, could 'never put off his wonder completely every morning [...] in remembering that he had dined yesterday with a Duchess',[100] in *Saint Oscar* he is very aware of his status as, in Kiberd's words, an 'uninvited guest', and he reacts to this by being a guest who leaves his hosts hostages to their own fortune.[101] His exaggerated, flamboyant mask of a parasite drags the parasitism of his hosts into the limelight; as Wallace reflects: 'I've never been sure whether you're an upper-class wastrel or a feckless Mick. Maybe there's not much difference' (*SO*, p. 32). By performing as such a cracked looking-glass, Wilde, as Kiberd noted, undermined the colonial stereotype of the 'stage Irishman': '[t]he ease with which [he] effected the transition from stage Irishman to stage Englishman was his ultimate comment on the hollowness of the antithesis, on the emptiness of both notions.'[102]

Mimesis is eschewed in this operation, as built into that mode is a privileging of that which is being reproduced; instead, to adopt Lloyd's terms, Wilde produced 'per-version[s]' of these types through forms of 'parodic' translations.[103]

> Parodic translation strips away the 'outward mask and semblance' of poetry precisely by emphasizing the radical disjunction between the surface

> and the interior which, supposedly, it masks in order to reveal. So radical is that disjunction, indeed, that what is found when the original is unmasked is the absence of substance, the subsistence of 'incongruity' and 'anomaly' rather than the subjective integrity of the original poet.[104]

Such a mode of translation fractures the relationship between nature and representation, which has been based on seeing the latter as dependent on and serving the needs of the former – a hierarchical structure that is replicated in other binaries, such as man/woman and coloniser/colonised. Instead of inverting this hierarchy, through the processes of Wilde's parodic, perverse translations, the signifier is sacrificed in a way that it asserts its independence from the limits of this relationship with the signified, and can thus form new associations with different, unsanctioned signifiers.[105] In this precisely metaphorical fashion, Wilde exposes what Eagleton calls the 'ceaseless sameness-within-difference' that drives hegemonic discourses and suppresses the play of difference-within-apparent-sameness that constitutes the self.[106]

Interpreting roles: *The Madame MacAdam Travelling Theatre*

> What is tragic is not the impossibility but the necessity of repetition.
>
> — Jacques Derrida[107]

While the 1995 production of McGuinness's *Uncle Vanya* was the last Field Day tour, it constituted, in effect, a belated curtain call as the company's annual tours came to a close in 1991 with *The Madame MacAdam Travelling Theatre*. As is recounted in the Epilogue, from the late 1980s, there was a gradual dissolution of the energies necessary to maintain a touring company. Yet, although all the directors recognised that the theatrical side of the enterprise was in great trouble, there was no uniformity about how to proceed, and the indecision and confusion within the board wounded *Madame MacAdam*'s tour. Surveying the state of the company at the end of 1991 (a fateful year for Field Day), Deane acknowledged, in a letter to all the directors, that the board had not supported the play properly and that 'FDay [had] let Tom Kilroy down in many important respects'; a failure that 'is all the sadder because the theatre has been FDay's flagship for so long'.[108] Pelletier has argued that a meta-theatrical 'ironic self-consciousness' informed Kilroy's decision to offer this play to Field Day; this irony is redoubled and assumes a darker hue in light of the fact that a drama about the difficulties of touring should have been dogged by severe production problems.[109] So frustrated did the cast and crew become, that they sent a group letter of complaint to the Field Day directors, cataloguing a list of problems with almost every aspect of the project: the director was inexperienced; Kilroy's presence at rehearsals was intimidating; the tour's planning and itinerary was haphazard; the set was impractical; advance publicity was poor; the negative reviews were disillusioning; and even the handling of the dog was not properly organised.[110] Considering this level of disorganisation and discontent, it is not surprising that the tour was a failure, and the *Madame MacAdam* experience hastened the end of Field Day's touring.

Touring companies – the so-called fit-up companies – played a crucial role in bringing theatre to audiences outside Dublin in the early- and mid-twentieth century. The most important of these was Anew McMaster's company, which toured Ireland – playing almost exclusively in small towns – from 1925 to 1960. It was, in Morash's description, 'as much like a travelling circus as a theatre', but had a remarkable impact on Irish theatrical life: 'over the years small Irish towns saw a line-up that in retrospect reads like an Equity *Who's Who*: Harold Pinter, Milo O'Shea, Patrick Magee, T. P. McKenna, Coralie Carmichael and Kenneth Haigh', with the result that amateur theatre thrived in the towns it visited regularly.[111] McMaster also had a formative influence on Kilroy: invited to his school, he talked Kilroy's class through *Hamlet*, playing all the roles, including a 'simpering Ophelia'. The intimate insight into theatrical performance that he provided – the fascinating and somewhat estranging sight of a man sitting in a classroom in glistening make-up – left the young Kilroy with the feeling that theatre could be created in any context.[112]

Rea has recalled that it was a 'pleasure to go to towns where people would say: "You're the first professional production we have had since Anew McMaster was here thirty years ago."'[113] In some respects, Field Day saw itself as operating in the tradition of the fit-ups and for some time had the idea of producing a play based on the experiences of these companies. On 28 August 1987, Friel sent a letter to Harold Pinter (whose 'first proper job on the stage' was touring Ireland for two years in the early 1950s with McMaster),[114] in which he introduced Field Day as an 'enterprise [...] not unlike Anew McMaster's', and asked if Pinter could do 'something for our company – ideally a play that had to do with McMaster, or your Irish experience with McMaster, or just about fit-ups generally'.[115] While Pinter could not do this, the programme for the first performances of *The Madame MacAdam Travelling Theatre*, which is very loosely based on McMaster's enterprise, included an extract from Pinter's 'Mac', an affectionate evocation of his time with McMaster.

> The people came down to see him. Mac travelled by car, and sometimes some of us did too. But other times we went on the lorry with the flats and props, and going into Bandon or Cloughjordan would find the town empty, asleep, men sitting upright in dark bars, cow-pads, mud, smell of peat, wood, old clothes. We'd find digs; wash basin and jug, tea, black

> pudding, and off to the hall, set up stage on trestle tables, a few rostra, a few drapes, costumes out of the hampers, set up shop, and at night play, not always but mostly, to a packed house (where had they come from?); people who listened, and who waited to see him, having seen him before, and been brought up on him.[116]

In its tours, Field Day also enjoyed similar, essentially improvised experiences: Rea, for instance, has recounted how, in certain towns, the curtain only rose when the audience had found its way to the venue – after the work on the farms had been completed and tea had been had – rather than at the scheduled time.[117]

Suitably for a play based on a touring company, the set of *Madame MacAdam* is dominated by a van, which also – like the use of a video screen in *Double Cross* – registers Kilroy's 'resistance to the limiting effects of stage space'.[118] While the letter from the cast and crew indicates that this set proved to be cumbersome and impractical, it embodies the meta-theatrical significance of the drama: it is constructed so that it becomes a stage upon which the characters perform both the company's plays and their lives. Therefore, audiences, as Hayley notes, 'are forced to question the distinction between those worlds [...] [and] are never allowed to forget that [they] are in the theatre'.[119] According to Murray, the interplay of these two worlds caused critical confusion, and contributed to the hostile reviews of the play 'from the Dublin critics, who [...] refused to see, to accept, what was in front of them[:] [...] a world not of realism but of deliberate make-believe'.[120] Rather than viewing theatre as distinct from the world at large, the play proposes three different, often conflicting, understandings of how theatre provides a paradigm for life.

Cut adrift in the newly post-colonial Ireland during the Second World War, Madame MacAdam's company is running out of steam. Things have changed greatly since the time when Lyle Jones emerged from Wales as a sort of dashing Richard Burton-type to sweep the English Madame MacAdam off her feet; now, he is a fading, camp old-timer. Madame also seems to be beyond romance, and has settled into her role as a practical, mothering type. The distance between the two is reflected in their divergent views on the actor's task. Both believe that what transpires in a theatre is a mysterious miracle; however, while Lyle Jones holds that an 'actor [...] is forever exposed' (*MMTT*, p. 27), for Madame, actors 'create through concealment' (*MMTT*, p. 43). Madame sees donning the actor's mask as a prophylactic that protects the identity of this mystery, knowledge of which should be available only to those who have been initiated into the thespian rites. Lyle Jones, on the other hand, thinks that the miracle of theatre occurs through others' perception of the mask: 'we actors are the creation of our audiences. They create us nightly. We exist only in their imaginations. When they walk out of the theatre we cease to exist' (*MMTT*, p. 17). This is the credo of *Saint*

Oscar: the self attains its highest expression through what Hayley has called 'self-denial'.[121] Like Wilde, Lyle Jones does not restrict the illusion of theatre to the stage, as is most evident in his participation in a greyhound racing betting fraud – a conspiracy that not only calls upon his thespian skills, but also requires the dog to be masked, to appear as another dog. This appals and exasperates Madame, who believes it is dangerous to transgress the boundary between the stage and real life: actors 'can only play behind the lines. Beyond is the pit of darkness' (*MMTT*, p. 30). Such a view also betrays an ethics of 'self-denial', but one that is manifestly different from Lyle Jones's: by foregrounding the lines, Madame sees her role as that of a transparent medium through which the original text is relayed. In opposition to Madame's impulses towards literal translation, Lyle Jones domesticates the text he interprets. However, as their interactions indicate, these positions are neither fixed nor are they fully other to each other.

Madame unveils the play as a banal repetition of tired, conventional scenarios: another helping of 'the usual fare' (*MMTT*, p. 1). With its focus trained on the meaning of the original, her sort of literal translation resorts to stock phrases in the receiving language that fail to revive the source text – '[a] love story. A lost child. Villainy at large' (*MMTT*, p. 1). This sort of repetition gives Lyle Jones 'the willies' (*MMTT*, p. 13); yet, he, as Madame notes, has proclaimed each audience he's played before for twenty-five years to be the best. Without some level of repetition, as Lyle Jones claims, 'one is unable to go on' (*MMTT*, p. 12). However, what is clear is that this must be continually questioned; when it becomes habit, it devitalises the performance. While Lyle Jones justifies his domestication of the text by prioritising the audience, he nonetheless domesticates the audience through repetition: the different audiences he has played to for twenty-five years are an undifferentiated entity, as is evident in his rather stereotypical opinion of Irish audiences.

Tableaux drawn from Irish nationalist history are the principal ingredients of the company's fare. Re-enacting the capture and trial of Robert Emmet, these British actors get their most enthusiastic reactions for lines such as: 'I know thy dream, to free Ireland from the Saxon yoke' (*MMTT*, p. 15). These statements are divested of meaning for the company; the actors regard them as little more than rhetorical exercises which they don't confuse with history: 'the Irish simply adore these Hibernian melodramas' (*MMTT*, p. 17). 'The Irish', as the Wilde of *Saint Oscar* puts it, 'have no history. It belongs to the British' (*SO*, p. 25). This proprietorship persists in the post-colonial Ireland of *Madame MacAdam* because of the linguistic hegemony of English. As was testified to by Lombard in *Making History*, nationalism clad its heroes in the rhetoric of the British empire, which, at times, sounds like a schoolboy's literal translation of a Latin text: '[o]h, Providence, guide me this night. Hark! Who's there?' (*MMTT*, p. 15). Lyle Jones, indeed, explicitly frames Emmet's revolution in terms of Cromwell's takeover of Westminster. Rather than expressing any real meaning, what is important is the clichéd heroic music of this rhetoric, as is evident when Lyle Jones strays from Emmet's speech from the dock – one of the central texts in Irish nationalism – to Sydney Carton's thoughts on the scaffold in *A Tale of Two Cities*, and the divergence is not immediately noticed by the audience. Such a lack of discrimination is a symptom of Lyle Jones's form of domestication and his need to please an audience: an indispensable aspect of the fluency he aims for is achieved through a certain 'now-ness', which involves eliding or smoothing out the past into a more readily digestible form. On the other side of the footlights, the audiences' reactions to these performances provide insights into the 'discursive values and institutional limits', to use Venuti's terms, of mid-twentieth-century Ireland.[122]

Madame finds Ireland a rather menacing place because, like Lyle Jones, the Irish do not discriminate between the stage and real life. However, the stage upon which the Irish play out their lives is both fantastical and restricted through forms of self-imposed domestication. The paucity of imaginative freedom available in the Irish Free State is seen in the reactions of Marie Therese and Jo (teenagers from the town) to the arrival of the theatre company. For both, it represents a form of escape: Marie Therese sees it as a release into unreality, away from the quotidian boredom of rural Ireland; it offers Jo, on the other hand, a getaway into reality, away from the unreality of Ireland during the Second World War. While Marie Therese cultivates romantic illusions that she maintains as illusions because she needs the security of constriction, Jo embraces reality by meeting and having sex with Rabe, an actor with the company. When she tries to talk about this experience to Marie Therese, she discovers that her Irish Catholic upbringing has failed to supply her with the appropriate words, whereas it compels Marie Therese to insist that the event belongs entirely in the realm of language – that it could not have happened. Neither, in short, can connect language to experience, and this is portrayed as a symptom of the somewhat unreal, disconnected nature of neutral Ireland, where the Second World War even assumed a different guise: as 'the Emergency'.

Madame MacAdam is the second play in this chapter that is set during the Second World War; however, unlike *Double Cross*, the action of this piece takes place far from the crucibles of the London Blitz and the fall of Berlin. While neutrality was, as Elizabeth Bowen reported, 'Eire's first *free* self-assertion', it was also an act of restriction and, on some levels, a repetition of British practices and rhetoric.[123] Wills has pointed out that the term 'Emergency' came from the 1939 Emergency Powers Bill, which 'was modelled on the British Act' that was passed a week before it, and both had the same restrictive objectives of 'maintaining public order and suppressing dissent, and [...] regulating and controlling essential supplies'.[124] Through a strict system of censorship, Ireland was to be particularly inoculated against the parasitical germs that thrived in communication, and this helped induce what Keogh has described as a 'womb-like sense of security' in the country.[125] As Marie Therese illustrates, the ignorance that this fostered was translated into an expression of Catholic innocence, which was, at the same time, being promoted as an expression of Irish identity; in Wills's words: '[p]olitical censorship not only compounded the existing moral and religious censorship – it made it appear patriotic, and so all the harder to criticise.'[126] While there is little argument over the fact that, on a practical level, neutrality was Ireland's only option – it had no army, navy or air force to talk of – neutrality, nonetheless, generated a form of moral vacuum, which was most tellingly manifested in de Valera's reprehensible signing of the book of condolences on Hitler's death (an act that might be said to be a far too literal, post-colonial translation of the British ideal of 'fair play').

In the play, this failure of moral authority is manifested in the insecure manner in which the Irish assume their roles as hosts of the British theatre company. At its worst, this finds expression in the violent hostility of Bun Bourke, the town baker and leader of the Local Defence Force unit. His insecurities are particularly visited upon Rabe, who, as a Jew, is the most other figure in the British troupe. Bourke resents his sexual freedom, and pounces on Rabe while he is making love with Jo. A girl has been kidnapped – yet another lost child in the Field Day *oeuvre* – and Bourke accuses Rabe of having stolen her. In fact, Jo – a lost child herself – had taken her as a form of emotional compensation for her absent parents. However, as Young Maher reports, Bourke beat a confession out of Rabe and wants 'to send for the hangman beyond in England, Mr Pierpoint to come and hang him' (*MMTT*, p. 55). Salmon, in a *realpolitik* reading, has argued that Irish neutrality essentially depended upon British tolerance, as Ireland was not capable of protecting itself.[127] In a not entirely dissimilar manner, Bourke has not shaken off the culture of dependence fostered by colonialism: he is incapable of taking responsibility for the consequences of his actions as host, and automatically expects the execution of justice to come from England.

Bourke's right to nominate himself as host is challenged by the Sergeant – a decent, if somewhat corrupt, policeman – and their power struggle finds expression in their different attitudes to the theatre. When Bourke and his men appear onstage dressed as Nazis, they offer a literal enactment of Benjamin's dictum that fascism practises '*the aestheticizing of politics*'.[128] Such an act is an attempt to limit and harness a power that fascinates and repels – Bourke also fears the shape-changing capacity of the theatre, which he denounces as ungodly and irrational. His deformed enchantment with the theatre is most fully expressed when he swaggers onstage at the end of the play, proclaiming himself king, because he is wearing a crown he has stolen, along with the rest of the costumes, from the now-departed troupe. He further believes that in possessing the costumes, he has captured the actors: '[w]e have 'em now, boys. Where'll they be now without their gee-gaws? Naked fuckers. Stripped' (*MMTT*, p. 77). The actors are their masks, and his views of the theatre can be seen as a distorted, inverted reflection of Lyle Jones's, that warn of the political dangers when this attitude is stripped of its playfulness. It is at this too-late juncture that the Sergeant takes a decisive stand against Bourke, and stops him from setting fire to the costumes. As Bourke and his gang trail off, he examines this garb, which he finds 'poor looking enough in the broad daylight! Would you credit it that people could be so fooled now?' (*MMTT*, p. 78). While the Sergeant acknowledges the magic of theatre, he, like Madame, holds that there is a rigid difference between it and real life. According to Madame, the rending of such an ethical *cordon sanitaire* between reality and fantasy is inherently dangerous: '[o]nce one puts on a uniform one is in danger of unleashing one's violence' (*MMTT*, pp. 66–7). Nonetheless, despite insisting on the necessity of discriminating between theatre and reality, she views the world precisely in theatrical terms. This ties her to Lyle Jones, and it is the blending of her lucid ethical sense of the danger in theatre and the self-conscious playfulness of Lyle Jones that has kept the company on the road.

A third mode of theatrical interpretation is also performed in this play through the character of Rabe. In this, the danger of theatre takes on an intensely physical, non-verbal, form, which is also an expression of, and response to, the political context of the play. Rabe is suffering in Ireland: he is on the precipice of a breakdown and fears his identity is splitting – that he has 'begun to crack up again' (*MMTT*, p. 39). His previous experience of being placed in a psychiatric institution is recalled as the physical pain he endured trying to crash through its walls, and these symptoms have reappeared as he and the company are trying to break out of Ireland. Having constructed itself as an asylum from the war, Ireland has turned into another form of repressive asylum. In the Introduction, it was argued that an enabling understanding of the 'fifth province' as a refuge and sanctuary is predicated on forms of interaction with the other: this is what neutral Ireland worked to prevent.

Rabe's experiences have correspondences with those of Antonin Artaud, who claimed that he 'travelled to "CHRISTIAN IRELAND" in search of political asylum' in 1937.[129] Like Rabe, he did not encounter much in the way of Irish hospitality; instead, Ireland was the site of his decisive psychological breakdown. He was deported from the country as 'a destitute and undesirable alien' and, on arrival back in France, he was taken in a straitjacket to Le Havre hospital and thence to an insane asylum in Rouen.[130]

Through his notion of '[d]anger', Artaud, according to Kilroy, 'in his own demented way went closer than anyone else to communicat[ing] [the] environment [and] impact' of what might be termed non-literal theatre.[131] Danger, for Artaud, takes expression through the actor who offers her/himself as a scapegoat – a self-sacrifice – through whom the dangers of society can be released and confined to the theatre: 'theatre [...] is a revelation, urging forward the exteriorisation of a latent cruelty through which all the perversity of which the mind is capable [...]. Theatre [...] unravels conflicts, liberates powers, releases potential and if these and the powers are dark, this is not the fault of the plague or theatre, but life.'[132] Rabe echoes this, declaring that he wants 'a theatre which can hold—danger. [...] Where danger can detonate upon a stage. [...] if theatre can do that, there will be less—danger left in the world' (*MMTT*, pp. 24–5). He has witnessed firsthand the release of the danger that is always latent in humanity: he stood up to Mosley's fascist thugs in the East End of London and, in response, they burned down his father's shop. He feels responsibility and, simultaneously, contempt for his father's passivity in the face of this barbarity. Recalling it, he commands his father to perform – to '[d]ance' (*MMTT*, p. 24) – to act out the danger, to become an active rather than a passive sacrifice, and so drain Rabe of his confused sense of guilt, which drives his need for a theatre of danger.

What Artaud termed the 'cruelty' of theatre is a casting off of masks; a process that reveals not an authentic, whole self, but the 'lies' and 'aimlessness' that constitute identity.[133] Rabe's description of the play he wants to stage similarly involves a radical move behind any possible mask, which nonetheless unveils yet another performance.

> Curtain goes up, a man's skull opens up and we see the inside. All these creatures parade and dance. Such costumes, you've never seen the like, raging colours, material like metal and fur, all pouring out of this skull. Then it all ends, this play or whatever, this play ends in a great red hunt. The hunting down of these creatures, actors carrying long, pointed spears, see. And sound, yes, drums but also strange instruments like horns or bugles but different, whistles and conches. [...] I want to play the hunter and the hunted at the same time. (*MMTT*, p. 40)

Such a performance, in going beyond representation, goes beyond interpretation, and operates on the same level as the medium of radio does in McLuhan's understanding; to recall from the section on *Double Cross*: '[t]he subliminal depths of radio are charged with the resonating echoes of tribal horns and antique drums'.[134] Artaud also evoked the image of such primal sounds to indicate a form of theatre that exceeds representation: a theatre 'aimed at the system by exact means, the same means as the sympathetic music used by some tribes'.[135] As Derrida noted, an impossibility marked the unlimited ambition of Artaud's project: '[t]he theater of cruelty is not a *representation*. It is life itself, in the extent to which life itself is unrepresentable. Life is the nonrepresentable origin of representation.'[136] Artaud's theatre would be autochthonous: it is to be where life has its origin. As such, it would have no history, no previous examples to guide it. In order to attain this state in the twentieth century, Artaud first had to destroy, and his targets are precisely structures of representation: 'the profound essence of Artaud's project', as Derrida put it, is that he '*wanted to erase repetition in general*'.[137] As language depends on repetition for comprehension, Artaud's project is an attack on language itself, and in place of spoken language in the theatre, he advocates a 'physical language, aimed at the senses and independent of speech'.[138] Such a language does not represent as such, but rather gestures to a transcendental realm, beyond language: '[a]rt is not the imitation of life, but life is the imitation of a transcendental principle which art puts us into communication with once again'.[139] Setting off as a Wildean aphorism, Artaud's statement ends up evoking Benjamin's argument that translations do not imitate the original but recall the existence of a pure language beyond, and linking, all languages. The form of translation that Artaud looked to effect in his theatre would bring audiences into contact with a commensurate realm. In Benjamin, the impulse towards a pure language emanated from his Jewish experience on two levels: not only did it reflect, as Steiner notes, 'the Kabbalistic and gnostic tradition', but it was also, as Bush points out, 'informed by the jagged edges of historical experience'.[140] Similarly, Rabe's search for an original language beyond language is an expression of the 'problematic filiation' into which he has been hurled by the experience of violent anti-Semitism: sundered from the law and the traditional inheritance of the father, he seeks to uncover an alternative original – one that is always already an other. As Derrida commented, Artaud calls for '[t]heater as repetition of that which does not repeat itself, theater as the original repetition of difference', and it is precisely in this difference that intimations of an impossible universal original, such as the one Rabe seeks, might be apprehended.[141]

Grotowski noted that '[t]he paradox of Artaud lies in the fact that it is impossible to carry out his proposals. Does that mean that he was wrong? Certainly not.'[142] As Rabe, Artaud and Hölderlin illustrate, forms of schizophrenia cast shadows at the dangerous limits of representation: in the face of the inadequacy of language to express the self, identity undergoes a process of splitting and doubling. For it to describe a salutary process – one that deconstructs the primacy of an authoritative original – a vision such as Artaud's must be tempered through the masks of language. Indeed, in writing about Artaud's influence on his work, Kilroy fundamentally domesticates him by advocating a 'retreat or partial retreat from language' on the stage.[143] Rather than provide a blueprint, Artaud's theories, in Kilroy's version, prompt a discussion about modes of theatrical interpretation, and a commensurate three-way dialogue can be heard in *The Madame MacAdam Travelling Theatre*, where, through Madame, the three theatrical dispositions are put into communication with one another. While Lyle Jones inspired her to join the world of the theatre, she in turn brought Rabe into the troupe: '[s]he made it sound as if theatre could heal everything, make it whole again' (*MMTT*, p. 18). Her nurturing conception of wholeness, 'built upon human error and human frailty' (*MMTT*, p. 37), is based on an acceptance of imperfection, rather than an impossible attempt to get beyond actual life.

In other words, her conception of reality is rooted in an apprehension of the essential otherness of life, and this permits her to act as a moral compass for the stranded company. While she recognises Rabe's brilliance and his constant quest 'for the impossible, perfect performance' (*MMTT*, p. 65), she also appreciates Lyle Jones's essential role, despite her misgivings about him. For Rabe, Lyle Jones represents everything that is fake about the theatre, but Madame informs him that he could learn the crucial importance of '[p]layfulness' (*MMTT*, p. 39) from watching Lyle Jones. Indeed, ultimately, it is Lyle Jones's ability to bring things alive through his playful interpretations that rescues the company and gets them moving again. In the face of a threatening mob, he assumes the voice and pose of a *'Roman orator'* and quotes a speech from Shakespeare's *Coriolanus*. *'There is immediate total silence'* and, taken aback by his mask of authority, the crowd disperses.

> Knowest thou me yet?
> You common cry of curs, whose breath I hate
> As reek a' the rotten fens, whose loves I prize
> As the dead carcasses of unburied men
> That do corrupt the air—What would you have,
> You curs? Hence! Begone! (*MMTT*, p. 69)

Coriolanus's speech is a response to being called a traitor, and finishes with the insight known to every *traduttore traditore* – '[t]here is a world elsewhere' – which also marks the location where the attitudes to the theatre in this drama converge. For all three, bringing this other world to life is the function of theatre, which rather than simply representing life, instantiates the otherness that gives life its potential.

Epilogue. Afterlives

Things are not well in this house.

— *UV*, p. 17.

Tucked in the Field Day Archive is a memo from Friel instructing the company's administrator to pass on copies of *Dancing at Lughnasa* and invitations to its opening night (24 April 1990) to Kilroy, Heaney, Deane, Paulin and Hammond; all the directors, that is, with the exception of Rea.[1] Friel's decision to give *Dancing at Lughnasa* to the Abbey rather than to Field Day splintered his relationship with Rea – a fracture that was no doubt exacerbated by the outstanding success of the play. As Woodworth noted, *Dancing at Lughnasa* 'would have brilliantly restored the flagging fortunes of Field Day'; instead, it hastened the demise of the theatre company.[2] While the successful running of the enterprise's projects depended upon the sustained dedication of all the directors, Friel and Rea had primary responsibility for the theatre company throughout the 1980s (while Deane took charge of the publishing), and the enterprise never developed a structure that could withstand any diminution of interest on their parts. As was stressed in the Introduction, Field Day was bound together by ties of friendship more than anything else, and this informality created organisational problems as the company expanded, as can be heard in a letter from Julie Barber, the company's administrator, to Friel (3 October 1985), in which she informed him that 'on occasions I have felt completely unsure as to what is wanted or expected of me'.[3] In this light, Friel and Rea's falling out over *Dancing at Lughnasa* brought to a head underlying problems that had been eroding the sustainability of the theatre company for some years.

In an attempt to respond to these, Friel sent a letter to Rea (31 July 1989) outlining proposals to streamline the organisation of Field Day by setting up committees to deal with its different activities, so that the directors would be '[f]reed from the tyranny of board approval for every small decision'. Field Day, as he put it, 'is a very different operation from the old one when you and I—with lots of drive and enthusiasm and love and with the dedication of the true amateur—produced a play a year. Those relatively innocent days are over.' In this new dispensation, Rea would assume the role of 'artistic administrator', which Friel thought was 'right and proper; and you will do it wonderfully well'. Friel suggested this because he 'cannot—indeed must not—have any administrative role nor any day-to-day role': he has had a couple of serious health scares and '[t]he doctor insists the cause is over-work and overanxiety and the cure the shedding of a lot of activities'. He finishes this warm letter to Rea with the exhortation 'Long Live Field Day!'[4] The problem, though, was that Rea was not in a position to assume these duties either. In common with the other directors, his career took off in the 1980s, culminating in his nomination for an Oscar for his role in Neil Jordan's *The Crying Game* (1992). Though he remained the most passionate defender of the theatre company (with the support of Deane, who believed that, if it were closed, 'the heart would be taken out of the enterprise'),[5] it is notable that many of his submissions on its behalf in the early 1990s are letters, faxes and transcriptions of telephone conversations forwarded to meetings he could not attend because he was engaged in different projects abroad. David Hammond was the only director to have his permanent residence in Northern Ireland throughout this period; the others were, for the most part, based either in the Republic or England. However, as their work gained increasing international recognition, it became virtually impossible to gather them all in one place at the same time. As the headline on John Burns's article on Field Day's decision to go on sabbatical in 1993 had it, the company was a 'Casualty of Success', and Burns cited an unnamed board member's opinion that '[p]eople grew apart in the way showbands and pop groups do. The individual members became bigger than Field Day could ever be.'[6]

The deleterious consequences of this lack of cohesion were becoming increasingly evident: the 1990 production of *The Cure at Troy* 'had not been a success', while 1991's *Madame MacAdam* was a 'total disaster', as the minutes of the directors' meeting, 4 September 1992, reported.[7] Deane's diagnosis was unequivocal: 'the weakness that has afflicted the last two tours and productions has nothing to do with the plays themselves but with the loss of Stephen Rea as a central power in FD theatre.' In the past, his and Friel's theatrical abilities had camouflaged Field Day's underlying structural problems and the fact that 'FD theatre is always too expensive, too rushed, too panicked. We can't afford the 3 month tour.'[8] In Rea's absence, neither the Heaney nor the Kilroy play had been ready for its Derry première and, as the company's manager Gary McKeone wrote to the directors (9 March 1992), this was resented in Derry, where the feeling was that, instead of being the key performance, Derry had been used as a rehearsal for later performances.[9] Moreover, while Field Day was almost permanently short of cash and could never compete with the rates of pay provided by Dublin theatre companies, its reputation ensured that good actors were interested in working with the company. However, with the disengagement of Friel and Rea from the day-to-day management of the tours, this artistic compensation was lacking, with the result that more accomplished actors were not prepared to work for less than they could earn in Dublin – a central factor, Kilroy believed, in the failure of *Madame MacAdam*.[10] Rather than setting a standard for theatrical excellence – a primary motivation behind the founding of Field Day – the company, as Friel wrote to Deane (27 December 1991), had 'lost artistic respect in theatre circles. We are now looked on as a third-rate—and disagreeable to work with—group.'[11] The sorry state of the theatre company, and the failure to conceive of a practical way of remedying it, had persuaded most directors by early 1992 that that aspect of Field Day's operations should be at least mothballed for a period, if not wound up entirely. Rea and Deane, however, continued to fight for the theatre company and the other directors were not willing to deny their friends the possibility of extending its life, even if they were not convinced of its feasibility and were not prepared to take active roles in it. Neither fully alive nor dead, from 1992 the theatre company entered into a limbo-like state of abeyance, from which it emerged four times, in different forms, in the following six years.

As the theatre operations declined, Field Day's publishing activities grew; but these too brought problems for the company. 1991 was not only the year of *Madame MacAdam*, it was also a crucial year in the company's publishing history. A collection of essays, *Revising the Rising*, was produced, which gathered diverse voices from different disciplines to debate the manifold significances of 1916 and how it has come to be understood in Ireland. To a large extent, this volume rehearses the debate between revisionists and anti-revisionists, and is notable for strident contributions from Kiberd, Deane and Edna Longley, who broadens the focus of 1916 from the Easter Rising to encompass the ways in which understandings of the Battle of the Somme have contributed to the construction of Irish identities. While this book has come to be recognised as an important intervention, its immediate impact was overshadowed by *The Field Day Anthology of Irish Writing*, which was published in the same year with much fanfare and to extremely hostile reviews. More than any other, this episode contributed to the perception that Field Day was a group of unreconstructed Northern Irish patriarchal nationalists – a view that still holds some sway. Critics immediately noted the egregious lack of women's writing in the *Anthology*'s 4,044 double-columned pages, as well as the fact that not one of the editors of the forty-four different sections was a woman. Furthermore, the sections devoted to political writings, which are for the most part edited by Deane, omit the story of Irish women: the only mention of feminism in the section 'Political Writings and Speeches, 1900–1988' is Francis Sheehy Skeffington's 1914 article, 'War and Feminism'. Deane, who had taken on the Herculean task of general editor of the three volumes, did not shirk from his culpability: in a letter to the other directors, he took full responsibility for the 'serious weakness' of the *Anthology*, but was also angry about the way in which the issue of gender was 'exploited mercilessly' by revisionist critics.[12] Indeed, as Meaney noted, certain revisionists, including Roy Foster, seemed to have undergone a Damascene conversion to feminism in 1991.[13] However, the significance of this blindness to gender cannot be confined to skirmishes at the borders of revisionism; instead, a question such as the one posed by Spivak as a starting point of her critique of Foucault and Deleuze, has to be directed at the *Anthology*: '[w]hy', she asks, 'should such occlusions be sanctioned in precisely those intellectuals who are our best prophets of heterogeneity and the Other?'[14] To judge by his response to Nuala O'Faolain on a 1991 RTÉ television programme, Deane would appear to have been asking himself a version of this question: '[t]o my astonishment and dismay, I have found that I myself have been subject to the same kind of critique to which I have subjected colonialism ... I find that I exemplify some of the faults and erasures which I analyze and characterize in the earlier period.'[15] The general critical verdict on this failure has seen it as a symptom of Field Day's focus on the Northern crisis and consequent disregard for the social issues that played such central roles in the political life of the Republic. Mulhern characterises this in topographical terms: in the *Anthology*, he believes, '"Dublin" is overwritten as "Derry"' – 'the spiritual centre' of Field Day and 'the symbolic capital of the Northern crisis'.[16] Viewing Ireland from this perspective elides, in particular, the history of the conservative force of the Catholic Church in the Republic. As is noted in Chapter 3, Field Day's critical and artistic output lacks a sustained critique of the Catholic Church, and while this is undoubtedly a major element in the *Anthology*'s blindness to gender, another crucial factor was the manner in which these volumes were informed by a postcolonial reading of the history of translation in Ireland that occluded Irish 'domestic' experience.

In *Heroic Styles* (1984), Deane announced the idea of an anthology that would deconstruct the 'prevailing idea of what it is that constitutes the Irish reality'.[17] As it is developed in the *Anthology*, this 'idea' is revealed as a myth that has been generated by an 'axis of translation'.[18] Instead of functioning as an act and trope through which the unassimilated contradictions of Irish identity might be recognised and reimagined, translation, in Deane's account, becomes the site that has fostered an 'Irishness' rooted in concepts of originality and authenticity. As Irish culture from the eighteenth century onward became increasingly Anglophone, the allure of a lost Irish 'original' became stronger and assumed an almost mythical form, precisely because it was becoming more unknowable in its 'pure', 'authentic' guise. This prompted a recovery of Gaelic literature through translation, which, Deane contends, had the effect of entombing it and bequeathing Gaelic culture a future as a spectre of authenticity that has haunted Irish cultural and political life. Furthermore, by both ennobling and rendering inert Irish culture, this translational internment functioned 'as a means of cultural conciliation' between the Anglo-Irish Ascendancy and the Catholic population of Ireland, and between Ireland and Britain.[19] According to Deane, Anglo-Irish translators such as Charlotte Brooke and Samuel Ferguson sought 'in culture a reconciliation of those forces and interests that remained steadfastly opposed in politics': translation became the means through which political problems were transformed into cultural or artistic quandaries that were, in a circular manner, theoretically amenable to a form of resolution through translation.[20]

Thomas Moore's *Irish Melodies*, which Deane reads as acts of 'transmogrification', consolidated this style of cultural translation.[21] Hugely popular in Ireland and Britain, the *Melodies* translated the corybantic Gaelic tradition into the cool elegance of English verse, in a manner that not only provided evidence of a Gaelic cultural reservoir that the Irish might draw upon, but which also provided evidence for British palates that this heritage was amenable to modernisation and civilisation. Deane reads this domesticating 'translation of something wild and savage into something regulated and polished', as 'a metaphor of the translation of catholic into protestant, of native and antique authenticity into modern and equally native civility'.[22] He argues that this civilising process that allowed for – indeed, thrived on – unreconstructed notions of authenticity and loss, was adopted by Irish nationalism, with the result that Irish cultural nationalism was little more than a translation of a colonial discourse. While the translational link this establishes between colonialism and post-colonial nationalism is convincing (and has been argued for in this study), the inherent irony involved in mounting a sustained critique of Irish nationalism for its cultural, rather than political, bias through an anthology of major Irish writers points to a problem at the heart of Deane's argument.

What Deane describes is a domesticating mode of translation, and his focus on this – implicitly seeing it as the only form of translation that operated in Ireland – is based on a restricted conception of culture; one that, in effect, views it in terms of the coloniser. Working to foreignise this domesticating mode of translation, his deconstructive narrative does not so much challenge the terms of this colonial translation of Ireland as invert them. Moore, he notes, prepared his verses for accompaniment by the pianoforte, and Deane marks out the drawing rooms where his melodies were sung as one of the sites where *bourgeois* Irish cultural nationalism blossomed.[23] The 'axis of translation' ultimately hinges on just such a limited and privileged conception of culture, as it is only in such a *bourgeois* and Anglophone sphere that translation could be construed as a mode of reconciliation. Elsewhere in nineteenth-century Ireland, translation was a register of the cultural disjunctions wrought by colonialism; as *Translations* showed, it was a necessary response to a landscape that was being shifted under the feet of the country's Irish speakers. Surveying the *Anthology*, Breandán Ó Buachalla pointed to its Anglophone bias and the relative paucity of Irish language material, and argued that these volumes constituted a colonising account of Irish experience:

> any foreigner perusing FDA (1991) could not but conclude that English was always the primary medium of writing in Ireland, whereas the status of Irish was ever ephemeral and marginal. [...] the end product – the 'meta-narrative' – is, whatever the intention, a classic colonising exercise in appropriation and marginalisation.[24]

Rather than being simply marginalised, however, Irish assumes the role of a spectre in the *Anthology*, and this reflects the influence of Said's *Orientalism* on Deane's attitude to Irish and to translation. Said conceived of Renan's Orientalist endeavours in Frankenstein-like terms: in his 'philological laboratory' he dissected what are presented as the dead and ossified Semitic languages in such a way that this action provided, through a form of oppositional comparison, an aura of life for European languages.[25] Likewise, for Deane, Celticist 'philological scholarship'[26] (to which Renan also contributed) looked to transmit the essential energy of the expiring Irish language into English, thereby effecting an 'ultimate reconciliation between the English language and that essential spirit'.[27] The latent Gothic note here is even more explicit in his 1985 *Celtic Revivals*, where Deane visualises the translation activities that set the stage for the Revival as acts of vampirism: acts through which the English language revived itself 'on the corpse of the Irish language it destroyed'.[28] As the implicit passivity, or incipient inertness, of Irish in these constructions suggests, such views of colonial translation operations essentially occlude domestic agency.

Clifford has pointed out that, in *Orientalism*, Said uses European and Enlightenment modes of thinking to examine, not the condition of the Oriental, but the construction of the West, and this, Clifford argued, is symptomatic of a 'condition of homelessness'.[29] As has been shown, homelessness was the chosen physical condition of Field Day, and is a central thematic concern of the dramas. It is also, as Deane noted, a prerequisite for a mature criticism; in his reading of Friel's drama, he has pointed out the 'ultimate perception' of Friel's protagonists 'is that fidelity to the native place is a lethal form of nostalgia, an emotion that must be overcome if they are [...] to grow up'.[30] In the *Anthology*, this nostalgia is seen as a colonial construction and as the principal mode through which the Irish engage with the concept of home. Like Said, Deane is only really concerned with the condition of the colonised on the plane of their discursive formation by the coloniser, and his postcolonial objective is to write against this construction. In place of the productive tensions between a homeless critical intelligence and the demands created by political situation of 'home' that mark much of Deane's critical work, in the *Anthology* he is almost exclusively occupied with dismantling the construction of 'Irishness' by rendering the concept foreign. Pursuing this project through an anthology that also has the objective of reclaiming Irish writers from the English canon fashions a text that attempts, as Longley put it, 'to deconstruct and canonise in the same gesture'.[31] Such a disjunction is a symptom of the inherent paradox of the very idea of a postcolonial national anthology. Said has argued that anthologies were one of the classical means employed by Orientalists to portray the Orient: they were compiled in the conviction that only a 'small set of powerful examples' was required for the Western reader to understand the Orient and, therefore, constituted a form of colonial, domesticating, translation.[32] Deane attempts to forestall the domesticating implications of compiling extracts from representative Irish figures by framing the fragments in terms of a foreignising narrative of Ireland. But in doing so, he repeats, to some extent, the motions of the colonial, domesticating, translation that he critiques: in his desire to reinscribe foreignness into the concept of 'Irishness', he fails to provide a critical account of the Irish 'domestic', which is left unexamined and open to essentialising interpretations. Deane never outlines how the 'domestic' might be reconstructed through the *Anthology*'s re-translation of 'the Irish reality', as his Introductions throughout the three volumes do not read the texts in ways that acknowledge domestic agency in Irish experience. This gives a curious aura of the tomb to the first three volumes, as the extracts chosen from the perspective of a rather elitist view of 1,500 years of Irish culture form a formidable cryptic carapace, at the centre of which resides a troubled spectre of the Irish 'domestic' and the Irish language.

Volumes IV and V were published in 2002 and are devoted to Irish women's experiences. They supplement the first three volumes by problematising the would-be plenitude of the first instalment in the same moment as they supply the whole *Anthology* project with a postcolonial version of the 'domestic'.[33] More explicitly postcolonial than the first three volumes, rather than Said, they turn to the work of Subaltern Studies scholars in a manner which does not allow for, in Spivak's terms, 'the positivist inclusion of a monolithic collectivity of "women" in the list of the oppressed whose unfractured subjectivity allows them to speak for themselves against an equally monolithic "same system"'.[34] No space is granted in these pages to the possibility of a unified subjectivity that can be labelled 'women' and consigned to the archive; instead, as the 'Preface' claims, 'a many-voiced narrative' is articulated, which ranges from the works of feminists who have established public voices to innumerable texts from women who have lived their lives entirely within the confines of the domestic.[35] Whereas the first three volumes ultimately did not interrogate the construction of private Irish identities from the perspective of the public expressions of 'Irishness', these constantly question official articulations of Irish identity through what are, in canonical terms, marginal texts: voices from hospitals, prisons, workhouses, convents, the pages of children's literature, and from the travelling community are heard here. As Meaney argues in volume V:

> [i]f it is simply a case of filling in the gaps and producing independent anthologies and histories of women's writing, until an inclusive canon which is no longer prejudiced by patriarchal culture comes into existence, then feminist history's attitude to literature as such is not fundamentally radical. If, however, the function of specific anthologies of women's writing is to put into question the whole notion of literary canons constituted by reference to a universal standard of literary greatness, then the project of feminist literary history is far more wide ranging and far more radical.[36]

The extracts in these two volumes are compiled from the other side of history, and this is presented as an uneven site of difference. In other words, the domestic becomes the location where difference might be encountered in Irish experience; instead of representing the known and the deadeningly comfortable (which is how it is implicitly dismissed in Deane's 'Introduction' to the section on Joyce in volume III), the domestic is shown to be that which is unknown and jagged in Ireland, precisely because it has developed in the deep shadow cast by various public, male-dominated, (post-)colonial discourses which have essentialised it, *inter alia*, as a symbol of uncomplicated nationhood.[37]

These volumes have the subtitle 'Irish Women's Writing and Traditions', and from them emerges a repositioning of the political and cultural valence of the domestic by means of a rethinking, and thinking together, of the marginal and the traditional through, amongst other things, performance. As Angela Bourke explains in her 'Introduction' to the 'Oral Traditions' section: '"[t]raditional" in this context does not mean sternly conservative or resistant to change [...] A "traditional singer" [...] is one who possesses a particular combination of musical, lyrical and personal sensibilities, recognized by other singers and by the audience.'[38] Echoing Kearney's open-ended definition of tradition that was cited in the reading of *The Communication Cord*, the traditional here is that which is always variable: it achieves a provisional form through the interactions of the performer and the audience. Such a fluid concept of the traditional introduces a radical flux into a construction of an Irish canon, contaminating it with 'foreign bodies' that have always been present – such as women and travellers – in a way that the radical alterity embodied in their subaltern experiences unmoors the notion of Irish identity. 'Irishness' becomes a site of translation, and not just a foreign concept that has been translated on to Irish experience. Whereas Deane's analysis halts at the point of deconstructing how 'Irishness' became a mark of difference in the colonial situation, an exploration of the modalities of difference and otherness that that context has produced in Irish experience is presented in the second instalment. Indeed, Deane's foreignising operation is impossible with texts that have no definitive form, no stable meaning and no proprietorial author, which is frequently the case in these volumes. Such texts are always somewhat foreign and must seek out forms of temporary domesticity; their meanings depend, it might be said, on the sort of hospitality that takes place in performance.

The emphasis on performance and hospitality in the second instalment, and the ways in which it supplements and problematises the view of translation in volumes I–III, might have brought the *Anthology* into dialogue with the enterprise's theatrical side; however, by 2002 the theatre company was no longer operating. The sporadic signs of life in the 1990s never amounted to a resurrection of the company as a sustained, ongoing, venture. Throughout the theatre company's existence there were many projects that failed to get off the ground. There were, for instance, several attempts to stage a musical – one based on Joyce's *The Dead* with a libretto by Paul Muldoon, and another based on the life of the popular nineteenth-century balladeer Johnny Patterson. Offers of commissions to write plays were also made in 1987 to Derek Mahon, John McGahern and Eugene McCabe; and while McGahern did produce a version of Tolstoy's *The Power of Darkness*, Friel had to turn it down, telling him that although the company loved the play, 'there is absolutely no possibility of Field Day doing any play with 13+ characters'.[39] In the early 1990s, however, there was an air of desperation about the projects that were suggested; ideas were being canvassed 'without a motive or focus except to show signs of life', as Heaney observed in a fax to the directors.[40] There was also the fact that the *Anthology* had left the company massively in debt, and consequently projects that would generate income were sought.[41] Amongst other ideas, Arthur Miller was invited to direct *The Crucible* for the company in 1991.[42] Nothing came of this or of other ventures that got beyond the initial stages, but foundered in the uncertain currents within the Field Day boardroom. The company, for instance, entered into a 'memorandum of agreement' with Donal O'Kelly to produce his play *The Dogs* on 13 January 1992.[43] However, two months later, O'Kelly wrote in extreme frustration to the company: he felt that Field Day had lost confidence in him and he had lost confidence in it. His major problem was with its haphazard organisation, noting that 'communications are extremely fuzzy, both in terms of who to communicate with, and how'.[44] In the same year, Derry turned its back on the company, when the City Council decided not to fund a production of *The Freedom of the City* that was to be directed by Rea for its major arts festival 'Impact 92'. Friel addressed this with unequivocal anger: 'Field Day has effectively put Derry on the theatre map and is now being disregarded during this hugely important year in the city.'[45] Barely drawing breath at this juncture, the company warmed to Judy Friel's proposal that they produce 'high quality rehearsed readings' of plays, as it would at least indicate a presence.[46] While the thrust of this idea was that the plays would be by first-time writers, the only plays to get this treatment were David Rudkin's 1974 piece for BBC radio, *Cries From Casement as His Bones are Brought to Dublin,* which was produced in Derry, Belfast and Dublin in 1992, and, in 1994, John Berger and Nella Bielski's *Goya's Last Portrait.*

A play about a process of translation – in the sense of moving remains from one burial site to another – Rudkin's piece has many correspondences with other Field Day dramas. Like Bracken and Joyce in *Double Cross,* this Casement is a spectral (radio) voice. As a traitor, he is both doubled – 'Casement's voices, mine, all mine' (*CFC,* p. 16) – and split: for his inextricable other within, Crippen, he is 'HalfPaddy' (*CFC,* p. 9). Neither his career investigating slave conditions nor his involvement in Irish nationalism is credited as his main achievement; instead, his major accomplishment is that, like Wilde, '[t]hrough horror, sickness, danger, sodomy, farce, he hacks out a new definition of himself' (*CFC,* p. 24). A man who lived without a fixed address, he is presented as being in permanent exile, in search of a voice: '[w]here is my speech? [...] I—am—unvoiced ... Three times an exile. Irishman in England. Ulsterman in Ireland. Queer in the world' (*CFC,* p. 30). Even the posthumous translation that he undergoes will not bring him home, as rather than being buried in Antrim, as he wished, he is given a hero's re-burial in Glasnevin cemetery in Dublin. On discovering this, he protests and insults the officiating cardinal, who brusquely tells him to take his place in the Irish nationalist pantheon: '[l]ie down. Be a good hero, shut your mouth. Be a good patriot, lie down' (*CFC,* p. 76).

While Casement has a place in Field Day's roll of contradictory traitors, a rehearsed reading of a radio play was not going to revitalise the enterprise and, in these entropic conditions, the board of directors drifted apart. The last director in was the first out, with Kilroy resigning in 1992. He felt that the company was losing its focus and relevance in Ireland,[47] and that he was 'unable to contribute in any substantive way to the evolution of Field Day', to a large extent, 'because it had already achieved its own finished form' before he became a part of it.[48] This is overly self-critical: as was argued in the last chapter, from this distance it is apparent that Kilroy's work contributed greatly to the theatre company. Alongside *Translations, Double Cross* stands as Field Day's most enduring dramatic achievement, and these were the two Field Day plays to be canonised by the company itself in the *Anthology.* Of the other plays, despite the questions that have been raised about *Making History*'s dramatic qualities, it, *Pentecost* and *The Cure at Troy* have attracted most critical attention (and will continue to do so) because of the challenging political perspectives they articulate, and because of the significance of Friel, Heaney and Parker in contemporary Irish literature.

A year after Kilroy's resignation, in April 1993, the board decided to take a sabbatical; news that most commentators greeted with obituary-like articles.

Within the group, however, Rea and Deane continued to act as if Field Day had a future as a theatre company, and this led to a serious split with the other four directors, who were not happy with the fact that the duo were making decisions under the banner of Field Day without the consent of the board. When Friel, Hammond, Heaney and Paulin heard that Rea had made a funding submission for *Uncle Vanya*, they issued a joint memo (22 November 1993), stating that this action

> overrides the wishes of four of the six board members and places them in a difficult position. We must ensure that board decisions are respected and that this application for funding is withdrawn either by SR [Rea] or by the board. If it is not withdrawn, the board of FD no longer has a function and must seriously consider disbanding.[49]

Besides wanting to keep the theatrical side of the enterprise alive, Rea also wanted to avail of the £25,000 Prudential award that was to be granted to Field Day's next production. He argued that the company should not let this opportunity pass and, on these grounds, the board decided, with little enthusiasm, to go ahead with *Uncle Vanya* on 21 January 1994.[50] This was to be Friel's last involvement with Field Day: three days later he sent a resignation letter to all the directors.[51] Throughout this drawn-out dissolution, the friendships that had bound the group were put under considerable strain. In an affectionate letter to Friel, Deane acknowledged that feelings had been hurt and registered his sadness that Friel should have left in such circumstances; he finished by noting that 'FD both enriched and interfered with our friendship; it will do neither any longer. The friendship stays.'[52] Friel had been Field Day's lynchpin and without his involvement, Heaney, Paulin and Hammond were increasingly sidelined, and they drifted away from the company in the second half of the decade.

McGuinness's version of *Uncle Vanya* is similar to Friel's *Three Sisters* in that it keeps the setting of Chekhov's play and inserts 'Irishisms' – '[g]ive my head peace' (*UV*, p. 21); 'been on the tear' (*UV*, p. 22) – into a text that does not radically deviate from a British English rendition. Thematically, it also shares *Three Sisters*'s concern with the fate of a house in the Russian provinces into which a parasite has been introduced. Whereas the takeover of the house by Natasha takes place on the stage of *Three Sisters*, Serebriakov is well ensconced at the start of *Uncle Vanya*. There is a need to remove this infirm parasite from the house, as he is sucking the life out of the place; the ambivalent parasitical compensations that Natasha introduced into the sisters' house are absent, or, rather, are dissipated. As Vanya explains to Elena, this parasite was willingly endured at first because he was considered a lustrous addition to their lives, and the proceeds of the years Vanya and Sonia spent toiling on the estate were all sent, uncomplainingly, to support Serebriakov's academic life in the city.

> We were the right fools. We starved ourselves. We turned the pennies into pounds and we made him a fortune. We sent it all off to him. I was proud of him. I was proud of his learning. He was my life's breath. Brilliant. All he wrote, all he said – fool's gold. Now he's handed in his cards and we can see what it amounts to. Not one page of his work will survive him. (*UV*, p. 20)

As in several other Field Day dramas, the characters are caught up in the dynamics of a self-sacrificing loyalty to people who do not deserve this, who sap their lives, leaving them 'ghost[s]', which is how Vanya describes himself (*UV*, p. 18).

Their efforts were not enough for Serebriakov to afford a house in the city in his retirement, and so he has come, like a metropolitan colonist, and lodged himself in this house in the provinces. He has no feeling for this place beyond an egotistical disgust that he should have to live there and looks to exploit it in order to get away from it: he proposes that they sell the estate, so that he can use the money to buy a villa in Finland. Appalled by the audacity of this scheme, Vanya is pushed into the position of protecting the inherited rights of those who live on the estate: '[m]y dead father bought this estate as a dowry for my sister. Up to now I've been simple enough to believe we were not living under Turkish law and I thought this estate would pass from my sister to Sonia' (*UV*, p. 44). Adhering to the authority of the dead father is not, as it was in *High Time*, a confirmation of patriarchal rule, rather it subverts its assumptions. A defence of the rights of women was also a message that Field Day found congenial in the aftermath of the *Anthology* controversy; as Rea indicated in a 1993 letter to Friel, a concern about the representation of women and about 'raising any political heckles at the moment', informed his choice of play.[53] However, rather than representing a new type in Field Day's dramas, Sonia finds a place alongside the other women characters who unearth forms of hope in reality, such as *Translations*'s Maire, *Making History*'s Mabel and Lena from *Boesman and Lena*. While, like Antigone, Sonia invokes religion in her final speech, she grounds it in such a manner that it becomes a perspective from which a fidelity to mundane temporality is articulated. Her speech – 'a pragmatic, female message', as Longley called it – refigures her and Vanya's self-sacrifice in a manner that echoes the conclusion of *Pentecost*.[54]

> We will live, Uncle Vanya, we will live. Live through day after long day. Night after long night. Fate waits to send us trials. We'll have patience, we'll get through them. So, we work for other people. Work hard, work harder, and we grow old. Then, come the day and hour, we die quietly. Look up into its face, death, and there it is, the grave. When we rise from it, we'll tell how we suffered – how we sorrow. God will show us mercy. Full of mercy, God. Uncle Vanya, you and me, we will see a life that is full

> of beauty, full of light, of grace. Rejoice. We will. The days of our misfortune we will look upon with tender eyes. We will smile. And we will rest. Uncle, I believe this. (*UV*, p. 63)

McGuinness characterised this speech as 'a lament for what has gone before and an illogical hope for peace and rest in the future', and pointed out that one could not 'listen to that [...] in Derry in 1995 and not feel some connection'.[55] In August 1994, the IRA called its first ceasefire in twenty-five years; compared to the tours of the 1980s, this final Field Day tour, which ran from February to April 1995, set off in a radically different political atmosphere. There was, in brief, a glimmer of hope that the quiet, quotidian concerns that Sonia evoked would come to dominate life in Northern Ireland.

Uncle Vanya had a resonance in these political circumstances, and has links with other Field Day dramas through themes such as loyalty, fidelity and even mapping – Astrov draws up an alternative, anti-positivist map that challenges the empty promises of modernisation. As such, it was hardly breaking new ground and was instead a safe option for what remained of Field Day – a perfunctory revisiting of territory that had already been charted by the company. Moreover, while Friel's *Three Sisters* fashioned an anti-hegemonic response to the English domestication of Chekhov in 1981, by 1995 Wardle could comment with some validity that 'Chekhov over the past 20 years has become an honorary Irish author.'[56] O'Toole further argued that this Irish domestication of Chekhov, which 'may look like a profound connection with Europe, [...] is in fact a way of avoiding any such connection', because 'there has been [...] almost no relationship with any European dramatists other than Chekhov and Ibsen'.[57] It was not just critics who had reservations about this production; there was a palpably different atmosphere at its première at the Guildhall, marked by the absence of stalwart Field Day first-nighters such as John Hume, Martin McGuinness and, most notably, Friel. As Mary Holland observed:

> [t]he days when a Field Day first night in Derry's imposing Guildhall seemed like an act of cultural defiance by the entire Nationalist community are gone. In the Eighties it was quite common to see terrorists and poets and priests mingling together in the audience. Peace, which has brought new hope and visible prosperity to the much-bombed city, has changed everything. People no longer feel such a desperate need for the kind of resonances that were immediately evident in plays like *Translations* [...] or the plea for peace that rang out from *The Cure at Troy*.[58]

Deane criticised Field Day in 1991 for not having 'established a new audience'; by 1995, even its loyal audience was deserting it, not only because of the changing political environment, but also because *Uncle Vanya* was barely recognised as a Field Day play as, amongst other reasons, it was well known that most of the original directors doubted the continued viability and salience of the theatre company.[59] In these conditions, it would have required a *Translations*-like phenomenon to re-energise the company, not a tour of a translation of Chekhov that had the added misfortune of coinciding with Louis Malle's film *Vanya on 42nd Street.*

In the 2006 documentary, *The Story of Field Day,* Rea refused to write Field Day off entirely as a theatrical venture, and suggested that there might be future productions along the lines of the company's final project: its November 1998 co-production with the Belfast company Tinderbox of Stewart Parker's 1984 *Northern Star*. Should there not be another production, there could hardly have been a more appropriate play with which to bring the curtain down on the Field Day Theatre Company. As is hinted at by the employment of polemical pamphlets, Field Day's critical approach has had an eighteenth-century flavour (it should not be forgotten that Deane's doctoral thesis concerned the reception of French Enlightenment in England).[60] More importantly, with its non-sectarian, republican, anti-colonial philosophy, the United Irishmen, perhaps more than any other group in Irish history, represented an ideal for Field Day. The failure of the 1798 rebellion was 'the loss of the one genuine opportunity Ireland has had to benefit from the advances of the European Enlightenment', as Deane put it in a proposal for a film documentary on 1798 – *A Moment of Hope* – which he was to script in 1998. He goes on to outline, in a manner that echoes his comments in the programme for *Northern Star,* how his involvement in this project had little to do with commemorating the rebellion's bicentenary; rather, he wanted

> to show that this particularly tragic sequence of events is still unfinished, that we are still inheritors and victims of it and that it is important to recognise this so that we may better control and govern forces that have gained government over us. The worst of those forces is sectarianism; that is the caricature of the ideal shared and equal citizenship which was basic to the United Irish movement. [...] The decade of the 1790s is a moment in which we can see some of the complexities of our present situation take initial shape [...] and in which we can see the chance and the means to transcend them. [...] The commemoration of 1798 is, for me, as much a salute to the possible future as to the complex past.[61]

As ever in Field Day thinking, history is never completed, but continues to take shape as it moulds subsequent generations. As Whelan has noted, '[t]he 1798 rebellion was fought twice: once on the battle fields and then in the war of words which followed in those bloody footprints'.[62] *Northern Star* participates in the ongoing second battle in the manner suggested by Deane: it turns back to history in order to knock it out of the groove within which it has been lodged for 200 years.

By translating the radical thinking that was being nurtured in France and in the United States into Ireland in the 1790s, the United Irishmen sought to overcome a history of colonialism and sectarianism; as McCracken sardonically puts it in the play, they are the 'logical men' who wish to escape the 'dungeon' of history (*NS*, p. 68). But the characters onstage are already ghosts: their failure to achieve this objective has condemned them and Ireland to '*the continuous past*' (*NS*, p. 3). This continuity is a stasis, and is embodied in the Phantom Bride and her basilisk-like gaze. She, as noted in Chapter 3, is the spectral progenitor of the other squabbling ghosts of Irish Protestant history in Parker's trilogy – who are only quietened when Marian asks for Lily's forgiveness in *Pentecost*. *Northern Star*, however, does not confine the United Irishmen to the unchanging temporality of despair; instead, it bounces their history off different temporal planes, and in this manner keeps their message in play: the continual past is performed in the continual present of the stage, which is, at the same time, the site of a meta-theatrical history of Anglo-Irish drama. Field Day's 1998 production was not a tour, but was staged in the First Presbyterian Church on Belfast's Rosemary Street, which further heightened these reverberations between the past and the present, as this was where many of the protagonists of 1798 attended religious services. Outside this building is a plaque that holds out the promise of an open hospitality that points towards an enabling politics:

> [t]he members of this congregation believe in the freedom of individuals to form their own opinions and fashion their own faith, free from compulsion or coercion from any ecclesiastic or civil authority. We do not seek to exclude persons because of their beliefs nor do we seek to impose ours upon them. We welcome all who are seeking to work out and understand the purpose of life, with freedom of thought and without constraint of uniformity.[63]

Billington noted that, '[s]hort of seeing Hamlet at Elsinor, you could hardly find a play that comes closer to home'.[64] Playing in this site of liberal, non-subscribing Presbyterian thinking is also a philosophical and thematic homecoming of sorts for Field Day. As is argued in Chapter 3, Northern Presbyterianism holds more intellectual and dramatic interest for the company than Catholicism: at no point is it suggested in Field Day's output that Catholicism might supply an enabling model in the way that liberal Presbyterianism provided a humane and rational blueprint for the United Irishmen.

Standing outside the Guildhall in 1980, Friel, as can be seen in the documentary *The Story of Field Day*, discussed the somewhat surreal backdrop to the rehearsals for *Translations*: how the cast and crew were negotiating bomb scares, roadblocks and the military. Founded in this context, Field Day sought to fashion a refuge where voices might be heard that were not ensnared in the insistent clatter of militarised logic that was inducing a fatal sclerosis in what were already hardened modes of identifications. Field Day's plays brought out the movement that was being occluded in these formations. On its stages, identity never goes unquestioned; rather, by being brought into a critical dialogue with itself and with the other, the self is shown to be always split and doubled. Kilroy's description of Field Day as 'process' can be best understood in terms of this fundamental, translational dynamic. Inimical to univocality, this is a mode of operation that resists movement towards definable political results, and this, ironically, came to frustrate Kilroy, who has written that,

> one of the reasons why I eventually resigned from the board was because I wanted to push Field Day towards more overt political gestures. I realised that this could not happen because Field Day had no single ideology; it consisted of disparate, even conflicting, elements and its true energy was one of process, not movement towards fixed goals.[65]

While this is undoubtedly the case, Field Day's indirect intervention into the 'Troubles' need not be so readily discounted, as the terms through which it articulated this translational process have found echoes in the Northern Irish peace process. 1998, the bicentenary of the United Irishmen's failure and the year of the last Field Day drama, was also the year in which the Belfast Agreement was signed. This has proven to be a durable basis upon which Northern Ireland can negotiate its future, and running through its text are elements of the language and themes that find expression in the Field Day plays. The Agreement acknowledges, for instance, that the only way of honouring those who died in the 'Troubles' is to work towards 'reconciliation, tolerance, and mutual trust'.[66] These terms, which describe motions of Steiner's model of translation, speak of a need to re-imagine and re-translate the past without eliding it. Fundamental, therefore, to the Agreement is a recognition and registering of difference as a starting point rather than as dead end. Instead of seeing Northern Ireland as a fixed site, different understandings of what constitutes the Northern Irish home are noted and, so that it might prove to be more hospitable than it has been, its future form is left open, in the hands of those who live there. No direct Field Day influence on the drafters of the Agreement is in any manner being suggested here; however, the reverberations in the Agreement of the rhetoric and themes of the plays illustrate that while the processes of difficult accommodation that Field Day staged throughout the 1980s did not propose political solutions, they nevertheless articulated the contradictory conditions necessary for the functioning of politics in Northern Ireland.

Notes

Introduction

1. Cited, Geoffrey Stokes, 'Field Day Tours America', *An Gael* (Summer/Samhradh 1987), pp. 29–31, p. 31.
2. Transcript of Interview with Kilroy, FDP, MS 47,308/3.
3. Ibid.
4. Francis Mulhern, *The Present Lasts a Long Time: Essays in Cultural Politics* (Cork: Cork University Press in association with Field Day, 1998), p. 151.
5. Jacques Derrida, 'Des Tours de Babel', in J. F. Graham (ed. and trans.), *Difference in Translation* (Ithaca and London: Cornell University Press, 1985), pp. 165–207, p. 171.
6. For an overview see José Mariá Rodríguez García, 'Introduction: Literary into Cultural Translation', *diacritics*, 34: 3–4 (Fall–Winter 2004), Special issue on 'Literary into Cultural Translation', pp. 3–30.
7. Emily Apter, *The Translation Zone: A New Comparative Literature* (Princeton: Princeton University Press, 2006), p. 6.
8. Sandra Bermann, 'Introduction', in Sandra Bermann and Michael Wood (eds), *Nation, Language and the Ethics of Translation* (Princeton: Princeton University Press, 2005), pp. 1–10, p. 5.
9. Gayatri Chakravorty Spivak, *Death of a Discipline* (Calcutta and New Delhi: Seagull Books, 2004), p. 12.
10. Shane Weller, 'In Other Words: On the Ethics of Translation', *Angelaki*, 10: 3 (December 2005), pp. 171–87, p. 172.
11. J. Hillis Miller, *Others* (Princeton: Princeton University Press, 2001), pp. 273–4.
12. Thomas Keenan, *Fables of Responsibility: Aberrations and Predicaments in Ethics and Politics* (Stanford: Stanford University Press, 1997), pp. 1–5 and 7–42. See especially pp. 30–1, where Keenan invokes and develops J. Hillis Miller's description of the moment of 'the impossibility of reading, where reading nevertheless takes place' (p. 31).
13. Spivak, *Death of a Discipline*, p. 10. The embedded citation is from Derrida's *Given Time: I. Counterfeit Money*.
14. Jacques Derrida, *Monolingualism of the Other; or, The Prosthesis of Origin*, trans. Patrick Menash (Stanford: Stanford University Press, 1998).
15. Algeria 'too had been incorporated within a metropolitan system, and yet had inevitably also remained a colony of France. This intensified the bitterness of the separation [...]. It also intensified the difficulty of achieving independence and perhaps overstimulated the colonized country's desire for and dream of autonomy.' Seamus Deane, 'Dumbness and Eloquence: A Note on English as We Write It in Ireland', in Clare Carroll and Patricia King (eds), *Ireland and Postcolonial Theory* (Notre Dame: University of Notre Dame Press, 2003), pp. 109–21, p. 111.
16. Thomas Kinsella, *The Dual Tradition: An Essay on Poetry and Politics in Ireland* (Manchester: Carcanet, 1995), p. 64.
17. Pascale Casanova, *The World Republic of Letters*, trans. M. B. DeBevoise (Cambridge, MA: Harvard University Press, 2004), pp. 303–23. For a critique of Casanova's

work on Ireland, see Joe Cleary, 'The World Literary System: Atlas and Epitaph', *Field Day Review*, Vol. 2 (2006), pp. 196–219.

18. Stokes, 'Field Day Tours America', p. 29.
19. Deane, Lecture at Magee College Derry, 24 January 1985, FDP, MS 47,116/1.
20. Aside from the translations in *The Field Day Anthology* (many of which are by Deane), the only other translation from Irish published by Field Day is: Seamus Heaney, *Sweeney Astray* (Derry: The Field Day Theatre Company, 1983). The translations of Sophocles, Chekhov and Molière are rendered, to varying extents, in different forms of Hiberno-English. However, the only specific engagement with Hiberno-English is to be found in Tom Paulin's pamphlet, *A New Look at the Language Question*, in *Ireland's Field Day* (London: Hutchinson, 1985), pp. 3–22. As Hildebidle notes, 'Field Day is [...] an exercise in the repossession of language, fought on the terrain of English rather than Irish.' John Hildebidle, 'Translation or Retranslation? Field Day and the Reversal of Cultural Colonizations', in Patricia Lynch, Joachim Fischer and Brian Coates (eds), *Back to the Present: Forward to the Past. Irish Writing and History since 1798*, Vol. II (Amsterdam and New York: Rodopi, 2006) pp. 165–74, p. 174.
21. Paulin, *A New Look at the Language Question*, p. 3.
22. Tejaswini Niranjana, *Siting Translation: History, Post-Structuralism, and the Colonial Context* (Berkeley and Los Angeles: University of California Press, 1992); Eric Cheyfitz, *The Poetics of Imperialism: Translation and Colonisation from The Tempest to Tarzan* (Oxford: Oxford University Press, 1991); Vicente L. Rafael, *Contracting Colonialism: Translation and Christian Conversion in Tagalog Society Under Early Spanish Rule* (Durham: Duke University Press, 1993); Michael Cronin, *Translating Ireland: Translation, Languages, Cultures* (Cork: Cork University Press, 1996).
23. Susan Bassnett and Harish Trivedi, 'Introduction: Of Colonies, Cannibals and Vernaculars', in Susan Bassnett and Harish Trivedi (eds), *Post-Colonial Translation: Theory and Practice* (London and New York: Routledge, 1999), pp. 1–18, p. 2.
24. Edward W. Said, 'The Public Role of Writers and Intellectuals', in Bermann and Wood (eds), *Nation, Language and the Ethics of Translation*, pp. 15–29, p. 24.
25. Anthony Roche, *Contemporary Irish Drama: From Beckett to McGuinness* (Dublin: Gill and Macmillan, 1994), p. 208.
26. 'Proposal by Stephen Rea Regarding the Setting up of a Theatre Company', FDP, MS 47,119/1.
27. FDP, MS 47,116/1; same letter to all four.
28. *The Story of Field Day* (British Broadcasting Corporation, Northern Ireland, broadcast 18 October 2006).
29. FDP, MS 47,113/2-3.
30. Dennis O'Driscoll, *Stepping Stones: Interviews with Seamus Heaney* (London: Faber and Faber, 2008), p. 420.
31. Letter, 5 May (no year), FDP, MS 47,118.
32. Deane's recollection in 2009, given in personal correspondence, is that two Friel plays (most probably *Fathers and Sons* and his version of Macklin's *A True-Born Irishman*) were offered to the company as stopgaps and were not rejected as such, but plans to produce them were shelved when other plays became available. Nonetheless, in 1992 Friel said he 'was still hurting about' the manner in which these plays were treated (Minutes, directors' meeting, 4 September 1992, FDP, MS 47,115/3). On the other hand, the directors clearly turned down Paulin's *All the Way to the Empire Room*, as it was 'not theatrical enough' (Letter from McKeone to Paulin, 1 February 1991, FDP, MS 47,123/4). Friel believed that Paulin began to

withdraw from Field Day 'because of our indelicate handling of his plays' (Letter to Deane, 13 June 1992, FDP, MS 47,118).

33. Cited, Stephen Regan, 'Ireland's Field Day', *History Workshop Journal*, 33 (1992), pp. 25–37, p. 26.
34. Patrick Quilligan, 'Field Day's New Double Bill', *The Irish Times*, 18 September 1984, p. 10.
35. BBC, *The Story of Field Day*.
36. Brian Friel, *Essays, Diaries, Interviews: 1964–1999*, ed. Christopher Murray (London: Faber and Faber, 1999), p. 113.
37. Deane, 'Field Day Pamphlet Series. A Comment' (n.d.), FP, MS 37,181/1.
38. FP, MS 37,181/7.
39. Richard Kirkland, *Literature and Culture in Northern Ireland Since 1965: Moments of Danger* (London and New York: Longman, 1996), pp. 128–9.
40. Deane, Lecture at Magee College, FDP, MS 47,116/1.
41. Ibid.
42. Shaun Richards 'Breaking the "Cracked Mirror": Binary Oppositions in the Culture of Contemporary Ireland', in Colin Graham and Richard Kirkland (eds), *Ireland and Cultural Theory: The Mechanics of Authenticity* (Basingstoke: Macmillan, 1999), pp. 99–118, p. 108.
43. Seamus Deane, 'Introduction', in Terry Eagleton, Fredric Jameson and Edward W. Said, *Nationalism, Colonialism and Literature* (Minneapolis: University of Minnesota Press, 1990), pp. 3–19, p. 6.
44. Edward W. Said, *Orientalism* (London and New York: Penguin Books, 2003), p. 353.
45. Edward W. Said, *Culture and Imperialism* (New York: Vintage Books, 1993), p. 200.
46. Ibid., p. 206.
47. Seamus Deane, 'Preface', in *Ireland's Field Day*, pp. vii–viii, p. vii. For historical accounts of these stereotypes, see L. Perry Curtis, *Apes and Angels: The Irishman in Victorian Caricature* (Washington and London: Smithsonian Institution Press, 1997); and Robert Fitzroy Foster, *Paddy and Mr Punch: Connections in Irish and English History* (London: Allen Lane, 1993).
48. Seamus Deane, *Civilians and Barbarians*, in *Ireland's Field Day*, pp. 33–42, p. 42.
49. Said, *Culture and Imperialism*, p. 56.
50. Deane, *Civilians and Barbarians*, p. 33.
51. Joseph J. Lee, *Ireland, 1912–1985: Politics and Society* (Cambridge: Cambridge University Press, 1989), p. 454.
52. Deane, *Civilians and Barbarians*, p. 42.
53. David Lloyd, *Ireland After History* (Cork: Cork University Press in association with Field Day, 1999), p. 60. At the same time, the importance of the symbolism of religion and martyrdom to the emotional impact of the strikes cannot be denied. Perhaps the best recent negotiation of the contending discourses brought into play by the hunger strikes has been Steve McQueen's 2008 film *Hunger*. It registers the mythical imagery evoked by the dirty protests and the hunger strikes, but never lets it overwhelm the human experience of these, nor does it at any point admit a stereotypical 'civilian and barbarian' reading of the interaction of the forces of the state and the prisoners.
54. Denis Donoghue, 'Afterword', in *Ireland's Field Day*, p. 113.
55. Paulin, *A New Look at the Language Question*, p. 10.
56. Cited, ibid., p. 4.

57. Douglas Robinson, *Translation and Empire: Postcolonial Theories Explained* (Manchester: St. Jerome Publishing, 1997), p. 55.
58. Cronin, *Translating Ireland*, p. 71.
59. Lawrence Venuti, *The Translator's Invisibility: A History of Translation* (London and New York: Routledge, 1995), pp. 17–39.
60. George Steiner, *After Babel: Aspects of Language and Translation*, 2nd edn. (Oxford and New York: Oxford University Press, 1992), p. 49.
61. Helen Lojek, 'Brian Friel's Plays and George Steiner's Linguistics: Translating the Irish', *Contemporary Literature*, 35: 1 (Spring 1994), pp. 83–99, p. 90.
62. Jacques Derrida, *Positions*, trans. Alan Bass (London: Athlone Press, 1981), p. 20. Essentially this describes the process of metaphor, which I. A. Richards conceived of in the analogous concepts of tenor and vehicle: I. A. Richards, *The Philosophy of Rhetoric* (Oxford: Oxford University Press, 1979), pp. 89–138. *The New Princeton Encyclopedia of Poetry and Poetics* highlights metaphor's imaginative, disruptive potential, defining it as 'a trope, or figurative expression, in which a word or phrase is shifted from its normal uses to a context where it evokes new meanings' (A. Preminger and T. V. F. Brogan (eds), *The New Princeton Encyclopedia of Poetry and Poetics* (Princeton: Princeton University Press, 1993), p. 760). For an extended insight into the problematic nature of metaphor see Paul Ricoeur, *The Rule of Metaphor: Multi-Disciplinary Studies of the Creation of Meaning in Language*, trans. Robert Czerny, Kathleen McLaughlin and John Costello, SJ (London: Routledge & Kegan Paul, 1978); and the review of it, Dominick LaCapra, 'Who Rules Metaphor?', *diacritics*, 10 (December 1980), pp. 15–28. The essays in the collection, Sheldon Sacks (ed.), *On Metaphor* (Chicago: University of Chicago Press, 1979), also give a good overview of different perspectives on this trope.
63. Paul de Man, *The Resistance to Theory* (Manchester: Manchester University Press, 1986), p. 83.
64. As Huw Griffiths put it: 'inadequacy – the inability of language to be consistent with itself – is perhaps the condition for language as metaphor.' Huw Griffiths, 'Translated Geographies: Edmund Spenser's "The Ruines of Time"', *Early Modern Literary Studies*, 4: 2–3 (September 1998), pp. 1–26. Available at: http://extra.shu.ac.uk/emls/04-2/griftran.htm
65. Seamus Heaney, *Preoccupations: Selected Prose, 1968–1978* (London: Faber and Faber, 1980), p. 56.
66. Spivak, *Death of a Discipline*, p. 12.
67. Peter Hallward, *Absolutely Postcolonial: Writing Between the Singular and the Specific* (Manchester: Manchester University Press, 2001), p. xx.
68. Paul Ricoeur, 'Rhetoric – Poetics – Hermeneutics', in W. Jost and M. J. Hyde (eds), *Rhetoric and Hermeneutics in Our Time: A Reader* (New Haven: Yale University Press, 1997), pp. 60–72, p. 70.
69. Steiner, *After Babel*, p. 23.
70. A. T. Q. Stewart, *The Narrow Ground: The Roots of Conflict in Ulster* (Aldershot: Gregg Revivals, 1993), p. 56.
71. Arjun Appadurai, *Modernity at Large: Cultural Dimensions of Modernization* (Minneapolis: University of Minnesota Press, 1996), p. 58.
72. Friel, *Essays, Diaries, Interviews*, p. 114. Dipesh Chakrabarty, *Provincializing Europe: Postcolonial Thought and Historical Difference* (Princeton: Princeton University Press 2000).
73. Luke Gibbons and Kevin Whelan, 'In Conversation with Stephen Rea, 2 February 2001, Yale University', *The Yale Journal of Criticism*, 15: 1 (2002), pp. 5–21, p. 9.

74. Transcript of interview with Rea, FDP, MS 47,308/7.
75. Seamus Deane, 'The Pathos of Distance: Edward Said and the Intellectual Class', *Suitcase: A Journal of Transcultural Traffic*, 2: 1–2 (1997), pp. 44–55, p. 55. Deane's reading of Said is discussed in: Conor McCarthy, 'Edward Said and Irish Criticism', *Éire-Ireland*, 42: 1–2 (Spring/Summer 2007), pp. 311–35, pp. 325–33.
76. Neil Smith and Cindi Katz, 'Grounding Metaphor: Towards a Spatialized Politics', in M. Keith and S. Pile (eds), *Place and the Politics of Identity* (London and New York: Routledge, 1993), pp. 67–83, p. 68.
77. Scott Boltwood, *Brian Friel, Ireland, and The North* (Cambridge: Cambridge University Press, 2007), pp. 139–43.
78. Kirkland, *Literature and Culture in Northern Ireland*, p. 131.
79. Heaney employed these terms when speaking of Field Day's objectives in *History Boys on the Rampage* ('Arena', British Broadcasting Corporation, 1988).
80. Quilligan, 'Field Day's New Double Bill'.
81. Seamus Deane, *Heroic Styles: The Tradition of an Idea*, in *Ireland's Field Day*, pp. 45–60, p. 58.
82. Seamus Deane, 'Field Day's Greeks (and Russians)', in Marianne McDonald and J. Michael Walton (eds), *Amid Our Troubles: Irish Versions of Greek Tragedy* (London: Methuen, 2002), pp. 148–64, p. 148.
83. Richard Kearney, 'Editorial I: Endodermis', *The Crane Bag*, 1: 1 (Spring 1977), pp. 3–5, p. 4.
84. Alwyn Rees and Brinley Rees, *Celtic Heritage: Ancient Traditions in Ireland and Wales* (London: Thames & Hudson, 1961), p. 118. T. F. O'Rahilly, *Early Irish History and Mythology* (Dublin: The Dublin Institute for Advanced Studies, 1946), pp. 171–83. I am grateful to Prof. Mícheál MacCraith for discussing with me the references to the five provinces in O'Rahilly.
85. Kearney, 'Editorial I', p. 4.
86. Mark Patrick Hederman, 'Poetry and the Fifth Province', *The Crane Bag*, 9: 1 (1985), pp. 110–19, pp. 110–11.
87. Michel Foucault, *The Order of Things: An Archaeology of the Human Sciences* (London and New York: Routledge, 2002), p. xviii.
88. As Dauenhauer points out through Ricoeur, it is precisely the non-presence of utopias that gives them an ethical socio-political significance: Bernard P. Dauenhauer, *Paul Ricoeur: The Promise and Risk of Politics* (Lanham: Rowman & Littlefield, 1998), pp. 217–18. The historiographer Reinhart Koselleck has also pointed out that utopias, although 'condemned to be wrong', exist precisely because of the possibility of alternative histories. Reinhart Koselleck, *The Practice of Conceptual History: Timing History, Spacing Concepts*, trans. Todd Presner (Stanford: Stanford University Press, 2002), pp. 84–99, p. 99.
89. George M. Logan and Robert M. Adams, 'Introduction', in Thomas More, *Utopia*, ed. Logan and Adams (Cambridge: Cambridge University Press, 1989), pp. xi–xxviii, p. xi.
90. Hederman, 'Poetry and the Fifth Province', p. 111.
91. Edna Longley, 'A Reply', *The Crane Bag*, 9: 1 (1985), pp. 120–2, p. 120.
92. For an overview of this criticism, and in particular of Field Day's use of the 'fifth province', see Michael Böss, 'The Postmodern Nation: A Critical History of the "Fifth Province" Discourse', *Etudes Irlandaises*, 27: 1 (Spring 2002), pp. 139–59, pp. 148–9.
93. Richard Kearney, *Postnational Ireland: Politics, Culture, Philosophy* (London and New York: Routledge, 1997), p. 100.

94. As Derrida has noted, 'the theme of a transcendental signified took shape within the horizon of an absolutely pure, transparent, and unequivocal translatability.' Derrida, *Positions*, p. 20.
95. Richard Kearney, *Transitions: Narratives in Modern Irish Culture* (Dublin: Wolfhound Press, 1988), pp. 123–60, p. 125.
96. Homi K. Bhabha, *The Location of Culture* (London and New York: Routledge, 1994), p. 4.
97. Ibid., pp. 1–2.
98. Richard Kearney, 'Introduction', in Richard Kearney (ed.), *Paul Ricoeur: The Hermeneutics of Action* (London: Sage, 1996), pp. 1–2, p. 1.
99. F. C. McGrath, *Brian Friel's (Post)Colonial Drama: Language, Illusion, and Politics* (Syracuse: Syracuse University Press, 1999), p. 209.
100. Seamus Heaney, 'A Field Day for the Irish', *The Times*, 5 December 1988.
101. Friel, *Essays, Diaries, Interviews*, p. 101.
102. Edward W. Said, *The World, The Text, and The Critic* (Cambridge, MA: Harvard University Press, 1983), p. 116. Said differentiated between 'instinctual filiation' and 'affiliation', which expresses the social forms through which modes of filiation are represented (p. 24).
103. Cited, John Gray, 'Field Day Five Years On', *Linenhall Review*, 2: 2 (Summer 1985), pp. 4–10, p. 7.
104. Deane, Lecture at Magee College, FDP, MS 47,116/1.
105. Seamus Deane, *Reading in the Dark* (London: Vintage, 1997).
106. Deane, Lecture at Magee College, FDP, MS 47,116/1.
107. Seamus Deane, 'Derry: City Besieged Within the Siege', *Fortnight*, 198 (October 1983), pp. 18–19, p. 18.
108. Shaun Richards, 'To Bind the Northern to the Southern Stars: Field Day in Derry and Dublin', *The Irish Review*, 4 (Spring 1988), pp. 52–8, p. 52.
109. As Deane has noted: '[b]oth Irish communities have, in fact, cultural traditions which glamorise the position of being under siege from a world hostile to their unique and treasured beliefs.' Seamus Deane, 'Remembering the Irish Future', *The Crane Bag*, 8: 1 (1984), pp. 81–92, p. 85.
110. Robert Fitzroy Foster, *Modern Ireland, 1600–1972* (London: Penguin, 1988), p. 146.
111. Alasdair MacIntyre, *Whose Justice? Which Rationality?* (London: Duckworth, 1988), p. 378.
112. Eamonn McCann, *War and an Irish Town*, 3rd edn. (London: Pluto Press, 1993), p. 287.
113. Thirteen people were killed immediately and one died subsequently in hospital. The British government launched a new inquiry into 'Bloody Sunday' chaired by Lord Saville in April 1998. Its findings (published 15 June 2010) found that all the dead were innocent and prompted an official apology from the British Prime Minister, David Cameron. Richard Norton-Taylor's drama, *Bloody Sunday: Scenes from the Saville Inquiry* (London: Oberon Books, 2005), splices together depositions to this inquiry.
114. Foster, *Modern Ireland, 1600–1972*, p. 588.
115. See McCann, *War and an Irish Town*, pp. 79–80.
116. Christopher Morash, *A History of Irish Theatre, 1601–2000* (Cambridge: Cambridge University Press, 2002), p. 233. In his account of the first night of *Translations* (pp. 233–41), Morash further notes that '[s]taging *Translations* in Derry […] generated the kind of local, community passion usually associated with the amateur

theatre (and, indeed, amateur companies would quickly make the play a staple of the festival circuit)' (p. 238).

117. Brian Friel, *The Freedom of the City* (Oldcastle, Meath: Gallery Press, 1992), p. 47.
118. Gibbons and Whelan, 'In Conversation with Stephen Rea', p. 9.
119. Clifford Geertz, *The Interpretation of Cultures: Selected Essays* (New York: Basic Books, 1973), p. 335.
120. BBC, *The Story of Field Day*.
121. Elmer Andrews, *The Art of Brian Friel: Neither Reality nor Dreams* (Basingstoke: Macmillan, 1995), pp. 72–3.
122. Judith Butler, *Gender Trouble: Feminism and the Subversion of Identity* (New York and London: Routledge, 1999), p. 177.
123. Ibid., p. 173
124. Ibid., p. 175.
125. Ibid.
126. McGrath, *Brian Friel's (Post)Colonial Drama*, p. 7.
127. Ibid., p. 6.
128. Bhabha, *The Location of Culture*, p. 67.
129. David Lloyd, *Nationalism and Minor Literature: James Clarence Mangan and the Emergence of Irish Cultural Nationalism* (Berkeley: University of California Press, 1987), p. xi.
130. Bhabha, *The Location of Culture*, p. 228.
131. Antoine Vitez, Georges Banu, Alain Girault and François Rey, 'The Duty to Translate: An Interview with Antoine Vitez', in Patrice Pavis (ed.), *The Intercultural Performance Reader* (New York and London: Routledge, 1996), pp. 121–30, pp. 126–7.
132. Susan Bassnett, *Translation Studies*, 3rd edn. (London and New York: Routledge, 2002), pp. 119–20.
133. Roche, *Contemporary Irish Drama*; Christopher Murray, *Twentieth-Century Irish Drama: Mirror Up to Nation* (Manchester: Manchester University Press, 1997); Morash, *A History of Irish Theatre*; Nicholas Grene, *The Politics of Irish Drama: Plays in Context from Boucicault to Friel* (Cambridge: Cambridge University Press, 1999); Lionel Pilkington, *Theatre and the State in Twentieth-Century Ireland: Cultivating the People* (London and New York: Routledge, 2001).
134. David Cairns and Shaun Richards, *Writing Ireland: Colonialism, Nationalism and Culture* (Manchester: Manchester University Press, 1988); Kirkland, *Literature and Culture in Northern Ireland Since 1965*; Andrews, *The Art of Brian Friel*; Richard Pine, *The Diviner: The Art of Brian Friel* (Dublin: University College Dublin Press, 1999); McGrath, *Brian Friel's (Post)Colonial Drama*; Boltwood, *Brian Friel, Ireland, and The North*.
135. Andrews, *The Art of Brian Friel*, pp. 172 and 179.
136. Colm Tóibín, 'On The Literary Wing', *Times Literary Supplement*, 28 April 1995, p. 10.
137. Seamus Deane, 'Hot House Flowers or Anthologica Hibernica', *Gown: Literary Supplement* (Spring/Summer 1990), p. 19.
138. Edna Longley, 'Poetry and Politics in Northern Ireland', *The Crane Bag*, 9: 1 (1985), pp. 26–40, p. 33.
139. Longley, 'A Reply', pp. 120–1.
140. Kirkland, *Literature and Culture in Northern Ireland*, p. 90.
141. Edna Longley, *The Living Stream: Literature and Revisionism in Ireland* (Newcastle: Bloodaxe Books, 1994), p. 174.

142. BBC, *The Story of Field Day*.
143. Conor McCarthy, *Modernisation: Crisis and Culture in Ireland, 1969–1992* (Dublin and Portland, OR: Four Courts Press, 2000), p. 228.
144. Longley was suggested as a possible pamphleteer by Field Day in 1983–4, with Paulin, in particular, urging that she be invited (FP, MS 37,181/1). Nothing came of this, or of the 1986 proposal that she might edit the Yeats section of the *Anthology*, which Deane objected to on the grounds that she might use it merely as an opportunity to pursue her hostilities against Field Day. Meeting, 14 June 1986, FP, MS 37,181/2.
145. Marilynn J. Richtarik, *Acting Between the Lines: The Field Day Theatre Company and Irish Cultural Politics, 1980–1984* (Oxford: Clarendon Press, 1994). Carmen Szabó, *"Clearing the Ground": The Field Day Theatre Company and the Construction of Irish Identities* (Newcastle: Cambridge Scholars Publishing, 2007).
146. Szabó, *"Clearing the Ground"*, p. 81.
147. Ibid., p. 12.

1. Translating the Irish Past

1. Draft for 1984 essay on Field Day, FDP, MS 47,166.
2. Georg H. W. Hegel, 'Preface', *The Philosophy of Right*, trans. S.W. Dyde (New York: Cosimo Inc., 2008), p. xxi.
3. Grene, *The Politics of Irish Drama*, p. 40.
4. John C. Greene, Patricia L. Haberstroh and Adrian Frazier, 'Wealth, Gender, Politics: Three Views on *The Field Day Anthology of Irish Writing*', *Éire-Ireland*, 27: 2 (Summer 1992), pp. 111–31, p. 130.
5. Pine, *The Diviner*, p. 1.
6. Ibid., pp. 2–3.
7. Ibid., p. 184.
8. Seamus Heaney, *An Open Letter*, in *Ireland's Field Day* (London: Hutchinson, 1985), pp. 23–32, p. 26.
9. Heaney, 'A Field Day for the Irish'.
10. Spivak, *Death of a Discipline*, p. 15.
11. As Chakrabarty has shown, for colonial historiography, native cultures were essentially belated. See Chakrabarty, *Provincializing Europe*, pp. 6–11.
12. Notes, 24 November 1978, FP, MS 37,085/1.
13. Paul Ricoeur, *Time and Narrative*, Vol. I, trans. Kathleen Blamey, Kathleen McLaughlin and David Pellauer (Chicago: University of Chicago Press, 1990), p. 176. An alternative, somewhat distorted, version of this passage is cited in Brian Friel, John Andrews and Kevin Barry, 'Translations and A Paper Landscape: Between Fiction and History', *The Crane Bag*, 7: 2 (1983), pp. 118–24, p. 118.
14. Cited, David Woodward, 'Cartography and the Renaissance: Continuity and Change', in David Woodward (ed.), *The History of Cartography*, Vol. III, Part I (Chicago: University of Chicago Press, 2007), pp. 3–24, p. 22.
15. Positivism's roots are in Saint-Simon's and Comte's searches for standard scientific methods that might be used to evaluate social life. Later, 'logical positivists' supplemented this by arguing 'that sensory knowledge was the most certain and that any concept not directly about sensory experiences should be "translatable" into observational concepts. Those concepts that could not be so translated, primarily metaphysical concepts, were rejected as meaningless.' Harold Kincaid,

'Positivism in the Social Sciences', in Edward Craig (ed.), *Routledge Encyclopedia of Philosophy*, Vol. VII (New York and London: Routledge, 1998), pp. 558–61, p. 559. For a brief discussion of the link between empiricism and positivism, see Alan Lacey, 'Positivism', in Ted Honderich (ed.), *The Oxford Companion to Philosophy* (Oxford and New York: Oxford University Press, 1995), pp. 705–6. Britain, according to Kail, is 'empiricism's spiritual home'. Peter Kail, 'Empiricism', in Anthony Grayling, Andrew Pyle and Naomi Goulder (eds), *The Continuum Encyclopedia of British Philosophy*, Vol. II (London and New York: Thoemmes Continuum, 2006), pp. 981–3, p. 981.

16. Derek Gregory, *Geographical Imaginations* (Oxford: Blackwell Publishers, 1993), p. 171.
17. Bernhard Klein, 'Partial Views: Shakespeare and the Map of Ireland', *Early Modern Literary Studies*, 4: 2–3 (September 1998), pp. 1–20. Available at: http://extra.shu.ac.uk/emls/04-2/kleipart.htm
18. Frantz Fanon, *The Wretched of the Earth*, trans. C. Farrington (London: Penguin Books, 2001), p. 77.
19. The 'civilians and barbarians' stereotype can be heard in embryonic form in Gerald of Wales's twelfth-century text: see Giraldus Cambrensis [Gerald of Wales], *The History and Topography of Ireland [Topographia Hiberniae]*, trans. John J. O'Meara (Portlaoise: Dolmen Press, 1982). For an analysis of these texts that underlines the need to question the expectation of a certain type of impartial, scientific veracity from a 'geography', see Joep Leerssen, *Mere Irish and Fíor-Ghael: Studies in the Idea of Irish Nationality, Its Development and Literary Expression prior to the Nineteenth Century* (Cork: Cork University Press, 1996), pp. 32–76.
20. Cited, Paul Binding, *Imagined Corners: Exploring the World's First Atlas* (London: Headline Books, 2003), p. 222.
21. Cheyfitz, *The Poetics of Imperialism*, p. 15.
22. Kearney, *Transitions*, p. 156.
23. Mary Hamer, 'Putting Ireland on the Map', *Textual Practice*, 3: 2 (Summer 1989), pp. 184–201, p. 188.
24. J. H. Andrews, *A Paper Landscape: The Ordnance Survey in Nineteenth-Century Ireland* (Oxford: Clarendon Press, 1975).
25. Cited, Patricia Boyne, *John O'Donovan (1806–1861): A Biography* (Kilkenny: Boethius, 1987), p. 9.
26. A. S. Green, cited, J. H. Andrews, *History in the Ordnance Map: An Introduction for Irish Readers*, 2nd edn. (Kerry: David Archer, 1993), p. 2.
27. Joep Leerssen, *Remembrance and Imagination: Patterns in the Historical and Literary Representation of Ireland in the Nineteenth Century* (Cork: Cork University Press, 1996), p. 102.
28. Cited, Andrews, *A Paper Landscape*, p. 148.
29. Mary Louise Pratt, *Imperial Eyes: Travel Writing and Transculturation* (London and New York: Routledge, 1992), pp. 201f.
30. For interpellation see Louis Althusser, *Lenin and Philosophy, and Other Essays*, 2nd edn., trans. Ben Brewster (London: New Left Books, 1977), pp. 121–73, pp. 160–70.
31. Thomas Davis, *Essays, Literary and Historical* (Dundalk: Dundalgan Press, 1914), p. 382.
32. Thomas Davis, *Our National Language* (Dublin: Gaelic League, n.d.), p. 1.
33. Leslie P. Thiele, 'The Agony of Politics: The Nietzschean Roots of Foucault's Thought', *American Political Science Review*, 84: 3 (September 1990), pp. 907–25, p. 921.

34. Kearney, *Transitions*, pp. 155–7. Elmer Andrews has translated Kearney's concepts into an opposition between '[t]he "Gaelic" view of language' and the '"English" view of language'; however, things are never so clear-cut in *Translations*. Andrews, *The Art of Brian Friel*, p. 170.
35. Ordnance Survey, *Memoir of the City and North Western Liberties of Londonderry, Parish of Templemore (1837)*, Facsimile edition (Derry: North-West Books, 1990). The project was cancelled in 1840 because of its cost. Indeed, the Board of Ordnance took a dim view of the memoir scheme in general as 'they saw the Survey's ventures into non-cartographic scholarship as a perversion of its rightful purpose'. Andrews, *History in the Ordnance Map*, p. 2.
36. Ordnance Survey, *Memoir*, pp. 208–10.
37. Ibid., p. 209.
38. Martin Heidegger, *Poetry, Language, Thought*, trans. and intro. Albert Hofstadter (New York: Harper & Row, 1975), p. 154.
39. Kearney, *Transitions*, p. 126.
40. Heidegger, *Poetry, Language, Thought*, p. 132.
41. *The Metrical Dindshenchas*, Vols. I–XII, trans. and ed. Edward Gwynn (Dublin: Hodges, Figgis, & Co., 1903–35).
42. Pine, *The Diviner*, p. 199.
43. Andrews links Hugh with the fathers in *The Playboy of Western World* and *Juno and the Paycock*, noting that 'Hugh has more dignity and more self-knowledge than either of these characters, but he is no more capable of positive action.' Andrews, *The Art of Brian Friel*, p. 180.
44. Heaney, *Preoccupations*, p. 145. The colonial significance of this sequence is discussed in Michael Wiley, *Romantic Geography: Wordsworth and Anglo-European Spaces* (Basingstoke: Macmillan Press, 1998), pp. 79–120. I am grateful to Martina Domines Veliki for assisting me with these references to Wordsworth's poetry.
45. 25 September 1979, FDP, MS 47,119/1.
46. 'Sunday', (n.d.) (1979), FDP, MS 47,119/1.
47. For a complete biography, see Boyne, *John O'Donovan*.
48. Paddy John Browne, 'Wonderful Knowledge: The Ordnance Survey of Ireland', *Éire-Ireland*, 20: 1 (Spring 1985), pp. 15–27, p. 23.
49. Colby's instructions, cited, Andrews, *A Paper Landscape*, p. 120.
50. Ibid.
51. Brian Friel, 'Extracts from a Sporadic Diary', in Tim Pat Coogan (ed.), *Ireland and the Arts* (London: Namara, 1986), pp. 56–61, p. 58.
52. Friel et al., 'Translations and A Paper Landscape', p. 123.
53. Steiner, *After Babel*, p. 236. McGrath provides a full exploration of this idea, as he locates Friel's work in a tradition he terms 'the Irish Art of Lying'. McGrath, *Brian Friel's (Post)Colonial Drama*, pp. 13–48.
54. Pilkington, *Theatre and the State in Twentieth-Century Ireland*, p. 214.
55. 'Then they said, "Come let us build ourselves a city, and a tower with its top in the heavens, and let us make a name for ourselves; otherwise we shall be scattered abroad upon the face of the whole earth."' Genesis 11:4, *The New Revised Standard Version Bible*, in *The Complete Parallel Bible* (Oxford and New York: Oxford University Press, 2000), p. 20. Kiberd offers a reading of *Translations* in terms of the Tower of Babel; see Declan Kiberd, *Inventing Ireland* (London: Jonathan Cape, 1995), pp. 636–7.
56. Notes, 22 November 1978, FP, MS 37,085/1.
57. Steiner, *After Babel*, pp. 312–16.

58. Samia Mehrez, 'Translation and the Postcolonial Experience: The Francophone North African Text', in Lawrence Venuti (ed.), *Rethinking Translation: Discourse, Subjectivity, Ideology* (London and New York: Routledge, 1992), pp. 120–38, p. 121.
59. Niranjana, *Siting Translation*, p. 68.
60. Steiner, *After Babel*, p. 428.
61. Ibid., pp. 39–40 and 476.
62. Robert S. Smith, 'The Hermeneutic Motion in Friel's *Translations*', *Modern Drama*, 34 (1991), pp. 392–409, p. 407.
63. Albert Hofstadter, 'Introduction', in Heidegger, *Poetry, Language, Thought*, pp. ix–xxii, pp. xix–xx.
64. Pine, *The Diviner*, p. 202.
65. Friel, 'Extracts from a Sporadic Diary', p. 57.
66. Patrick John Dowling, *The Hedge Schools of Ireland* (Dublin and Cork: The Talbot Press, 1935), p. 108.
67. Heidegger, *Poetry, Language, Thought*, p. 132.
68. Steiner, *After Babel*, p. 413.
69. Friel, *Essays, Diaries, Interviews*, p. 87.
70. Friedrich Nietzsche, *The Gay Science*, trans. Walter Kaufmann (New York: Random Books, 1974), p. 82.
71. Murray dismisses it as 'the closest Friel ever came to writing a pamphlet'. Christopher Murray, 'Palimpsest: Two Languages as One in *Translations*', in Donald E. Morse, Csilla Bertha and Mária Kurdi (eds), *Brian Friel's Dramatic Artistry: 'The Work Has Value'* (Dublin: Carysfort Press, 2006), pp. 93–108, p. 94.
72. Fintan O'Toole described it as 'a hesitant move into unknown territory', and in common with other critics found the play too static and lacking in drama. See Fintan O'Toole, 'A Hesitant Move into Unknown Territory', *The Irish Times*, 'Weekend', 24 September 1988, p. 5; Brian Brennan, 'The Reinvention of Hugh O'Neill', *The Sunday Independent* (Ireland), 25 September 1988, p. 15; Michael Billington, 'History Kisses the Blarney Stone', *The Guardian*, 7 December 1988; Paul Taylor, 'Stranger than Faction', *The Independent*, 7 December 1988. However, transcripts of the brief interviews with audience members after the first performances of the play in the North conducted for the BBC documentary, *History Boys on the Rampage*, would tend to dispute the dominant critical view of the play, as the audiences show a great appreciation of the complexities of the drama. While some queried elements of the play by noting its parallels with O'Faoláin, most saw it as a salutary act of historical demythologising and reflected on its connections with the 'Troubles' (FDP, MS 47,308/2). Mary Holland also records that the play inspired an intense historiographical and political debate in a house on the Bogside: Mary Holland, 'A Field Day for Irish Theatre', *The Observer Magazine*, 30 October 1988, pp. 60–3, pp. 60–1.
73. McGrath, *Brian Friel's (Post)Colonial Drama*, p. 231.
74. McKeone also points to the similarities between *Faith Healer* and this play: Gary McKeone, 'Fact and Faction', *Theatre Ireland*, 17 (December 1988/March 1989), pp. 7–8, p. 7.
75. Ronan Fanning, 'Nationalist Perspectives on the Past: A Symposium. I. The U.C.D. Debate (a) The Meaning of Revisionism', *The Irish Review*, 4 (Spring 1988), pp. 15–19, p. 16. Myth, indeed, became a historiographical category of sorts; see, for instance, T. W. Moody, 'Irish History and Irish Mythology', in Ciaran Brady (ed.), *Interpreting Irish History: The Debate on Historical Revisionism, 1938–1994* (Dublin: Irish Academic Press, 1994), pp. 71–86.

76. See T. W. Moody and R. D. Edwards, 'Preface to *Irish Historical Studies*', in Brady (ed.), *Interpreting Irish History*, pp. 35–7. Although coming from different perspectives, overviews of revisionism in Ireland would not contest this statement; see, for instance, Nancy J. Curtin, '"Varieties of Irishness": Historical Revisionism, Irish Style', *The Journal of British Studies*, 35: 2 (April 1996), pp. 195–219; Stephen Howe, 'The Politics of Historical "Revisionism": Comparing Ireland and Israel/ Palestine', *Past and Present*, 168: 1 (2000), pp. 227–53; Kevin Whelan, 'The Revisionist Debate in Ireland', *boundary 2*, 31: 1 (2004), pp. 179–205.
77. Robert Fitzroy Foster, *The Irish Story: Telling Tales and Making It Up in Ireland* (London and New York: Allen Lane, The Penguin Press, 2001). Controversially, Gibbons proposed that 'the spectre of nationalism conjured up by revisionism is largely of its own making'. Luke Gibbons, 'Challenging the Canon: Revisionism and Cultural Criticism', in Seamus Deane (ed.), *The Field Day Anthology of Irish Writing*, Vol. III (Derry and New York: Field Day and Norton, 1991), pp. 561–8, p. 568.
78. For Longley, '*Translations* refurbishes an old myth.' Longley, 'Poetry and Politics in Northern Ireland', p. 29.
79. Sean Connolly, 'Dreaming History: Brian Friel's *Translations*', *Theatre Ireland*, 13 (1987), pp. 42–4, p. 44. Likewise, Fenton declared the play historically inaccurate because the peasants seemed more sophisticated than the crude English soldiers: 'I would have believed more readily in the historical accuracy of the picture had the English officers been less oafishly unlatined.' James Fenton, 'Ireland: Destruction of an Idyll', *The Sunday Times*, 28 September 1980, p. 40.
80. Friel et al., 'Translations and A Paper Landscape', p. 122.
81. Walter Benjamin, 'On the Concept of History', in *Selected Writings, Vol. 4, 1938–1940*, ed. H. Eiland and M.W. Jennings, trans. E. Jephcott et al. (Cambridge, MA and London: The Belknap Press of Harvard University Press, 2003), pp. 389–400, p. 395.
82. Friel et al., 'Translations and A Paper Landscape', p. 122.
83. Seamus Deane, 'Wherever Green is Read', in Máirín Ní Dhonnchadha and Theo Dorgan (eds), *Revising the Rising* (Derry: Field Day, 1991), pp. 91–105, p. 101. Ó Tuathaigh put this more diplomatically, suggesting that Irish historians 'have, on the whole, been less courageous – or at least less enterprising – in looking to critical theory for insights into how the activity of historical writing/representation is conducted'. M. A. G. Ó Tuathaigh, 'Irish Historical "Revisionism": State of the Art or Ideological Project', in Brady (ed.), *Interpreting Irish History*, pp. 306–26, p. 325.
84. Howe, *Ireland and Empire*, pp. 110–20. In the course of this critique, he touches on the trope of translation in Field Day's productions, seeing it correctly as a symbol of postcolonial ambivalence. However, he does not develop this, but rather dismisses it by wondering if the persistence of this trope meant that Field Day was committed to a critical examination of Irish nationalism or if it simply replaced 'bad' colonial translations with new enabling ones. He insinuates, without actually putting forward an argument, that the latter is the case, after which the subject of the ambivalent trope of translation is dropped (pp. 114–15).
85. Ibid., p. 147. Ultimately Howe cannot sustain his argument that Ireland was not a colony, and in his conclusion doubles back and posits that Ireland did have a 'colonial past [...] though one that took unique hybrid forms, involving extensive integration and consensual partnership as well as exploitation and coercion' (p. 232). Joe Cleary provides a comprehensive overview of the debate about Ireland's status as a colony in '"Misplaced Ideas"? Colonialism, Location, and Dislocation

in Irish Studies', in Clare Carroll and Patricia King (eds), *Ireland and Postcolonial Theory* (Notre Dame: University of Notre Dame Press, 2003), pp. 16–45. Revised as 'Irish Studies, Colonial Questions: Locating Ireland in the Colonial World', in Joe Cleary, *Outrageous Fortune: Capital and Culture in Modern Ireland* (Dublin: Field Day Publications, 2007), pp. 14–46.

86. Lloyd, *Ireland After History*, p. 82.
87. Ranajit Guha, 'On Some Aspects of the Historiography of Colonial India', in Ranajit Guha and Gayatri Chakravorty Spivak (eds), *Selected Subaltern Studies* (Oxford: Oxford University Press, 1988), pp. 37–44, p. 40.
88. Connolly, 'Dreaming History', pp. 43–4.
89. Pilkington, *Theatre and the State in Twentieth-Century Ireland*, p. 219.
90. Andrews offered the *faux-naïf* objection that it is 'a play about late twentieth-century Ireland'. Friel et al., 'Translations and A Paper Landscape', p. 120.
91. Pilkington, *Theatre and the State in Twentieth-Century Ireland*, pp. 219–20.
92. Joe Cleary, 'Introduction: Ireland and Modernity', in Joe Cleary and Claire Connolly (eds), *The Cambridge Companion to Modern Irish Culture* (Cambridge: Cambridge University Press, 2005), pp. 1–21, p. 3.
93. Lloyd, *Ireland After History*, pp. 77–88.
94. McCarthy, *Modernisation, Crisis and Culture in Ireland, 1969–1992*, pp. 30 and 22.
95. Benjamin, 'On the Concept of History', p. 391.
96. Hiram Morgan, 'Making History: A Criticism and a Manifesto', *Text and Context* (Autumn 1990), pp. 61–5, p. 61.
97. Foster, *Modern Ireland, 1600–1972*, p. 4.
98. Ibid., p. 3n.
99. Kathleen Hohenleitner, 'The Book at the Center of the Stage: Friel's *Making History* and *The Field Day Anthology of Irish Writing*', in Stephen Watt, Eileen Morgan and Shakir Mustafa (eds), *A Century of Irish Drama: Widening the Stage* (Bloomington and Indianapolis: Indiana University Press, 2000), pp. 239–55, p. 241.
100. Chakrabarty, *Provincializing Europe*, p. 17.
101. Steven Ellis, 'Nationalist Historiography and the English and Gaelic Worlds in the Late Middle Ages', *Irish Historical Studies*, 25 (1986–7), pp. 1–18.
102. Brendan Bradshaw, 'Nationalism and Historical Scholarship in Modern Ireland', *Irish Historical Studies*, 26: 104 (November 1989), pp. 329–51, pp. 337–46.
103. Ibid., p. 346.
104. Ibid., p. 348.
105. Lloyd, *Ireland After History*, p. 17.
106. Nicholas Canny, 'Review of Making History', *The Galway Advertiser*, 27 October 1988.
107. Roy Foster, '"We Are All Revisionists Now"', *The Irish Review*, 1 (1986), pp. 1–5, p. 2.
108. Foster, *The Irish Story*, p. 5. Ricoeur, however, has argued that Propp's model is ultimately unique to the Russian folktale, and that it 'is not transposable, for example, to the tales of Grimm or of Andersen, still less to narrative structures more removed from the Russian folktale and from the form of the folktale in general'. Paul Ricoeur, *Hermeneutics and the Human Sciences: Essays on Language, Action and Interpretation*, trans. and ed. John B. Thompson (Cambridge and Paris: Cambridge University Press and Editions de la Maison des Sciences de l'Homme, 1981), p. 282.

109. Hayden White, 'Historical Discourse and Literary Writing' (text of paper delivered at the conference, *Literature and Its Others*, University of Turku, Finland, 9 May 2003).
110. McGrath, *Brian Friel's (Post)Colonial Drama*, p. 227.
111. For a concise account, see Tom Dunne, 'A Polemical Introduction: Literature, Literary Theory and the Historian', in Tom Dunne (ed.), *The Writer As Witness: Literature As Historical Evidence* (Cork: Cork University Press, 1987), pp. 1–9, p. 3.
112. Michel de Certeau, *The Writing of History*, trans. Tom Conley (New York: Columbia University Press, 1988), p. xxvii.
113. Ricoeur, *Hermeneutics and the Human Sciences*, pp. 293–4.
114. Ibid., p. 294.
115. LaCapra, for instance, finds that Ricoeur's assertions 'rely on a very attenuated or overly expansive notion of narrative and are interesting only when they enable a far-ranging inquiry into the different possibilities and modalities of narrative'. Dominick LaCapra, *Writing History, Writing Trauma* (Baltimore and London: Johns Hopkins University Press, 2001), p. 64.
116. Aristotle, 'The Poetics', in Aristotle, *In Twenty-Three Volumes: XXIII. The Poetics. "Longinus", On the Sublime. Demetrius, On Style*, trans. W. Hamilton Fyfe (Cambridge, MA and London: Harvard University Press and William Heinemann, 1973), pp. 4–118, p. 35.
117. Ricoeur, *Hermeneutics and the Human Sciences*, p. 296.
118. Cited, Jonathan Bardon, *A History of Ulster: New Updated Edition* (Belfast: Blackstaff Press, 2001), p. 103.
119. Hiram Morgan, *Tyrone's Rebellion: The Outbreak of the Nine Years War in Tudor Ireland* (Woodbridge: Royal Historical Society and The Boydell Press, 1993), p. 217.
120. Brian Friel, 'Self-Portrait', *Aquarius*, 5 (1972), pp. 17–22, p. 18.
121. Brian Friel, 'Programme Note', in Field Day Theatre Company, *Making History by Brian Friel* (Programme for first performances), 1988, n.p.
122. As Morgan put it: 'we cannot allow Lombard to speak for historians because he is transparently a contemporary propagandist, nor can we allow O'Neill to interpret his own life because that would be subjective and equally flawed'. Morgan, 'Making History: A Criticism and a Manifesto', p. 62.
123. Luisa Passerini, *Autobiography of a Generation, Italy 1968*, trans. Lisa Erdberg (Hanover and London: Wesleyan University Press, 1996), p. 22.
124. As Robert Smith has noted, a similar motion underlines a hermeneutic understanding of interpretation. He reads Gadamer's metaphor of 'a "fusion of horizons" of the past and present' as denoting 'the ideal, if ever-elusive goal at which interpretation aims: the horizon recedes as the observer approaches it'. Smith, 'The Hermeneutic Motion in Friel's *Translations*', pp. 392–3.
125. Koselleck, *The Practice of Conceptual History*, p. 12.
126. Morgan, 'Making History: A Criticism and a Manifesto', p. 63. Byrne provides a brief biography of Lombard in Peter Lombard, Archbishop of Armagh, *The Irish War of Defence, 1598–1600 (Extracts from the 'De Hibernia Insula Commentarius')*, trans. and ed. Matthew J. Byrne (Dublin and Cork: Cork University Press, 1930), p. viii.
127. Morgan, *Tyrone's Rebellion*, p. 216. For Fogarty, the relationship between O'Neill and Mabel represents an attempt on Friel's part to imagine 'a utopian counter-version of Irish history'. Anne Fogarty, 'The Romance of History: Renegotiating the Past in Thomas Kilroy's *The O'Neill* and Brian Friel's *Making History*', *Irish University Review*, 32: 1 (Spring/Summer 2002), pp. 18–32, p. 32.

128. Deane, 'Field Day's Greeks (and Russians)', p. 162n. McGrath argues that O'Faoláin 'portrays a much more complex, less sentimentalized O'Neill' than Friel's 'more two dimensional' character. McGrath, *Brian Friel's (Post)Colonial Drama*, p. 219.
129. Seán O'Faoláin, *The Great O'Neill: A Biography of Hugh O'Neill Earl of Tyrone, 1550–1616* (London, New York and Toronto: Longmans, Green and Co., 1942), pp. 36–49.
130. Morgan, *Tyrone's Rebellion*, p. 214.
131. Kiberd, *Inventing Ireland*, p. 633.
132. Hayden White, *Metahistory: The Historical Imagination in Nineteenth-Century Europe* (Baltimore and London: Johns Hopkins University Press, 1973), p. 21.
133. Deane, *Heroic Styles: The Tradition of an Idea*, p. 45.
134. Koselleck, *The Practice of Conceptual History*, p. 12.
135. Ibid., p. 13.
136. Hayden White, 'Foreword', in ibid., pp. ix–xiv, p. xi.
137. Freddie Rokem, *Performing History: Theatrical Representations of the Past in Contemporary Theatre* (Iowa City: University of Iowa Press, 2000), p. 6.
138. Seamus Deane, 'Brian Friel: The Name of the Game', in Alan J. Peacock (ed.), *The Achievement of Brian Friel* (Gerrards Cross: Colin Smythe, 1993), pp. 103–12, p. 111.
139. Bhabha, *The Location of Culture*, p. 235.
140. de Man, *The Resistance to Theory*, p. 83.
141. Koselleck, *The Practice of Conceptual History*, p. 7.
142. Hans-Georg Gadamer, 'The Problem of Historical Consciousness', in Paul Rabinow and William M. Sullivan (eds), *Interpretive Social Science: A Second Look*, trans. Jeff L. Close and Hans Fantel (Berkeley, Los Angeles and London: University of California Press, 1987), pp. 82–140, p. 136.
143. Ibid., p. 91.
144. Ibid., pp. 126–7 and 130–1.
145. Koselleck, *The Practice of Conceptual History*, p. 41.
146. Jacques Derrida, 'Circumfession', in Geoffrey Bennington and Jacques Derrida, *Jacques Derrida*, trans. Geoffrey Bennington (Chicago: University of Chicago Press, 1999), p. 315.
147. Cathy Caruth, *Unclaimed Experience: Trauma, Narrative, and History* (Baltimore and London: Johns Hopkins University Press, 1996), p. 3.
148. Ibid., p. 4.
149. LaCapra, *Writing History, Writing Trauma*, p. 41.
150. Ibid., p. 66.
151. Holding that both plays 'accept the inevitability of a hybrid Anglicized Ireland', McGrath reads the postcolonial historiography enacted in *Making History* in a commensurate fashion. McGrath, *Brian Friel's (Post)Colonial Drama*, pp. 212 and 231. Writing on *Translations*, in which he detects 'a strong nationalist strain', Grene notes the 'paradox [...] that a play asserting the irrevocable and uninterpretable nature of the pre-colonial past should have acted as such a readily available interpretation of the Irish experience for so many diverse audiences both in Ireland and outside'. Rather than seeing this, as Grene does, as a case of Friel 'hav[ing] it both ways', the contention here is that the disjunctions of the colonial situation bring to the fore the disjunctions of history itself, and that both *Translations* and *Making History* spoke to audiences on this level as well. Grene, *The Politics of Irish Drama*, pp. 45 and 43.

2. Translation, Home and Hospitality

1. Paul Ricoeur, 'Reflections on a New Ethos for Europe', in Kearney (ed.), *Paul Ricoeur: The Hermeneutics of Action*, pp. 3–13, p. 5.
2. While this is often cited as a direct quote, Craig told the Northern Ireland House of Commons on 24 April 1934: '[a]ll I boast is that we are a Protestant Parliament and Protestant State.' In the same place on 21 November 1934, he declared that he led 'a Protestant Government for a Protestant people'. See: http://cain.ulst.ac.uk/issues/discrimination/quotes.htm
3. Friel, *Essays, Diaries, Interviews*, p. 112.
4. Seán Ó Tuama recounts that one of the clearest signs of the changes that swept through Ireland in the aftermath of the Battle of Kinsale (1601) was that poets, who had held an exalted position in the old Gaelic order, were refused hospitality. Poets such as Ó Bruadair and MacCurta composed bitter verses on this theme. Seán Ó Tuama (ed.), *An Duanaire, 1600–1900: Poems of the Dispossessed*, trans. Thomas Kinsella (Dublin: The Dolmen Press and I gComhar na Gaeilge, 1981), p. xxvii.
5. James Joyce, *A Portrait of the Artist As a Young Man* (London: Grafton Books, 1986), p. 172.
6. For a thorough account of how Joyce's modernism relates to the fractured experience of colonial Ireland, see Emer Nolan, *James Joyce and Nationalism* (London and New York: Routledge, 1995).
7. For a brief account of the history of English in Ireland, see Terence Patrick Dolan (ed.), *A Dictionary of Hiberno-English: The Irish Use of English*, 2nd edn. (Dublin: Gill and Macmillan, 2004), pp. xxvii–xxix. While it has been almost entirely superseded in terms of scholarship, the pioneering study of Hiberno-English is P. W. Joyce, *English as We Speak It in Ireland* (Dublin: Wolfhound Press, 1988) (first published 1910).
8. Friel, *Essays, Diaries, Interviews*, p. 87.
9. Seamus Deane, 'General Introduction', in Deane (ed.), *The Field Day Anthology of Irish Writing*, Vol. I, pp. xix–xxvi, p. xxv.
10. Niranjana, *Siting Translation*, p. 167.
11. In essence, this is Howe's argument in *Ireland and Empire*. For a summary of his objections to this use of postcolonial theory in the Irish context, see: pp. 233f.
12. Longley, 'Poetry and Politics in Northern Ireland', p. 31; Howe, *Ireland and Empire*, p. 110.
13. Friel, *Essays, Diaries, Interviews*, p. 107.
14. Desmond Rushe, 'Derry *Translations*', *Éire-Ireland*, 15: 4 (Winter 1980), pp. 126–8, p. 128; Desmond Rushe, 'Friel at His Pleasing and Astonishing Best', *Irish Independent*, 25 September 1980, p. 9.
15. Lynda Henderson, 'A Fondness for Lament', *Theatre Ireland*, 17 (December 1988–March 1989), pp. 18–20, p. 18.
16. As Richtarik put it: '[w]ithin a few years the play had acquired the status of a classic'. Richtarik, *Acting Between the Lines*, p. 29.
17. Friel, *Essays, Diaries, Interviews*, p. 107.
18. Walter Benjamin, 'The Task of the Translator', in *Selected Writings, Vol. I, 1913–1926*, ed. Marcus Bullock and Michael W. Jennings (Cambridge, MA and London: The Belknap Press of Harvard University Press, 2004), pp. 253–63, p. 262.
19. Ibid., p. 258.

20. James Simmons, 'Brian Friel, Catholic Playwright', *The Honest Ulsterman*, 79 (1985), pp. 61–6, p. 66.
21. Emmanuel Levinas, *Entre Nous: On Thinking-of-the-Other*, trans. Michael B. Smith and Barbara Harshav (London: Athlone Press, 1998), p. 226.
22. Jacques Derrida, *The Ear of the Other: Otobiography, Transference, Translation. Texts and Discussions with Jacques Derrida*, trans. Peggy Kamuf (Lincoln and London: University of Nebraska Press and Schocken Books, 1985), p. 102.
23. Lloyd, *Ireland After History*, p. 81.
24. Gerald Dawe, 'Introduction', in Jean Lundy and Aodan MacPóilin (eds), *Styles of Belonging: The Cultural Identities of Ulster* (Belfast: Lagan Press, 1992), pp. 7–10, p. 7.
25. FP, MS 37,086/1.
26. Richard Kearney, *Myth and Motherland*, in *Ireland's Field Day* (London: Hutchinson, 1985), pp. 61–83, p. 63.
27. Ibid.
28. Colin Graham, '"...maybe that's just Blarney": Irish Culture and the Persistence of Authenticity', in Graham and Kirkland (eds), *Ireland and Cultural Theory*, pp. 7–28, p. 7.
29. Theodor W. Adorno, *The Jargon of Authenticity*, trans. Knut Tarnowski and Frederic Will (London: Routledge & Kegan Paul, 1973), p. 9.
30. Lloyd, *Ireland After History*, p. 81.
31. Fanon, *The Wretched of the Earth*, p. 180.
32. Lojek, 'Brian Friel and George Steiner's Linguistics', p. 86.
33. Homer, *The Odyssey*, trans. Robert Fagles (London and New York: Penguin Books, 1996), p. 256.
34. Paul Ricoeur, 'Memory and Forgetting', in Richard Kearney and Mark Dooley (eds), *Questioning Ethics: Contemporary Debates in Philosophy* (London and New York: Routledge, 1999), pp. 5–11, p. 7. See Sigmund Freud, 'Mourning and Melancholy', *The Standard Edition of the Complete Psychological Works*, Vol. XIV (1914–16), trans. and ed. James Strachey (London: The Hogarth Press and The Institute of Psycho-Analysis, 1957), pp. 239–58; and Sigmund Freud, 'Remembering, Repeating and Working-Through', *The Standard Edition of the Complete Psychological Works*, Vol. XII (1911–13), trans. and ed. James Strachey (London: The Hogarth Press and The Institute of Psycho-Analysis, 1958), pp. 147–56.
35. Ricoeur, 'Memory and Forgetting', p. 7.
36. Paul de Man, *Blindness and Insight: Essays in the Rhetoric of Contemporary Criticism* (London and New York: Routledge, 1996), p. 131.
37. Friel's research for this thesis included: Denis Donoghue's *Ferocious Alphabets*, Erving Goffman's *Forms of Talk* and George A. Miller's *Language and Speech*. FP, MS 37,093/1.
38. Rousseau's allegory of man's first encounter with another man – an 'originary' scene – and his naming of him as 'giant' forms an important element of de Man's reading of metaphor. Paul de Man, *Allegories of Reading: Figural Language in Rousseau, Nietzsche, Rilke and Proust* (New Haven and London: Yale University Press, 1979), pp. 149f.
39. See Derrida, *The Ear of the Other*, pp. 100–1.
40. In his notes for the play, Friel isolated this passage from Donoghue's *Ferocious Alphabets* as a possible ending for the play: '[e]ach person describes or tries to make manifest his own experience; the other, listening, cannot share the experience, but he can perceive it, as if at a distance. Complete proximity is impossible.

What makes a conversation memorable is the desire of each person to share experience with the other, giving and receiving. All that can be shared, strictly speaking, is the desire: it is impossible to reach the experience.' FP, MS 37,093/1; Friel's underlining.

41. Edwin Gentzler, *Contemporary Translation Theories* (London and New York: Routledge, 1993), p. 156.
42. Heidegger, *Poetry, Language, Thought*, pp. 215–16.
43. Kearney, *Myth and Motherland*, p. 80.
44. Irving Wardle, 'Seeking a Sense of Ireland', *The Times*, 5 October 1981, p. 11. Thomas Kilroy gives an account of the parallels he drew between Chekhov's Russia and Ireland in his 1981 version of Chekhov's *The Seagull*, in Thomas Kilroy, '"The Seagull": An Adaptation', in Vera Gottlieb and Paul Allain (eds), *The Cambridge Companion to Chekhov* (Cambridge: Cambridge University Press, 2000), pp. 80–90.
45. Wardle, 'Seeking a Sense of Ireland', p. 11.
46. Friel told Agnew that he turned to *After Babel* while working on *Three Sisters* to get some perspective on the type of translation he was preparing. Friel, *Essays, Diaries, Interviews*, p. 84.
47. Antoine Berman, *The Experience of the Foreign: Culture and Translation in Romantic Germany*, trans. S. Heyvaert (New York: State University of New York Press, 1992), p. 180.
48. Ibid.
49. This is discussed by S. Heyvaert, the text's translator: ibid., p. vii.
50. Lawrence Venuti, 'Translation, Community, Utopia', in Lawrence Venuti (ed.), *The Translation Studies Reader* (London and New York: Routledge, 2000), pp. 468–88, p. 468.
51. Emile Benveniste, *Indo-European Language and Society*, trans. J. Lallot (London: Faber and Faber, 1973), pp. 71–83.
52. Jacques Derrida and Anne Dufourmantelle, *Of Hospitality*, trans. Rachel Bowlby (Stanford: Stanford University Press, 2000), pp. 29 and 45.
53. See Venuti, *The Translator's Invisibility*, p. 20.
54. As Moss argues: 'Natasha and Andrei establish the main line of construction; their marriage is the network to which everything else attaches.' Howard Moss, 'Three Sisters', in Harold Bloom (ed.), *Anton Chekhov* (Philadelphia: Chelsea House Publishers, 1999), pp. 121–37, p. 122.
55. See Derrida, *Of Hospitality*, pp. 123–5.
56. Michel Serres, *The Parasite*, trans. Lawrence R. Schehr (Baltimore and London: Johns Hopkins University Press, 1982), p. 106.
57. As Senelick notes in a Serres-like fashion, '[o]nce Natasha gains a foothold, the indiscriminate feeding ends'. Laurence Senelick, *Anton Chekhov* (London: Macmillan, 1985), p. 113.
58. Moss, 'Three Sisters', p. 135.
59. Serres, *The Parasite*, p. 33.
60. Ibid., p. 79.
61. Ibid., p. 182.
62. Cited, Charles W. Meister, *Chekhov Criticism, 1880 Through 1986* (Jefferson and London: McFarland and Co., 1988), p. 246.
63. Moss, 'Three Sisters', p. 125.
64. However, as Gassner has pointed out, there is a great difference between English and US translations and performances of Chekhov, precisely because

the US, 'lacked the country houses that would afford a class basis for mournful Chekhovianism'. John Gassner, 'The Duality of Chekhov', in Robert Louis Jackson (ed.), *Chekhov: A Collection of Critical Essays* (Englewood: Prentice-Hall, 1967), pp. 175–83, p. 176.

65. Ulick O'Connor, 'Friel Takes Derry by Storm', *The Sunday Tribune*, 6 September 1981, p. 2.
66. Kenneth Tynan, *Tynan on Theatre* (London: Penguin Books, 1964), p. 273.
67. Muldoon noted that '[a]part from the occasional idiomatic phrase – "that's a wild big crowd", "as thick as poundies" – which seem incongruous, there's nothing remarkably "Irish", nothing obviously provincial.' Paul Muldoon, 'Lands and Languages', *Times Literary Supplement*, 9 October 1981 (FP, MS 37,092/4).
68. Diane Herron, 'Not a "Translation" in the Accepted Sense', *The Sentinel (Londonderry, Limavady, North-West)*, 16 September 1981, p. 10.
69. Friel, *Essays, Diaries, Interviews*, p. 84.
70. Donal O'Donnell, 'Friel and a Tale of Three Sisters', *The Sunday Press*, 30 August 1981, p. 19. Elsewhere, Friel expressed confidence in the ethics of his procedure in a way that indicated what he saw as the limits of a literal translation: '[t]he ideal condition would be to have a playwright who was fluent in Russian. But if you have to forgo the one, I think it's better for the translator to be a dramatist. There are bigger truths beyond that of the literal translation.' Friel, *Essays, Diaries, Interviews*, p. 100.
71. Ulf Dantanus, *Brian Friel: A Study* (London: Faber and Faber, 1988), p. 186.
72. Serres, *The Parasite*, pp. 42–3.
73. Yet, this ability still has a social force, as is underlined by Natasha's occasional, comic use of French to indicate her newly-minted sophistication (*TS*, pp. 56 and 104).
74. Peter Szondi, 'The Drama in Crisis: Chekhov', in Bloom (ed.), *Anton Chekhov*, pp. 169–74, p. 172.
75. Harold Bloom, 'Introduction', in Bloom (ed.), *Anton Chekhov*, pp. 1–7, p. 7.
76. Natasha is more garrulous in Friel's text compared to the somewhat clipped fashion in which Hingley renders her in the OUP translation; see Anton Chekhov, *Five Plays: Ivanov, The Seagull, Uncle Vanya, Three Sisters, and The Cherry Orchard*, trans. Ronald Hingley (Oxford and New York: Oxford University Press, 1980).
77. Cited, Bloom, 'Introduction', p. 1.
78. Hanna Scolnicov, 'Introduction', in Hanna Scolnicov and Peter Holland (eds), *The Play Out of Context: Transferring Plays From Culture to Culture* (Cambridge: Cambridge University Press, 1989), pp. 1–6, p. 1. Gorky recollects that Chekhov abhorred pretentious behaviour and associated this with speaking a foreign language, telling Gorky that '[e]very one should speak his own language'. Maxim Gorky, 'Fragments of Recollections', in Bloom (ed.), *Anton Chekhov*, pp. 13–23, p. 17.
79. Cronin, *Translating Ireland*, p. 143.
80. O'Donnell, 'Friel and a Tale of Three Sisters', p. 19.
81. Deane, 'Field Day's Greeks (and Russians)', p. 148.
82. Richard York, 'Friel's Russia', in Peacock (ed.), *The Achievement of Brian Friel*, pp. 164–77, p. 165. Tracy has envisioned Chekhov 'as a kind of presence in certain of Friel's plays, sharing, as he does, Friel's preoccupation with [...] language as shibboleth, non-communication, uneasy and conflicting memories of the past'. Robert Tracy, 'The Russian Connection: Friel and Chekhov', *Irish University Review*, 29: 1 (Spring/Summer 1999), pp. 64–77, p. 77.

83. Brian Friel, *Aristocrats: A Play in Three Acts* (Dublin: Gallery Press, 1980). Besides this play and *Three Sisters*, Friel has translated *Uncle Vanya* (Oldcastle: Gallery Books, 1998) and produced three short plays based on Chekhov: *The Yalta Game*, adapted from Chekhov's short story, 'The Lady with the Little Dog'; *The Bear*, a one-act Chekhov farce; and *Afterplay*, which bequeaths a translational afterlife to Andrey from *Three Sisters* and Sonya from *Uncle Vanya*. Brian Friel, *Three Plays After* (*Afterplay*; *The Bear, After Chekhov*; *The Yalta Game*) (Oldcastle: Gallery Press, 2002).
84. As Welch comments, 'at the play's end there is a curious air of jubilant relaxation'. Robert Welch, '"Isn't this your Job? – To Translate?": Brian Friel's Languages', in Peacock (ed.), *The Achievement of Brian Friel*, pp. 134–48, p. 142.
85. Seamus Deane, 'What is Field Day?', in *Field Day Theatre Company Presents Anton Chekhov's 'Three Sisters' in a New Translation by Brian Friel* (Programme for first productions), n.d., n.p.
86. Friel, *Essays, Diaries, Interviews*, p. 106.
87. Cited, Athol Fugard, *Notebooks 1960–1977*, ed. Mary Benson (London and Boston: Faber and Faber), p. 133.
88. The minutes from the Field Day board meeting, 5 June 1983, indicate the plays that were considered to replace *The Saxon Shore*: 'Afore Night Come, The Importance of Being Earnest, Anouilh's Antigone, The Colleen Bawn and Athol Fugard's The Island. SH [Seamus Heaney] also suggested Murder in the Cathedral. After much discussion, it was agreed that Fugard's play should be read by all concerned within the next 24–48 hours and a decision made. The urgency of the situation was stressed.' FP, MS 37,181/1.
89. Richtarik, *Acting Between the Lines*, p. 204. There were at least three different productions of *The Island* staged in the Republic of Ireland in the early- and mid-1980s, the most notable of which opened at The Gate in Dublin on 1 April 1986, and had John Kani and Winston Ntsbona (who had devised the play with Fugard) taking the main roles.
90. '"Boesman and Lena" for the Guildhall', *Derry Journal*, 2 September 1983, p. 7.
91. Pine, *The Diviner*, p. 234.
92. Cited, McCarthy, *Modernisation, Crisis and Culture in Ireland, 1969–1992*, p. 51.
93. Fugard, *Notebooks 1960–1977*, pp. 161 and 132.
94. Nadine Gordimer, 'English-Language Literature and Politics in South Africa', *Journal of South African Studies*, 2: 2 (April 1976), pp. 131–50, p. 144.
95. Brian Crow with Chris Banfield, *An Introduction to Post-Colonial Theatre* (Cambridge: Cambridge University Press, 1996), p. 97.
96. Cited, ibid., p. 97.
97. Cited, Loren Kruger, 'The Drama of Country and City: Tribalization, Urbanization and Theatre under Apartheid', *Journal of South African Studies*, 23: 4 (December 1997), pp. 565–84, p. 578.
98. Fugard, *Notebooks 1960–1977*, p. 160. Fugard had doubts about his approach: Walder reports that '[r]uminating upon the "social content" of *Boesman and Lena*, Fugard felt "nagging doubts", felt he was "opting out" and "not saying enough"; even though at one level the predicament of the "Coloured" couple "is an indictment", was this "explicit enough"?' Dennis Walder, *Athol Fugard* (Basingstoke: Macmillan, 1984), p. 74.
99. Kader Asmal, 'The Writer Under Apartheid', in The Field Day Theatre Company, *Athol Fugard's Boesman and Lena* (Programme for first productions), n.d., n.p.

100. *Beneukt*: 'unreasonable', 'crazy', 'fed up'. This and all further translations are taken from the Glossary inserted after the play: *BL*, pp. 251–5.
101. Walder, *Athol Fugard*, p. 13.
102. *Sinkplaat*: 'corrugated iron'; *klaar*: 'finished', 'ready'; *pondok*: 'hut', 'shack', 'little lean-to'; *baas*: 'boss'.
103. Nelson Mandela, *No Easy Walk To Freedom* (London: Heinemann, 1965), and *Long Walk to Freedom* (London: Little, Brown, 1994).
104. Bhabha, *The Location of Culture*, p. 66.
105. Bassnett and Trivedi, 'Introduction: Of Colonies, Cannibals and Vernaculars', pp. 12–13.
106. Kruger, 'The Drama of Country and City', p. 570n.
107. Athol Fugard, 'Antigone in Africa', in McDonald and Walton (eds), *Amid Our Troubles: Irish Versions of Greek Tragedy*, pp. 128–47, pp. 146–7n.
108. A Belfast *Sunday News* journalist felt that audiences 'needed a crash course in Swahili to have understood some of it'. Cited, Richtarik, *Acting Between the Lines*, p. 214.
109. Dennis Walder, 'Introduction', in Athol Fugard, *Selected Plays* (Oxford and Cape Town: Oxford University Press, 1987), pp. i–xxx, p. xii.
110. Fugard, *Notebooks 1960–1977*, p. 156.
111. One of the German entries quoted for '*heimlich*' gives as a possible use of the term: '[o]f animals: tame companionable to man. As opposed to wild'. Sigmund Freud, 'The "Uncanny"', in Sigmund Freud, *The Standard Edition of the Complete Psychological Works*, Vol. XVII (1917–19), trans. and ed. James Strachey (London: The Hogarth Press and The Institute of Psycho-Analysis, 1955), pp. 219–52, p. 222.
112. Freud, 'The "Uncanny"', p. 234. He develops this by giving an example of a dreamlike, repetitive revisiting of the same place, which produced 'the feeling of helplessness and of uncanniness' (p. 237).
113. Freud, 'Remembering, Repeating and Working-Through', pp. 150–1.
114. Ricoeur, 'Memory and Forgetting', p. 6.
115. *Kaffer nes*: 'a kaffir haunt'.
116. John Llewelyn, *Emmanuel Levinas: The Genealogy of Ethics* (London and New York: Routledge, 1995), p. 54.
117. Leslie Kane, *The Language of Silence: On the Unspoken and the Unspeakable in Modern Drama* (Cranbury, London and Mississauga: Associated University Presses, 1984), p. 24.
118. Ricoeur, 'Memory and Forgetting', p. 6.
119. *Manzi*: 'water'; 'Bhomboloza Outa, Bhomoloza': 'Cry out, old man, cry out'; 'Mlomo, ewe mlomo': 'Mouth, yes, mouth'; *dala*: 'old man'; *safa*: 'suffer'.
120. Bassnett and Trivedi, 'Introduction: Of Colonies, Cannibals and Vernaculars', p. 9.
121. Jacques Derrida, *Margins of Philosophy*, trans. Alan Bass (Brighton: Harvester Press, 1982), p. 315.
122. Ibid., pp. 328–9.
123. *Kaffertaal*: 'kaffir language'.
124. Freud, 'Remembering, Repeating and Working-Through', p. 155.
125. Cited, Hugh Raymend-Pickard, 'Derrida and the Fidelity to History', *History of European Ideas*, 28 (2002), pp. 13–20, p. 17. Hence, 'commitment to truth', as Raymend-Pickard notes, 'must take the form of "faith"' (p. 13).
126. Fugard, *Notebooks 1960–1977*, p. 191.

3. Loyal Translations: The Spirit and the Letter

1. Berman, *The Experience of the Foreign*, p. 4.
2. Cited, Tom Dunne, *Theobald Wolfe Tone, Colonial Outsider: An Analysis of his Political Philosophy* (Cork: Tower Books, 1982), p. 25.
3. Directors' meeting, 1–2 December 1984, FDP, MS 47,111/4.
4. W. J. McCormack, *The Battle of the Books: Two Decades of Irish Cultural Debate* (Mullingar: The Lilliput Press, 1986), p. 55.
5. This connection is heard in the list of possible themes for the third pamphlet series: 'the Irish constitution, the Failure of Socialism, ditto of Catholicism, The Rise of Feminism, the Negative reflection of Catholicism in literature, Nuclear Disarmament, Revisionism in Irish History'. FDP, MS 47,111/3.
6. Longley also noted an emphasis on Protestant theology and a failure to examine Catholicism in *The Field Day Anthology*. Longley, *The Living Stream*, p. 33.
7. This note is summed up in O'Toole's assessment: 'Field Day failed to produce a series of plays [...] which would have allowed other voices, sometimes saying very different things, to be heard.' Fintan O'Toole, 'Dissension in the Fifth Province', *The Irish Times*, 'Weekend', 1 May 1993, p. 5.
8. 'Frank McGuinness, *Carthaginians*: Comment', FDP, MS 47,116/2.
9. Both Heaney and Friel had major reservations, with Friel holding that it would require exceptional directing 'to conceal [...] its defects'. FDP, MS 47,120/2.
10. Memo from Julie Barber, 3 July 1987, FDP, MS 47,108/2.
11. Minutes of directors' meeting, 7–8 January 1984, FP, MS 37,181/1. As Friel wrote to McGuinness turning it down: 'were we an all-the-year-round company, like the Abbey, we would want to talk to you about the possibility of reworking those areas in the text that we felt weren't adequately realised. But we have one bite at one small cherry per annum; and the business of "developing" scripts isn't properly ours.' Letter, 16 January 1984, FDP, MS 47,146/7.
12. Letter from Friel to other directors, 21 September 1981, FP, MS 37,181/1.
13. David Rudkin, *The Saxon Shore* (London: Methuen, 1986). Richtarik notes that when the play was staged in the Almeida Theatre in London in 1986, '[h]ardly any of the critics made the connection with Northern Ireland'. Richtarik, *Acting Between the Lines*, p. 197.
14. David Rudkin, 'The Saxon Shore': http://www.davidrudkin.com/html/shore.html. As discussed in the Epilogue, Field Day staged a rehearsed reading of Rudkin's *Cries from Casement as his Bones are Brought to Dublin* in 1992.
15. Cited, Richtarik, *Acting Between the Lines*, p. 203. The links between the covenantal nature of these communities (and Israel) are discussed in Donald Harman Akenson, *God's Peoples: Covenant and Land in South Africa, Israel, and Ulster* (Ithaca and London: Cornell University Press, 1992).
16. Cited, Richtarik, *Acting Between the Lines*, p. 201.
17. Patrick Grant, *Literature, Rhetoric and Violence in Northern Ireland, 1968–98: Hardened to Death* (Basingstoke and New York: Palgrave Macmillan, 2001), p. 91.
18. These objections are reiterated in a letter from Deane to Friel (11 April 1983): 'I read the Rudkin synopsis very quickly and share the unease you obviously feel. It seems to be getting more and more like a baroque parable, settling somewhere between Gods and Fighting Men and Mercian Hymns. It may work, though; but hopes can't be too high.' FP, MS 37,181/7.
19. For minutes of these meetings see FP, MS 37,181/1. Perhaps the most succinct commentary on Field Day's attitude to Rudkin's play comes in a letter to Friel

from Kilroy. While this letter is undated, it is from this period, before Kilroy was on the board of directors. Of Rudkin he notes: 'you know how much I admire that man's work. He is a total original and aims for the top, although almost everything I know of his misses in some way. But they are the misses of a heavy-weight.' FP, MS 37,093/4.

20. Terence Rev. McCaughey, 'Synopsis of Address Given to Mark the Publication of the Third Set of Field Day Pamphlets, "The Protestant Idea of Liberty" on May 17 1985, in the Linenhall Library, Belfast.' 17 May 1985, n.p. (FDP, MS 46,974/2). McCormack has also lamented '[t]he absence of any extended self-critical reflection by an Irish Protestant writer of Carson's generation (or of any generation which grew up under Carson's sway)'. W. J. McCormack, '"The Protestant Strain": Or, A Short History of Anglo-Irish Literature from S. T. Coleridge to Thomas Mann', in Gerald Dawe and Edna Longley (eds), *Across A Roaring Hill: The Protestant Imagination in Modern Ireland. Essays in Honour of John Hewitt* (Belfast: Blackstaff Press, 1985), pp. 48–78, p. 76. Lynda Henderson is even more blunt, stating that '[t]he Ulster protestant Plantation culture, in its defensiveness, abjures poetry perhaps as Plato did.' Lynda Henderson, 'Not Even a Name: *The Saxon Shore* by David Rudkin', *Theatre Ireland*, 15 (May–August 1988), pp. 27–9, p. 29.
21. Meeting, 6 August 1986, FP, MS 37,181/2.
22. George Steiner, *Antigones* (Oxford: Oxford University Press, 1984), p. 128.
23. Stewart, *The Narrow Ground*, p. 83.
24. Fugard, '*Antigone* in Africa', p. 132.
25. In general, there was a much more favourable reaction to Mahon's portion of the double-bill. Fintan O'Toole, for instance, opined that, as the second part, '*High Time* [...] repeats as comedy what we have already seen as tragedy. But *High Time* gains considerably more from the association than *The Riot Act* does.' He concluded that *The Riot Act* was 'entirely unsatisfactory as a piece of theatre'. Fintan O'Toole, 'Field Day: On the Double', *The Sunday Tribune*, 27 September 1984, p. 16.
26. Christopher Murray, 'Three Irish Antigones', in Jacqueline Genet and Richard Allen Cave (eds), *Perspectives on Irish Drama and Theatre* (Gerrards Cross: Colin Smyth, 1991), pp. 115–29, p. 115. There were two other stage versions: Brendan Kennelly, *Sophocles' Antigone* (Newcastle upon Tyne: Bloodaxe Books, 1996) and Aidan Mathews's *Antigone: A Version* (Dublin, July 1984, unpublished). Besides these, Pat Murphy's film *Anne Devlin* can be considered a reworking of *Antigone*. See also: Brian Atkins, 'Irish Appropriations of Sophocles' *Antigone* and *Philoctetes*', in Michael Cronin and Cormac Ó Cuilleanáin (eds), *The Languages of Ireland* (Dublin and Portland, OR: Four Courts Press, 2003), pp. 167–78.
27. Minutes of Field Day board meeting, 5 June 1983, FP, MS 37,181/1.
28. Deane, 'Field Day's Greeks (and Russians)', p. 153.
29. In a letter to Friel (11 April 1983), Deane is adamant in his opposition to the idea: 'Anouilh's Antigone seems attractive. I would, overall, favour it but I really could not stand for CC O'B as translator. If he had any association with Field Day, I wouldn't. I don't think he would accept anyway, given that Tom Paulin is one of our number.' FP, MS 37,181/7.
30. This lecture was reproduced as: Conor Cruise O'Brien, 'Views', *The Listener*, 80: 2065, 24 October 1968, p. 526. Heaney's reflections on O'Brien's lecture, which he felt spoke to the fact that 'morality and legality are hardly identical at the moment', are in the same issue: Seamus Heaney, 'Old Derry's Walls', *The Listener*, 80: 2065, 24 October 1968, pp. 521–3, p. 523.
31. Conor Cruise O'Brien, *States of Ireland* (London: Panther, 1974), pp. 147–77.

32. Tom Paulin, *Ireland and the England Crisis* (Newcastle: Bloodaxe Books, 1984), pp. 23–38.
33. Ibid., p. 16.
34. Ibid., pp. 16 and 17. In an unwarrantedly dismissive fashion, O'Toole ascribed Paulin's change of political direction to 'rustic sentimentality'. Fintan O'Toole, 'What Paisley and Reggae Have in Common', *The Sunday Tribune*, 3 February 1985, p. 20.
35. Letter from Paulin to Friel, 23 February 1984, FP, MS 37,181/8.
36. '[W]hen my Country takes her place among the nations of the earth, then, and not till then let my epitaph be written.' Robert Emmet, 'Speech from the Dock', in *The Field Day Anthology of Irish Writing*, Vol. I, pp. 933–9, p. 938n.
37. See Séamus McKendry, *Disappeared: The Search for Jean McConville* (Dublin: Blackwater Press, 2000). Jean McConville's body was recovered and reburied in October 2003.
38. Derrida, *Of Hospitality*, pp. 109–11.
39. Cleary, *Outrageous Fortune*, p. 258.
40. Cited, Susan McKay, *Northern Protestants: An Unsettled People* (Belfast: Blackstaff Press, 2000), p. 296.
41. Tom Paulin, *Minotaur: Poetry and the Nation State* (London and Boston: Faber and Faber, 1992), p. 12. Tom Paulin, *The Day-Star of Liberty: William Hazlitt's Radical Style* (London: Faber and Faber, 1998). See also Barra Ó Seaghdha's review of Paulin's 2005 *Crusoe's Secret* in *Dublin Review of Books*: http://www.drb.ie/fa_Sentimental.html
42. Cleary, *Outrageous Fortune*, p. 258.
43. Paulin, *Ireland and the England Crisis*, p. 167.
44. Ibid., p. 170.
45. Ibid., pp. 163–4.
46. Akenson, *God's Peoples*, p. 126.
47. Cited, Linda Racioppi and Katherine O'Sullivan See, '"This We Will Maintain": Gender, Ethno-Nationalism and the Politics of Unionism in Northern Ireland', *Nations and Nationalism*, 7: 1 (2001), pp. 93–112, p. 98.
48. See, for instance, Elizabeth Shannon, *I am of Ireland: Women of the North Speak Out* (Boston, Toronto and London: Little, Brown and Co. 1989), p. 130.
49. Janice Holmes, 'Introduction' to 'The Century of Religious Zeal, 1800–74', *The Field Day Anthology of Irish Writing*, Vol. IV, pp. 537–41, p. 539.
50. For an account of the role of keening in Irish society see Angela Bourke, 'Introduction' to 'Lamenting the Dead', *The Field Day Anthology of Irish Writing*, Vol. IV, pp. 1365–7.
51. Roche, *Contemporary Irish Drama*, p. 259. Johnson called this speech the production's 'best moment [...] a very funny impersonation of a Humphrey Atkins/Brian Faulkner figure' (Robert Johnson, 'Antigone in Ulster Dialect', *Fortnight*, 227, 21 October 1985, p. 23). Murray sees Creon as a caricature of these roles: '[Paulin's] Antigone is no caricature, as Creon is.' Murray, 'Three Irish Antigones', p. 122.
52. Arthur Aughey, *Under Siege: Ulster Unionism and the Anglo-Irish Agreement* (London and New York: Hurst & Company and St. Martin's Press, 1989), p. 19.
53. Norman Porter, *Rethinking Unionism: An Alternative Vision for Northern Ireland* (Belfast: Blackstaff Press, 1998), p. 145.
54. Berman, *The Experience of the Foreign*, p. 172.
55. Steiner, *Antigones*, p. 105.
56. Benjamin, 'The Task of the Translator', p. 262.

57. Berman, *The Experience of the Foreign*, p. 18.
58. Ibid., p. 5.
59. Tom McCall, 'The Case of the Missing Body', *Le Pauvre Holterling: Blätter zur Frankfurter Ausgabe*, 8 (1988), pp. 53–73, p. 59.
60. Benjamin, 'The Task of the Translator', p. 254.
61. Paulin comments on the set design in Tom Paulin, '*Antigone*' in McDonald and Walton (eds), *Amid Our Troubles*, pp. 165–70, p. 168.
62. Johnson, 'Antigone in Ulster Dialect', p. 23.
63. Paulin, *A New Look at the Language Question*, p. 12.
64. Cleary places this play in a tradition of working-class domestic tragedies that show 'the influence of Sean O'Casey on contemporary political theatre in Ireland'. Cleary, *Outrageous Fortune*, p. 234.
65. Paulin, *A New Look at the Language Question*, p. 12.
66. Ibid., p. 11.
67. Derrida, *Of Hospitality*, pp. 25–7.
68. According to Paulin, in Hegel's reading, the conflict arises because of Creon's devotion to the 'daylight gods' of rational justice and Antigone's commitment to the gods of darkness, and the tragedy emerges from their 'one-sidedness'. Paulin has glossed this to emphasise that neither side alone bears the blame for the tragedy; rather, they are both culpable because of their commensurate intransigence: 'Hegel argued that [...] what is denied is the absoluteness of the claim of each' (Paulin, '*Antigone*', p. 167). This suggests a form of oppositional complementarity between the two dispositions: just as day only exists because of night, the attitudes of Creon and Antigone are, to some extent, dependent upon each other for their definition. For a fuller account of the intimate involvement of *Antigone* in Hegel's thought, see Steiner, *Antigones*, pp. 19–36.
69. Roche notes these elements in Creon's character, although he sees it in terms of a progression: from first being a 'Westminster functionary', Creon ends up 'sounding like [...] the hard man of violence, self-appointed leader of a gang of bullyboys'. Roche, *Contemporary Irish Drama*, pp. 259–60.
70. Akenson, *God's Peoples*, p. 134. As Paulin recounts, various Creons down the years, such as Terence O'Neill and Brian Faulkner, have dismissed Paisley as 'mindless' and as a 'demon doctor' (Paulin, *Ireland and the England Crisis*, p. 165). McCartney also repudiates Paisley's tradition by linking it with apartheid South Africa (Robert McCartney, *Liberty and Authority in Ireland* (Derry: Field Day Theatre Company, 1985), p. 23). Ironically, as the leader of a party that is still opposed to the Belfast Agreement – the United Kingdom Unionist Party – McCartney's 'liberal unionism' is clearly indissociable from the Antigone disposition it opposes.
71. J. S. McClelland, *A History of Western Political Thought* (London and New York: Routledge, 1996), p. 172.
72. Steve Bruce, *God Save Ulster: The Religion and Politics of Paisleyism* (Oxford: Clarendon Press, 1986), p. 258.
73. McCall, 'The Case of the Missing Body', p. 58.
74. Submissions by, respectively, Dr John Dunlop, Dr Godfrey Brown and Rev. Sydney Callaghan, cited, Andy Pollock (ed.), *A Citizen's Inquiry: The Opsahl Report on Northern Ireland* (Dublin: The Lilliput Press, 1993), p. 37. See also John Dunlop, *A Precarious Belonging: Presbyterians and the Conflict in Ireland* (Belfast: Blackstaff Press, 1995), pp. 83–93.
75. Akenson, *God's Peoples*, p. 150.

76. Marianne Elliott, *Watchmen in Sion: The Protestant Idea of Liberty* (Derry: Field Day Theatre Company, 1985), p. 7.
77. Akenson, *God's Peoples*, p. 350.
78. Cited, Bruce, *God Save Ulster*, p. 251.
79. Akenson notes that the prophetic element that underwrites this religious tradition depends on a limited version of the past: it can only be articulated after 'an agreed past has been established'. Akenson, *God's Peoples*, pp. 137–8.
80. Tom Paulin, 'Northern Protestant Oratory and Writing 1791–1985', in Deane (ed.), *The Field Day Anthology of Irish Writing*, Vol. III, pp. 314–18, p. 316. Paulin has also pointed out Paisley's 'highly theatrical' nature; see Paulin, *Ireland and the England Crisis*, p. 159.
81. Paulin, 'Northern Protestant Oratory and Writing 1791–1985', p. 316. The centrality of this biblical text to the puritan aesthetic is reasserted by Paulin in the final chapter of *The Day-Star of Liberty*, pp. 271–97.
82. Sigmund Freud, *The Interpretation of Dreams* (New York: Avon Books, 1998), pp. 294–8.
83. Ibid., pp. 312–13.
84. Adam Phillips, *Promises, Promises: Essays on Literature and Psychoanalysis* (London: Faber and Faber, 2000), p. 145.
85. While Antigone's impulse to bury speaks to a law beyond the law, she nonetheless states that she only defied Creon because it was her brother: '[b]ut if I'd a husband / or a child even / and they dead, dead but not buried, / I'd have laid no earth on them / if the law forbade me.' *RA*, p. 47.
86. Brown, *The Whole Protestant Community*, p. 20.
87. Antonio Gramsci, *Selections From the Prison Notebooks*, trans. Quintin Hoare and Geoffrey Nowell-Smith (London: Lawrence and Wishart, 1971), p. 276.
88. Kirkland, *Literature and Culture in Northern Ireland since 1965*, pp. 7–8.
89. Stewart Parker, *Dramatis Personae* (Belfast: John Malone Memorial Committee, 1986), p. 19.
90. Ibid., p. 6.
91. This background is evoked in Parker's 1974 radio play *The Iceberg*. For a discussion of the spectral links between this drama and *Pentecost*, see Richard Rankin Russell, 'Exorcising the Ghosts of Conflict in Northern Ireland: Stewart Parker's *The Iceberg* and *Pentecost*', *Éire-Ireland*, 41: 3 (Fall/Winter 2006), pp. 42–58. See also Brown's evocative portrait of post-Second World War Belfast: Terence Brown, 'Let's Go to Graceland: The Drama of Stewart Parker', in Jacqueline Genet and Elisabeth Hellegouarc'h (eds), *Studies on the Contemporary Irish Theatre* (Caen Cedex: Centre de Publications de l'Université de Caen, 1991), pp. 21–33, pp. 21–3.
92. Stewart Parker, 'State of Play', *The Canadian Journal of Irish Studies*, 7: 1 (June 1981), pp. 5–11, pp. 8 and 9.
93. Elmer Andrews, 'The Will to Freedom: Politics and Play in the Theatre of Stewart Parker', in Okifumi Komesu and Masaru Sekine (eds), *Irish Writers and Politics* (Gerrards Cross: Colin Smythe, 1989), pp. 237–69, p. 237.
94. Steiner, *Antigones*, p. 261.
95. Maguire, cited, Gerald Dawe, *The Rest is History* (Newry: Abbey Press, 1998), p. 61.
96. Lee, *Ireland, 1912–1985*, p. 445.
97. Ibid.
98. Bruce, *God Save Ulster*, p. 109.
99. Cited, Alvin Jackson, *Ireland, 1798–1998: Politics and War* (Oxford: Blackwell, 1999), p. 408.

100. Dawe, *The Rest is History*, p. 60.
101. Brown, 'Let's Go to Graceland', p. 27.
102. Roche, *Contemporary Irish Drama*, p. 221.
103. Jacques Derrida, 'The Parergon', trans. Craig Owens, *October*, 9 (Summer 1979), pp. 3–40, p. 20.
104. Andrews, 'The Will to Freedom', p. 266.
105. In the Preface to his 1987 television series, *Lost Belongings*, Parker also refers to his home as 'Northern Ireland, Ulster, the Six Counties, Lilliput, a place so fundamentally factional that it can't even agree on a name for itself'. Stewart Parker, *Television Plays*, ed. Clare Wallace (Prague: Litteraria Pragensia, 2008), p. 308.
106. Murray holds that Peter represents 'Parker's point of view' (Murray, *Twentieth Century Irish Drama: Mirror Up to a Nation*, p. 219). This connection is most evident in Peter's story of a civil rights confrontation on an American campus, as Parker had a similar formative experience.
107. Parker, *Dramatis Personae*, pp. 16 and 18.
108. Stewart Parker, 'Foreword', in Stewart Parker, *Plays: 2. Northern Star; Heavenly Bodies; Pentecost* (London: Methuen, 2000), pp. xiii–xiv, p. xiii.
109. Ibid., p. xiii.
110. Ibid., pp. xiii–xiv.
111. Lee, *Ireland, 1912–1985*, p. 447.
112. Elliott, *Watchmen in Sion*, pp. 9 and 8.
113. Ibid., p. 8.
114. Ibid., p. 7.
115. Cited, ibid., pp. 9–10.
116. Ibid., p. 15.
117. Lawrence Norfolk, 'Being Translated, or the Virgin Mary's Hair', *Barcelona Review* (September–October 2000). Available at: http://www.barcelonareview.com/20/e_ln.htm
118. Cited, Stephen Rea, 'Introduction' to Parker, *Plays: 2*, pp. ix–xii, p. xi.
119. Andrews, 'The Will to Freedom', p. 267.
120. Claudia W. Harris, 'From Pastness to Wholeness: Stewart Parker's Reinventing Theatre', *Colby Quarterly*, 27 (December 1991), pp. 233–41, p. 235.
121. René Girard, *Violence and the Sacred*, trans. Patrick Gregory (Baltimore: Johns Hopkins University Press, 1977), pp. 262–3.
122. Ibid., p. 8.
123. Ibid., p. 12.
124. Ibid., p. 49. Grant has also employed Girard's elaboration of the 'scapegoat' as a way of reading Northern Ireland: Patrick Grant, *Breaking Enmities: Religion, Literature and Culture in Northern Ireland, 1967–97* (Basingstoke and New York: Macmillan, 1999), pp. 98–9.
125. Girard, *Violence and the Sacred*, p. 39.
126. Ibid., p. 127.
127. Ibid., p. 128.
128. Ibid., pp. 127–8.
129. Ibid., p. 131.
130. Ibid., p. 134.
131. Richard Sieburth, 'The Guest: Second Thoughts on Translating Hölderlin', in Rosanna Warren (ed.), *The Art of Translation: Voices From the Field* (Boston: Northeastern University Press, 1989), pp. 237–43, p. 241.

132. Michael Parker, *Northern Irish Literature, 1975–2006*, Vol. II: *The Imprint of History* (Basingstoke and New York: Palgrave Macmillan, 2007), p. 99.
133. Derrida, *Of Hospitality*, pp. 123–5.
134. Grant, *Breaking Enmities*, p. 101.
135. Lee, *Ireland, 1912–1985*, p. 429.
136. Derrida, *Of Hospitality*, p. 125.
137. Cited, Paulin, *Ireland and the England Crisis*, p. 173.

4. Reciprocation and Resolution

1. Steiner, *After Babel*, p. 316.
2. Ibid., p. 319.
3. Hugh Denard, 'Seamus Heaney, Colonialism and The Cure: Sophoclean Re-visions', *PAJ: A Journal of Performance and Art*, 22: 66 (September 2000), pp. 1–18, p. 2.
4. See, for instance: Gerry Adams, *Hope and History: Making Peace in Ireland* (Kerry: Brandon Books, 2003), and Bill Clinton's 1995 speech at the Guildhall Square, Derry: http://cain.ulst.ac.uk/events/peace/docs/pres2.htm
5. Seamus Heaney, '*The Cure at Troy*: Production Notes in No Particular Order', in McDonald and Walton (eds), *Amid Our Troubles*, pp. 171–80, p. 176.
6. Ibid., pp. 173–4.
7. Seamus Heaney, *The Government of the Tongue: The 1986 T. S. Eliot Memorial Lectures and Other Critical Writings* (London: Faber and Faber, 1989), pp. 107–8.
8. W. J. McCormack, *From Burke to Beckett: Ascendancy, Tradition and Betrayal in Literary History* (Cork: Cork University Press, 1994), p. 54.
9. Deane, 'Field Day's Greeks (and Russians)', p. 148.
10. Seamus Heaney, '*The Cure at Troy*: A Note on the Play', in *The Cure at Troy by Seamus Heaney, after 'Philoctetes' by Sophocles* (Programme for first productions), 1990, n.p. He has also confessed that this choice was made because he 'simply had not the nerve to bring on a god two minutes from curtain'. Heaney, '*The Cure at Troy*: Production Notes in No Particular Order', p. 172.
11. Liliane Weissberg, 'Language's Wound: Herder, Philoctetes, and the Origin of Speech', *MLN*, 104: 3 (April 1989), pp. 548–79, p. 557.
12. Elizabeth Belfiore, '*Xenia* in Sophocles' *Philoctetes*', *The Classical Journal*, 89: 2 (1994), pp. 113–29, pp. 114–15.
13. Derrida, *Of Hospitality*, p. 125.
14. Cited, Belfiore, '*Xenia* in Sophocles' *Philoctetes*', p. 115.
15. '"Xenoi" ambiguously referred both to a "guest" and a "stranger" (or "foreigner"), although a stranger who might also be a Greek.' Lawrence Venuti, *The Scandals of Translation: Towards an Ethics of Difference* (London and New York: Routledge, 1998), p. 118.
16. Harsh expands on the particular significance of this bow: 'Heracles made civilization possible by achieving man's mastery over the beast and civilized man's superiority over the barbarian. [...] His most magnificent weapon [...] was the bow [...]. This bow, Apollo's gift to Heracles, symbolizes man's intelligence brought into action to guarantee man's domination of the earth.' Philip Whaley Harsh, 'The Role of the Bow in the Philoctetes of Sophocles', *The American Journal of Philology*, 81: 4 (October 1960), pp. 408–14, p. 412.
17. Belfiore, '*Xenia* in Sophocles' *Philoctetes*', p. 116.

18. The friend, for Derrida, is the one who 'already bears my death and inherits it as the last survivor'. Jacques Derrida, *Politics of Friendship*, trans. George Collins (London and New York: Verso, 1997), p. 13.
19. Marcel Mauss, *The Gift: The Form and Reason for Exchange in Archaic Societies*, trans. W. D. Halls (London and New York: Routledge, 2002), p. 16.
20. Belfiore notes that '[s]cholars have long debated the question of whether Philoctetes, his bow, or both are needed'. Belfiore, '*Xenia* in Sophocles' *Philoctetes*', p. 113n.
21. Weissberg, 'Language's Wound', pp. 557–8 and 550.
22. Ibid., p. 558.
23. See J. L. Austin, *How to Do Things with Words*, 2nd edn. (Oxford: Clarendon Press 1975), and J. L. Austin, 'Performative Utterances', in J. L. Austin, *Philosophical Papers*, 2nd edn., ed. J. O. Urmson and G. J. Warnock (Oxford: Oxford University Press 1970), pp. 233–52.
24. 'To witness the moment when pain causes a reversion to the pre-language of cries and groans is to witness the destruction of language; but conversely, to be present when a person moves up out of that pre-language and projects the facts of sentience into speech is almost to have been permitted to be present at the birth of language itself.' Elaine Scarry, *The Body in Pain: The Making and Unmaking of the World* (Oxford: Oxford University Press, 1985), p. 6.
25. Peacock links this with 'the endemic tensions in [Heaney's] poetry between private experience and public utterance, between poet as lyricist and poet as priest figure, between lyric and didactic poet'. Alan Peacock, 'Mediations: Poet as Translator, Poet as Seer', in Elmer Andrews (ed.), *Seamus Heaney: A Collection of Critical Essays* (London: Macmillan, 1992), pp. 233–55, p. 235.
26. In particular, he underlines that 'Philoctetes is not meant to be understood as a trimly allegorical representation of hardline Unionism' (Heaney, '*The Cure at Troy*: Production Notes in No Particular Order', p. 175). Brazeau has charted the confusion that is created if Philoctetes is aligned in this fashion: Robert Brazeau, 'Thomas Kinsella and Seamus Heaney: Translation and Representation', *New Hibernia Review*, 5: 2 (2001), pp. 82–98, p. 96. Heaney has acknowledged that these 'topical references were a mistake'. O'Driscoll, *Stepping Stones*, p. 421.
27. Cited, Peacock, 'Mediations: Poet as Translator, Poet as Seer', p. 235.
28. Jacques Derrida, 'What is a "Relevant" Translation?', trans. Lawrence Venuti, *Critical Inquiry*, 27 (Winter 2001), pp. 174–200, p. 188. See also Jacques Derrida, *On Cosmopolitanism and Forgiveness*, trans. Mark Dooley and Michael Hughes (London and New York: Routledge, 2001), p. 34.
29. Derrida, *On Cosmopolitanism and Forgiveness*, p. 33.
30. Derrida, 'What is a "Relevant" Translation?', p. 192.
31. John D. Caputo, 'Reason, History and a Little Madness: Towards an Ethics of the Kingdom', in Kearney and Dooley (eds), *Questioning Ethics*, pp. 84–104, p. 99.
32. Ibid., pp. 96–7.
33. David Lloyd, '"Pap for the dispossessed": Seamus Heaney and the Poetics of Identity', in Andrews (ed.), *Seamus Heaney: A Collection of Critical Essays*, pp. 87–116, p. 102.
34. Richard Kearney, *Strangers, Gods and Monsters: Interpreting Otherness* (London and New York: Routledge, 2003), p. 70.
35. Caputo, 'Reason, History and a Little Madness', p. 101.
36. Ibid.

37. For Deane, this is 'a play that looks for peace but cannot see how peace is going to come about'. It outlines a process of destruction through which peace might be arrived at: 'the arrival of justice – that first. Then war. Then peace. Troy has still to be sacked.' While his belief that 'Troy[ ] [...] refers to Northern Ireland' is too neat, Deane points to broader ethical and political problems that Heaney's translation seeks to elide: 'this prelude to the final battle, which seems to be about miraculous change, is not in any coherent sense really about an altercation that will bring reconciliation. Instead, it will bring victory to one side and defeat to the other.' Deane, 'Field Day's Greeks (and Russians)', p. 159.
38. Jacques Derrida, *Dissemination*, trans. Barbara Johnson (London: Athlone Press, 1981), p. 98.
39. Ibid., p. 110.
40. Heaney, '*The Cure at Troy*: Production Notes in No Particular Order', p. 175.
41. Heaney cites this line from Yeats's *A Vision* to describe *The Cure at Troy*. Ibid.
42. Terry Eagleton, *Nationalism: Irony and Commitment* (Derry: Field Day Theatre Company, 1988), pp. 7–8.
43. Conor Cruise O'Brien, *Writers and Politics* (London: Chatto & Windus, 1965), p. 105.
44. Belfiore, '*Xenia* in Sophocles' *Philoctetes*', p. 114.
45. Avital Ronell, *Dictations: On Haunted Writing* (Lincoln, NE and London: University of Nebraska Press, 1993), p. xii.
46. Earl of Roscomon, *An Essay on Translated Verse*, 2nd edn. (London: Jacob Tonson, 1685), pp. 7–8.
47. Norman Vance, *Irish Literature: A Social History. Tradition, Identity and Difference* (Oxford and Cambridge, MA: Basil Blackwell, 1990), pp. 62–3.
48. John Dryden, 'To the Earl of Roscomon, on his Excellent Essay on Translated Verse', in Roscomon, *An Essay on Translated Verse*, n.p.
49. Roscomon, *An Essay on Translated Verse*, p. 5.
50. Lori Chamberlain, 'Gender and the Metaphorics of Translation', in Venuti (ed.), *Rethinking Translation*, pp. 57–74, p. 57.
51. Dryden, 'To the Earl of Roscomon', n.p.
52. Chamberlain, 'Gender and the Metaphorics of Translation', p. 57.
53. FP, MS 37,181/1.
54. Bulletin, FP, MS 37,181/6.
55. FDP, MS 47,112/6.
56. FDP, MS 47,113/1.
57. Letter from Deane to Field Day directors, 14 December 1991, FDP, MS 47,116/4.
58. FDP, MS 47,114/2.
59. Steiner, *After Babel*, p. 319.
60. Chamberlain, 'Gender and the Metaphorics of Translation', p. 65.
61. Heaney, '*The Cure at Troy*: Production Notes in No Particular Order', p. 172.
62. Fugard, *Notebooks 1960–1977*, p. 198.
63. Daniel Tiffany, *Radio Corpse: Imagism and the Cryptaesthetic of Ezra Pound* (Cambridge, MA and London: Harvard University Press, 1995), p. 290.
64. 'Preface', *HT*, p. 7.
65. Ibid., p. 8.
66. Ibid., p. 7. In a letter to Deane, dated 20 March 1984, Friel informed him that he and Rea had seen the manuscript of *High Time*, 'and the difficulty we foresee is [...] the contrast—maybe even the conflict—between the Sixties setting and language and the seventeenth century behaviour and attitude of the women. Maybe the

audience will swallow this. I don't know. Maybe the difficulty can be resolved simply by costume. We'll see.' In a subsequent letter (21 June 1984), Friel is happier to report to Deane that 'Mahon's Molière has been updated from the Sixties to the Punk Eighties. Suggested by the 2 directors. Sounds a good idea to me—introduces the sense of threat and a reason for guardians/wards etc.' FDP, MS 47,118.

67. FDP, MS 46,895/3.
68. Fintan O'Toole, *Sunday Tribune*, 23 September 1984, p. 16.
69. Paul de Man, *Aesthetic Ideology*, ed. Andrzej Warminski (Minneapolis: University of Minnesota Press, 1996), p. 165.
70. Grant, *Literature, Rhetoric and Violence in Northern Ireland, 1968–98*, p. 62. Henderson, on the other hand, read the allegory as a warning to Ulster unionists about 'the destructive consequences of a siege mentality'. Cited, Richtarik, *Acting Between the Lines*, p. 235.
71. De Man, *Aesthetic Ideology*, pp. 178–9.
72. White, *Metahistory*, pp. 37–8.
73. There is, as McBride points out, 'an area of concordance between the views of the brothers'. Robert McBride, *Aspects of Seventeenth-Century French Drama and Thought* (London: Macmillan, 1979), p. 77. Harrison also contends that 'the two brothers employ different means to the same end'. Helen L. Harrison, *Pistoles/Paroles: Money and Language in Seventeenth-Century France* (Charlottesville: Rookwood Press, 1996), p. 138.
74. Paulin, *Ireland and the England Crisis*, p. 56.
75. Grant, *Literature, Rhetoric and Violence in Northern Ireland, 1968–98*, p. 61.
76. John Goodby, 'Reading Protestant Writing: Representations of the Troubles in the Poetry of Derek Mahon and Glenn Patterson's *Burning Your Own*', in Kathleen Devine (ed.), *Modern Irish Writers and the Wars* (Gerrards Cross: Colin Smythe, 1999), pp. 219–43, p. 228.
77. Paul Ricoeur, 'Imagination, Testimony and Trust: A Dialogue with Paul Ricoeur', in Kearney and Dooley (eds), *Questioning Ethics*, pp. 12–17, p. 17.
78. White, *Metahistory*, p. 38.

5. Masks: Men and History

1. Account of Field Day's activities and positions, March 1985, FP, MS 37,191/8.
2. Cited, Joe McMinn, 'Saint Oscar', *Theatre Ireland* (n.d.), pp. 9–13, p. 10 (FDP, MS 47,106/3).
3. Frantz Fanon, *Black Skin, White Masks*, trans. Charles Lam Markmann (New York: Grove Press, 1967), p. 228.
4. Aniket Jaaware, 'Of Demons and Angels and Historical Humans: Some Events and Questions in Translation and Post-colonial Theory', *The European Legacy*, 7: 6 (2002), pp. 735–45, p. 743.
5. Thomas Kilroy, 'Brian Friel: Playwright', *Connacht Tribune*, 31 October 1980, p. 15.
6. Thomas Kilroy, *The Secret Fall of Constance Wilde* (Oldcastle: Gallery Press, 1997); Thomas Kilroy, *The O'Neill* (Oldcastle: Gallery Press, 1995); Thomas Kilroy, *The Seagull: After Chekhov* (Oldcastle: Gallery Press, 1993). His interest in these split characters is discussed in the 2002 issue of the *Irish University Review* dedicated to his work. See, in particular, Thierry Dubost, 'Kilroy's Theatre of the Conflicted Self', *Irish University Review: A Journal of Irish Studies*, 32: 1 (Spring/Summer 2002), pp. 10–17, and Nicholas Grene, 'Staging the Self: Person and Persona in Kilroy's Plays', pp. 70–82.

7. Martine Pelletier, '"Against Mindlessness": Thomas Kilroy and Field Day', *Irish University Review: A Journal of Irish Studies*, 32: 1 (Spring/Summer 2002), pp. 110–25, p. 124.
8. Berman, *The Experience of the Foreign*, p. 35.
9. Gerald Dawe, 'An Interview with Thomas Kilroy', in Gerald Dawe and Jonathan Williams (eds), *Krino 1986–1996* (Dublin: Gill and MacMillan, 1996), pp. 230–5, p. 234.
10. Charles Edward Lysaght, *Brendan Bracken* (London: Allen Lane, 1979), p. 22.
11. Ibid., p. 26.
12. Ibid., p. 31.
13. Ibid., p. 34.
14. Cited, Andrew Boyle, *Poor, Dear Brendan: The Quest for Brendan Bracken* (London: Hutchinson, 1974), p. 101.
15. Lysaght, *Brendan Bracken*, p. 68.
16. Ibid.
17. Ibid., pp. 68–9.
18. Cited, ibid., p. 69.
19. Boyle, *Poor, Dear Brendan*, p. 89.
20. F. P. Lock, 'Burke, Ireland and India: Reason, Rhetoric and Empire', in Seán Patrick Donlan (ed.), *Edmund Burke's Irish Identities* (Dublin and Portland, OR: Irish Academic Press, 2006), pp. 154–70, p. 156.
21. Lysaght, *Brendan Bracken*, p. 82.
22. Seamus Deane, *Foreign Affections: Essays on Edmund Burke* (Cork: Cork University Press in association with Field Day, 2005), p. 90.
23. Edmund Burke, *An appeal from the new to the old Whigs, in consequence of some late discussions in Parliament, relative to the Reflections on the French revolution* (London, 1791), p. 130. Available at *Eighteenth Century Collections Online*: http://galenet.galegroup.com/servlet/ECCO
24. Kiberd stitches together excerpts from various speeches of Burke's on the subject of India to give an idea of these sentiments. '"The first men of that country", "eminent in situation", were insulted and humiliated by "obscure young men", pushing upstarts who "tore to pieces the most established rights, and the most ancient and most revered institutions of ages and nations".' Declan Kiberd, *Irish Classics* (London: Granta Books, 2000), p. 208.
25. Cited, Seamus Deane, *Celtic Revivals: Essays in Modern Irish Literature, 1880–1980* (London: Faber and Faber, 1985), p. 23.
26. Seamus Deane 'Introduction' to 'Edmund Burke', in Deane (ed.), *The Field Day Anthology of Irish Writing*, Vol. I, pp. 807–9, p. 809.
27. Luke Gibbons, *Edmund Burke and Ireland* (Cambridge: Cambridge University Press, 2003), pp. 174–5.
28. Lysaght, *Brendan Bracken*, p. 131.
29. Ibid., p. 132.
30. Terry Eagleton, 'Aesthetics and Politics in Edmund Burke', in Michael Kenneally (ed.), *Irish Literature and Culture* (Gerrards Cross: Colin Smythe, 1992), pp. 25–34, p. 29.
31. Cockett notes that Beaverbrook did publish this in the *Evening Standard* (23 June 1944), and an 'enraged Minister of Information used his powers to excise this passage from the later edition'. Richard Cockett (ed.), *My Dear Max: The Letters of Brendan Bracken to Lord Beaverbrook, 1925–1958* (London: The Historians' Press, 1990), p. 30.
32. Jacques Derrida, 'Foreword: *Fors*: The Anglish Words of Nicolas Abraham and Maria Torok', in Nicolas Abraham and Maria Torok, *The Wolf Man's Magic Word: A*

Cryptonymy, trans. Barbara Johnson (Minneapolis: University of Minnesota Press, 1986), pp. xi–xlviii, pp. xxx and xxi.

33. Ibid., p. xxi.
34. The nadir in their relationship was when Peter broke into Bracken's house and, tellingly, stole the Romney portrait of Burke, which he, unsuccessfully, held to ransom. Lysaght, *Brendan Bracken*, p. 96.
35. Jacques Derrida, *Specters of Marx: The State of Debt, the Work of Mourning and the New International*, trans. Peggy Kamuf (New York and London: Routledge, 1994), p. 39.
36. Barbara Johnson, 'Taking Fidelity Philosophically', in Graham (ed.), *Difference in Translation*, pp. 142–8, p. 143.
37. Ibid., p. 144.
38. Cited, Martin A. Doherty, *Nazi Wireless Propaganda: Lord Haw-Haw and British Public Opinion in the Second World War* (Edinburgh: Edinburgh University Press, 2000), p. 4.
39. Eric Hobsbawm, *Age of Extremes: The Short Twentieth Century, 1914–1991* (London: Abacus, 1995), p. 196.
40. Jeffrey Herf, *Reactionary Modernism: Technology, Culture, and Politics in Weimar and the Third Reich* (Cambridge: Cambridge University Press, 1984), p. 2.
41. Marshall McLuhan, *Understanding Media: The Extensions of Man* (London: Routledge & Kegan Paul, 1964), pp. 300 and 302.
42. Ibid., p. 299.
43. Cited, Doherty, *Nazi Wireless Propaganda*, p. 3.
44. For an analysis of the reactions in Britain to the Nazi radio broadcasts, see ibid., pp. 87–121.
45. J. A. Cole, *Lord Haw-Haw: The Full Story of William Joyce* (London and Boston: Faber and Faber, 1987), p. 158. Doherty relates that '[e]ven the War Cabinet was not immune', and the pre-Bracken Ministry of Information 'seem[ed] [...] to have convinced themselves that the German wireless was the real source of many rumours, while attempting to convince the public that it was not'. Doherty, *Nazi Wireless Propaganda*, pp. 111–12.
46. Cole, *Lord Haw-Haw*, p. 155.
47. F. T. Marinetti and Pino Masnata, 'La Radia (1933)', in Douglas Kahn and Gregory Whitehead (eds), *Wireless Imagination: Sound, Radio and the Avant-Garde*, trans. Stephen Sartarelli (Cambridge, MA and London: MIT Press, 1992), pp. 265–8, p. 267.
48. Doherty, *Nazi Wireless Propaganda*, p. 183. As Cole illustrates, the sobriquet could not originally have referred to Joyce, as it first appeared the day he received his contract as a broadcaster. Cole, *Lord Haw-Haw*, pp. 115–16.
49. Kearney, *Strangers, Gods and Monsters*, p. 154.
50. Derrida, 'Foreword: *Fors*', p. xvi.
51. Thomas Kilroy, 'Introduction', *DC*, pp. 11–15, p. 11.
52. Cited, Cole, *Lord Haw-Haw*, p. 22. Despite having distanced himself from the Fenians, the police turned down J. K. Bracken's application for a gun licence in 1894, stating that he was 'too dangerous a person to be granted one'. Boyle, *Poor, Dear Brendan*, p. 19.
53. Rebecca West, *The Meaning of Treason* (London: Virago, 1982), p. 13.
54. William Joyce, *Twilight Over England* (London: Imperial War Museum, 1992), p. 8.
55. For Fintan O'Toole, *Double Cross* 'is founded on the idea of the lives of Brendan Bracken and William Joyce [...] mirroring each other, of treason being just an exaggeration of loyalty rather than its opposite'. Fintan O'Toole, 'Double Vision: The Ambivalences of Thomas Kilroy', *Magill*, August 1988, pp. 57–9, p. 58.

56. Derrida, 'Foreword: *Fors*', p. xv.
57. As Fisk noted, during his time in the *Reichsrundfunkgesellschaft*, Joyce 'so disliked the Irish that he never came into their offices. He worked almost exclusively in the English section and rarely referred to Ireland, although he was remembered in Eire for some disparaging remarks about the Irish Naval Service's old gunboat *Muirchu*.' Robert Fisk, *In Time of War: Ireland, Ulster and the Price of Neutrality, 1939–45* (London: André Deutsch, 1983), pp. 346–7.
58. Cited, Doherty, *Nazi Wireless Propaganda*, pp. 188–9.
59. In his last recorded speech, a clearly drunk Joyce describes how he has warned Britain of the threat from Russia. He also foresaw a resurrection of Germany, the seeds of which were sown in the Battle of Berlin: 'the men who have died in the Battle of Berlin have given their lives to show that, whatever else happens, Germany will live.' William Joyce, 'Track 24: Joyce's last valedictory talk – unbroadcast. 30 April 1945', in Doherty, *Nazi Wireless Propaganda*, CD.
60. Ezra Pound, *The Cantos of Ezra Pound* (New York: New Directions, 1975), p. 425.
61. For an account of Joyce's trials, see West, *The Meaning of Treason*.
62. Oscar Wilde, 'The Truth of Masks', *Complete Works of Oscar Wilde* (Glasgow: HarperCollins, 1994), pp. 1156–73, p. 1173.
63. Terry Eagleton, 'Introduction', *SO*, pp. 1–9, p. 5.
64. Michael Riffaterre, 'Prosopopeia', *Yale French Studies*, 69 (1985), pp. 107–23, p. 108.
65. Cited, ibid., p. 107.
66. Richard Kearney, *The God Who May Be: A Hermeneutics of Religion* (Bloomington and Indianapolis: Indiana University Press, 2001), p. 126.
67. Cited, ibid., p. 127.
68. Patrick O'Connor and Seán Daffy, 'On the Importance of Not-Being Earnest: A Dialogue with Terry Eagleton', *Irish Studies Review*, 16: 1 (February 2008), pp. 55–69, p. 64.
69. Kiberd, *Irish Classics*, p. 328.
70. Oscar Wilde, 'The Soul of Man Under Socialism', *Complete Works of Oscar Wilde*, pp. 1174–97, p. 1179.
71. O'Connor and Daffy, 'On the Importance of Not-Being Earnest', p. 61.
72. Eagleton, 'Introduction', p. 8.
73. Cited, Kiberd, *Irish Classics*, p. 327.
74. *Saint Oscar by Terry Eagleton* (Programme for first performances, 1989), p. 4. Eagleton has recounted that '*St. Oscar* wasn't the original title. There was a more boring title, *The Fall of Oscar Wilde*.' However, the designer Bob Crowley 'couldn't get any inspiration out of that. So he homed in on the phrase "St. Oscar." Even then I didn't realise it was a sort of parody of *St. Genet*, Sartre's great worker – the saint and the sinner.' O'Connor and Daffy, 'On the Importance of Not-Being Earnest', p. 61.
75. Jean Baudrillard, *Symbolic Exchange and Death*, trans. Iain Hamilton Grant (London and Thousand Oaks: Sage Publications, 1993), p. 40.
76. Eagleton, 'Introduction', p. 3.
77. Nicholas Wroe, 'High Priest of Lit Crit', *The Guardian*, 2 February 2002. Available at: http://www.guardian.co.uk/books/2002/feb/02/academicexperts.highereducation
78. McMinn, 'Saint Oscar', p. 10 (FDP, MS 47,106/3).
79. Letter to Field Day administrator, 27 September 1991, FDP, MS 47,120/3.
80. Michael Billington, 'The Man in the Ironic Mask', *The Guardian*, 12 March 1990 (FDP, MS 47,124/4).

81. *The Cure at Troy*'s failure to match *Saint Oscar*'s appeal is noted at the directors' meeting, 7 December 1990, FDP, MS 47,114/1.
82. Eagleton, 'Introduction', p. 3.
83. Susan Sontag, *A Susan Sontag Reader* (London and New York: Penguin Books, 1983), pp. 428–9.
84. Terry Eagleton, *Heathcliff and the Great Hunger: Studies in Irish Culture* (London and New York: Verso, 1995), p. 334.
85. Roland Barthes, *Mythologies*, trans. Annette Lavers (New York: Hill and Wang, 1972), pp. 150–3.
86. Ibid., pp. 154–5.
87. Ibid., p. 129.
88. Ibid., p. 135.
89. Eagleton cited the first two sentences from this speech in a press release announcing that he was donating his royalties from the London run of *Saint Oscar* in the Hampstead Theatre to the Birmingham Six campaign. FDP, MS 46,921/1.
90. Eagleton, 'Introduction', p. 5.
91. Nolan, *James Joyce and Nationalism*, p. 15.
92. James Joyce, *Ulysses*, ed. and intro. Jeri Johnson (Oxford: Oxford University Press, 1993), pp. 6–7.
93. Oscar Wilde, 'The Picture of Dorian Gray', *Complete Works of Oscar Wilde*, pp. 17–159, p. 17.
94. Oscar Wilde, 'The Decay of Lying', *Complete Works of Oscar Wilde*, pp. 1071–92, p. 1082.
95. Ibid., p. 1086.
96. Joyce, *Ulysses*, p. 199.
97. See, for instance, ibid., pp. 6–7 and 196.
98. Cheyfitz, *The Poetics of Imperialism*, p. 34.
99. In personal correspondence, Eagleton confirmed that Wallace is an invention; a composite portrait of a type: 'a stereotypical working-class intellectual of the day'.
100. W. B. Yeats, *Autobiographies* (London: Macmillan, 1955), p. 138.
101. Declan Kiberd, 'The London Exiles: Wilde and Shaw', in Deane (ed.), *The Field Day Anthology of Irish Writing*, Vol. II, pp. 372–6, p. 375.
102. Ibid., p. 374.
103. Lloyd, *Nationalism and Minor Literature*, p. 115.
104. Ibid., p. 133.
105. De Man argued that an analogous process was at the heart of Barthes's work; see Paul de Man, 'Roland Barthes and the Limits of Structuralism', *Yale French Studies*, 77 (1990), pp. 177–90, pp. 180–1.
106. Eagleton, *Nationalism: Irony and Commitment*, p. 8.
107. Jacques Derrida, *Writing and Difference*, trans. Alan Bass (London and New York: Routledge, 1997), p. 248.
108. Letter, 14 December 1991, FDP, MS 47,116/4.
109. Pelletier, '"Against Mindlessness"', p. 124.
110. Letter, 9 November 1991, FDP, MS 46,936.
111. Morash, *A History of Irish Theatre, 1601–2000*, p. 177.
112. RTÉ, *Playwrights in Profile*, available at www.rte.ie/radio1/sundayplayhouse/rams/2007/4march.smil
113. Gibbons and Whelan, 'In Conversation with Stephen Rea', p. 9.
114. Harold Pinter, 'Mac', in *The Madame MacAdam Travelling Theatre* (Programme for first performances) (Derry: The Field Day Theatre Company, n.d.), pp. 6–7, p. 6.

115. FDP, MS 47,146/6.
116. Pinter, 'Mac', p. 7.
117. BBC, *The Story of Field Day.*
118. Thomas Kilroy, 'From Page to Stage', in Jacqueline Genet and Wynne Hellegouarc'h (eds), *Irish Writers and Their Creative Process* (Gerrards Cross: Colin Smythe, 1996), pp. 55–62, p. 61.
119. Barbara Hayley, 'Self-Denial and Self-Assertion in Some Plays of Thomas Kilroy: *The Madame MacAdam Travelling Theatre*', in Genet and Hellegouarc'h (eds), *Studies on the Contemporary Irish Theatre*, pp. 47–56, p. 51.
120. Christopher Murray, 'Thomas Kilroy's World Elsewhere', in Genet and Hellegouarc'h (eds), *Irish Writers and Their Creative Process*, pp. 63–77, p. 77.
121. Hayley, 'Self-Denial and Self-Assertion in Some Plays of Thomas Kilroy', p. 47.
122. Venuti, *The Translator's Invisibility*, p. 19.
123. Cited, Clair Wills, 'The Aesthetics of Irish Neutrality during the Second World War', *boundary 2*, 31: 1 (2004), pp. 119–45, p. 121.
124. Clair Wills, *That Neutral Island: A Cultural History of Ireland During the Second World War* (London: Faber and Faber, 2007), p. 46.
125. Dermot Keogh, *Twentieth-Century Ireland: Nation and State* (Dublin: Gill and Macmillan, 1994), p. 108.
126. Wills, *That Neutral Island*, p. 269.
127. Trevor C. Salmon, *Unneutral Ireland: An Ambivalent and Unique Security Policy* (Oxford: Clarendon Press, 1989), p. 110.
128. Walter Benjamin, 'The Work of Art in the Age of Its Technological Reproducibility: Third Version', in, *Selected Writings, Vol. 4*, pp. 251–83, p. 270.
129. '"An Absent-Minded Person of the Student Type": The Artaud File', *The Dublin Review*, 1 (Winter 2000–01), pp. 55–80, p. 64. This is what Artaud claimed in a letter after he was deported back to France. Before coming to Ireland, he described his purpose in going there as a search 'for the sources of a very ancient tradition' which required going to where Synge lived. He wanted to research the origins of a cane he had been given, which he was told was made from wood from the 'true cross' and which had also at one time been St. Patrick's staff (p. 56).
130. '"An Absent-Minded Person of the Student Type"', p. 57.
131. Thomas Kilroy, 'Two Playwrights: Yeats and Beckett', in Joseph Ronsley (ed.), *Myth and Reality in Irish Literature* (Ontario: Wilfrid Laurier University Press, 1977), pp. 183–95, p. 195.
132. Antonin Artaud, *The Theatre and Its Double*, trans. Victor Corti (London: Calder and Boyars, 1970), p. 21.
133. Ibid., p. 22.
134. McLuhan, *Understanding Media*, p. 299.
135. Artaud, *The Theatre and Its Double*, p. 63.
136. Derrida, *Writing and Difference*, p. 234.
137. Ibid., p. 245.
138. Artaud, *The Theatre and Its Double*, p. 27.
139. Cited, Derrida, *Writing and Difference*, p. 234.
140. Steiner, *After Babel*, p. 66. Peter Bush, 'Pure Language', in Mona Baker (ed.), *Routledge Encyclopedia of Translation Studies* (London and New York: Routledge, 1998), pp. 194–6, p. 194.
141. Derrida, *Writing and Difference*, p. 250.
142. Jerzy Grotowski, *Towards a Poor Theatre*, ed. Eugenio Barba (London: Methuen, 1975), p. 86.
143. Kilroy, 'Two Playwrights: Yeats and Beckett', p. 187.

Epilogue. Afterlives

1. FDP, MS 47,106/4.
2. Paddy Woodworth, 'Field Day Calls It a Night', n.d. (FDP, MS 47,107/3).
3. FP, MS 37,181/1.
4. FDP, MS 47,117/3.
5. Minutes of directors' meeting, 4 September 1992, FDP, MS 47,115/3.
6. *The Sunday Times*, 9 May 1993 (FDP, MS 47,107/3).
7. FDP, MS 47,115/3.
8. Letter from Deane to Field Day directors, 14 December 1991, FDP, MS 47,116/4.
9. FDP, MS 47,115/1.
10. Kilroy, fax to board of directors, 3 January 1992, FDP, MS 47,122/2.
11. FDP, MS 47,118.
12. Letter, 14 December 1991, FDP, MS 47,116/4.
13. Meaney 'was angry when [the *Anthology*] first appeared', but, as she recounts, 'what angered me even further was that the whole issue was then hijacked by those who wished to attack Field Day for other reasons. We saw critics and commentators who suddenly became born-again feminists, taking upon themselves the role of champion of women's writing for the first time ever. For example, I was galled by Roy Foster's attack on the lack of female writing, given that his history of Modern Ireland is a casebook of patriarchal history.' Cited, Eibhear Walshe, 'Women in the Annex?', *Irish Studies Review*, 2 (Winter 1992), pp. 13–14, p. 14.
14. Gayatri Chakravorty Spivak, 'Can the Subaltern Speak?', in Cary Nelson and Lawrence Grossberg (eds), *Marxism and the Interpretation of Culture* (Urbana and Chicago: University of Illinois Press, 1988), pp. 271–313, p. 272.
15. Cited, Catriona Crowe, 'Testimony to a Flowering', *The Dublin Review*, 10 (Spring 2003), p. 43.
16. Mulhern, *The Present Lasts a Long Time*, p. 155. Tóibín saw the issue in similar terms: 'the anthology's treatment of women is an imperative rather than a mistake. It arises from the view that Southern Ireland is contemptible and complacent, and thus the changes which have taken place here are nothing beside the great events which have rattled Northern Ireland.' Anthony Bradley, Edna Longley and Colm Tóibín, 'Review of *The Field Day Anthology of Irish Writing*, Vols I–III', *Canadian Journal of Irish Studies*, 18: 2 (December 1992), pp. 117–24, p. 122. See also Longley, *The Living Stream*, pp. 22–44.
17. Deane, *Heroic Styles*, p. 58.
18. Deane, 'General Introduction', p. xxv.
19. Ibid.
20. Ibid., p. xxiv.
21. Seamus Deane, 'Introduction' to 'Thomas Moore (1779–1852)', *The Field Day Anthology of Irish Writing*, Vol. I, pp. 1053–6, p. 1053.
22. Ibid., pp. 1055–6.
23. Ibid., pp. 1054–5.
24. Letter to the *Irish Times*, 19 November 2002, p. 15.
25. Said, *Orientalism*, p. 146.
26. Deane, 'General Introduction', p. xxv.
27. Deane, 'Introduction' to 'Poetry and Song 1800–1890', *The Field Day Anthology of Irish Writing*, Vol. II, pp. 1–9, p. 5.
28. Deane, *Celtic Revivals*, p. 13.
29. James Clifford, 'Review of *Orientalism*', *History and Theory*, 19: 2 (May 1980), pp. 204–23, p. 223.

30. Seamus Deane, 'Introduction', in Brian Friel, *Plays 1* (London: Faber and Faber, 1996), pp. 11–22, p. 13.
31. Longley, *The Living Stream*, p. 26.
32. Said, *Orientalism*, p. 125.
33. For 'supplement', see Jacques Derrida, *Of Grammatology*, trans. Gayatri Chakravorty Spivak (Baltimore and London: Johns Hopkins University Press, 1997), pp. 269–316.
34. Spivak, 'Can the Subaltern Speak?', p. 278.
35. 'Preface', *The Field Day Anthology of Irish Writing*, Vols. IV and V, pp. xxxii–xxxvi, p. xxxvi.
36. Gerardine Meaney, 'General Introduction' to 'Women and Writing, 1700–1960', *The Field Day Anthology of Irish Writing*, Vol. V, pp. 765–71, p. 766.
37. Seamus Deane, 'Introduction' to 'James Joyce (1882–1941)', *The Field Day Anthology of Irish Writing*, Vol. III, pp. 1–4.
38. Angela Bourke, 'General Introduction' to 'Oral Traditions', *The Field Day Anthology of Irish Writing*, Vol. IV, pp. 1191–7, p. 1196.
39. Letter from Friel to McGahern, 20 April 1988, FDP, MS 47,146/7.
40. Fax from Heaney to Field Day directors, 18 February 1993, FDP, MS 47,107/3.
41. The directors' meeting of 13 April 1991 reported that 'Marianne McDonald had guaranteed the 200,000 loan' that would bring *The Field Day Anthology of Irish Writing* to fruition. FP, MS 37,181/2.
42. Letter to Arthur Miller, 4 July 1991, FDP, MS 47,147/1.
43. FDP, MS 47,147/5.
44. Letter from O'Kelly to Kilroy, 31 March 1992, FDP, MS 47,107/1.
45. FDP, MS 47,145/1.
46. Memo, 30 April 1992, FDP, MS 47,107/1.
47. Fax from Kilroy to the Field Day board, 3 January 1992, FDP, MS 47,122/2.
48. Kilroy, letter of resignation from Field Day, 9 April 1992, FDP, MS 47,122/2.
49. FDP, MS 47,107/3.
50. Bulletin, 21 January 1994, FDP, MS 47,108/7.
51. Letter from Friel to directors, 24 January 1994, FDP, MS 47,117/4.
52. Letter from Deane to Friel, 6 February 1994, FP, MS 37,181/7.
53. Letter from Rea to Friel, 3 October 1993, FP, MS 37,181/3.
54. Edna Longley, 'Pamphlets and Professors', *Times Literary Supplement*, 17 March 1995.
55. Cited, Paul Ryan, 'A Field Day Once Again', *Irish Times*, 16 February 1995, p. 14.
56. Irving Wardle, 'Dandy, But Not Quite Fine', *Independent on Sunday*, 26 February 1995 (FDP, MS 47,213/1).
57. Fintan O'Toole, 'Chekhov without Control', *The Irish Times*, 7 March 1995, p. 10.
58. Mary Holland, 'The Trouble with Peace', *The Observer*, 26 February 1995 (FDP, MS 47,213/1).
59. Letter from Deane to Field Day directors, 14 December 1991, FDP, MS 47,116/4.
60. This thesis formed the basis for his book: Seamus Deane, *The French Revolution and Enlightenment in England, 1789–1832* (Cambridge, MA: Harvard University Press, 1988).
61. FDP, MS 47,150/3.
62. Kevin Whelan, *The Tree of Liberty: Radicalism, Catholicism and the Construction of Irish Identity* (Cork: Cork University Press in association with Field Day, 1996), p. 133.
63. Cited, Jane Coyle, 'Return of the United Irishmen', n.d. (FDP, MS 47,215).
64. Michael Billington, 'He Had a Dream', *The Guardian*, n.d. (FDP, MS 47,215).
65. Thomas Kilroy, 'Friendship', *Irish University Review*, 29: 1 (Spring/Summer 1999), pp. 83–9, p. 87.
66. Belfast Agreement, p. 2; available at: www.nio.gov.uk/agreement.pdf

Select Bibliography
Texts Published under the Field Day Imprint

Full bibliographic details of the plays produced by Field Day are to be found in the List of Abbreviations

Seamus Deane (General ed.), *The Field Day Anthology of Irish Writing*, Vols. I–III (Derry and New York: Field Day and Norton, 1991)
Angela Bourke et al. (eds), *The Field Day Anthology of Irish Writing, Vols. 4 and 5: Irish Women's Writing and Traditions* (Cork: Cork University Press in association with Field Day, 2002)

Field Day pamphlets

(Derry: The Field Day Theatre Company)

1983. Tom Paulin, *A New Look at the Language Question*; Seamus Heaney, *An Open Letter*; Seamus Deane, *Civilians and Barbarians*
1984. Seamus Deane, *Heroic Styles: The Tradition of an Idea*; Richard Kearney, *Myth and Motherland*; Declan Kiberd, *Anglo-Irish Attitudes*
1985. *The Protestant Idea of Liberty*: Terence Brown, *The Whole Protestant Community: The Making of an Historical Myth*; Marianne Elliott, *Watchmen in Sion: The Protestant Idea of Liberty*; Robert McCartney, *Liberty and Authority in Ireland*
1986. *Emergency Legislation*: Eanna Molloy, *Dynasties of Coercion*; Michael Farrell, *The Apparatus of Repression*; Patrick McGrory, *Law and the Constitution: Present Discontents*
1988. *Nationalism, Colonialism and Literature*: Terry Eagleton, *Nationalism, Irony and Commitment*; Fredric Jameson, *Modernism and Imperialism*; Edward Said, *Yeats and Decolonisation*

First six pamphlets collected and published as *Ireland's Field Day* (London: Hutchinson, 1985; Indiana: University of Notre Dame Press, 1986). Series 5 published as *Nationalism, Colonialism and Literature*, intro. Seamus Deane (Minneapolis: University of Minnesota Press, 1989)

Critical conditions: Field Day essays and monographs

(Cork: Cork University Press in association with Field Day)

1. Kevin Whelan, *Tree of Liberty: Radicalism, Catholicism, and the Construction of Irish Identity, 1760–1830* (1996)
2. Luke Gibbons, *Transformations in Irish Culture* (1996)
3. Joep Leerssen, *Mere Irish and Fíor-Ghael: Studies in the Idea of Irish Nationality, Its Development and Literary Expression prior to the Nineteenth Century* (1996)
4. Joep Leerssen, *Remembrance and Imagination: Patterns in the Historical and Literary Representation of Ireland in the Nineteenth Century* (1996)
5. Terry Eagleton, *Crazy John and the Bishop and Other Essays on Irish Culture* (1998)

6. Harry White, *The Keeper's Recital: Music and Cultural History in Ireland, 1770–1970* (1998)
7. Francis Mulhern, *The Present Lasts a Long Time: Essays in Cultural Politics* (1998)
8. Marc Caball, *Poets and Politics: Continuity and Reaction in Irish Poetry, 1558–1625* (1998)
9. David Lloyd, *Ireland After History* (1999)
10. Elizabeth Butler Cullingford, *Ireland's Others: Gender and Ethnicity in Irish Literature and Popular Culture* (2001)
11. Clare Carroll, *Circe's Cup: Cultural Transformations in Early Modern Writing about Ireland* (2001)
12. P. J. Mathews, *Revival: The Abbey Theatre, Sinn Féin, The Gaelic League and the Co-operative Movement* (2003)
13. Peter McQuillan, *Native and Natural: Aspects of the Concepts of 'Right' and 'Freedom' in Irish* (2004)
14. Clare O'Halloran, *Golden Ages and Barbarous Nations: Antiquarian Debate and Cultural Politics in Ireland, c.1750–1800* (2004)
15. Seamus Deane, *Foreign Affections: Essays on Edmund Burke* (2005)
16. William J. Smyth, *Map-making, Landscapes and Memory: A Geography of Colonial and Early Modern Ireland, c. 1530–1750* (2006)

Field Day files

(Dublin: Field Day Publications in association with the Keough-Naughton Institute of Irish Studies at the University of Notre Dame)

1. Joe Cleary, *Outrageous Fortune: Capital and Culture in Modern Ireland* (2006)
2. Marjorie Howes, *Colonial Crossings: Figures in Irish Literary History* (2006)
3. Kerby A. Miller, *Ireland and Irish America: Culture, Class, and Transatlantic Migration* (2008)
4. David Lloyd, *Irish Times: Temporalities of Modernity* (2008)
5. David Dwan, *The Great Community: Culture and Nationalism in Ireland* (2008)
6. Gregory Dobbins, *Lazy Idle Schemers: Irish Modernism and the Cultural Politics of Idleness* (2010)

Field Day music

(Dublin: Field Day Publications in association with the Keough-Naughton Institute of Irish Studies at the University of Notre Dame)

1. Séamas de Barra, *Aloys Fleischmann* (2006)
2. Patrick Zuk, *Raymond Deane* (2006)
3. Jeremy Dibble, *Michele Esposito* (2010)
4. Gareth Cox, *Seóirse Bodley* (2010)

Field Day filí

(Dublin: Field Day Publications in association with the Keough-Naughton Institute of Irish Studies at the University of Notre Dame)

1. Breandán Ó Buachalla, *Aogán Ó Rathaille* (2007)
2. Breandán Ó Conchúir, *Eoghan Rua Ó Súilleabháin* (2009)

Other publications

Seamus Heaney, *Sweeney Astray* (Derry: The Field Day Theatre Company, 1983)
Máirín Ní Dhonnchadha and Theo Dorgan (eds), *Revising the Rising* (Derry: Field Day, 1991)
October–December 2005: Field Day column in *The Village* magazine
Seamus Deane, Breandán Mac Suibhne and Ciaran Deane (eds), *Field Day Review*, Vols. 1–6 (Dublin: Field Day Publications in association with the Keough-Naughton Institute of Irish Studies at the University of Notre Dame, 2005–10)
Cormac Ó Duibhne (ed.), *The Field Day Archive* (Dublin: Field Day Publications in association with the Keough-Naughton Institute of Irish Studies at the University of Notre Dame, 2007)
Thomas Moore, *Memoirs of Captain Rock* (ed. and intro. Emer Nolan) (Dublin: Field Day Publications in association with the Keough-Naughton Institute of Irish Studies at the University of Notre Dame, 2008)
Niamh O'Sullivan, *Aloysius O'Kelly: Art Nation Empire* (Dublin: Field Day Publications in association with the Keough-Naughton Institute of Irish Studies at the University of Notre Dame, 2010)

Selected Critical Bibliography

Full bibliographic references are supplied in the endnotes. Key secondary texts used in this book are listed below. Texts published by Field Day are not included.

Irish drama, cultural and historical studies

Donald Harman Akenson, *God's Peoples: Covenant and Land in South Africa, Israel, and Ulster* (Ithaca and London: Cornell University Press, 1992)
Elmer Andrews, *The Art of Brian Friel: Neither Reality nor Dreams* (Basingstoke: Macmillan, 1995)
J. H. Andrews, *A Paper Landscape: The Ordnance Survey in Nineteenth-Century Ireland* (Oxford: Clarendon Press, 1975)
Scott Boltwood, *Brian Friel, Ireland, and The North* (Cambridge: Cambridge University Press, 2007)
Brendan Bradshaw, 'Nationalism and Historical Scholarship in Modern Ireland', *Irish Historical Studies*, 26: 104 (November 1989), pp. 329–51
Ciaran Brady (ed.), *Interpreting Irish History: The Debate on Historical Revisionism, 1938–1994* (Dublin: Irish Academic Press, 1994)
Steve Bruce, *God Save Ulster: The Religion and Politics of Paisleyism* (Oxford: Clarendon Press, 1986)
David Cairns and Shaun Richards, *Writing Ireland: Colonialism, Nationalism and Culture* (Manchester: Manchester University Press, 1988)
Clare Carroll and Patricia King (eds), *Ireland and Postcolonial Theory* (Notre Dame: University of Notre Dame Press, 2003)
Joe Cleary and Claire Connolly (eds), *The Cambridge Companion to Modern Irish Culture* (Cambridge: Cambridge University Press, 2005)
Seamus Deane, *Celtic Revivals: Essays in Modern Irish Literature, 1880–1980* (London and Boston: Faber and Faber, 1985)
Terry Eagleton, *Heathcliff and the Great Hunger: Studies in Irish Culture* (London and New York: Verso, 1995)

Robert Fitzroy Foster, *Modern Ireland, 1600–1972* (London: Penguin, 1988)

Brian Friel, John Andrews and Kevin Barry, '*Translations* and *A Paper Landscape*: Between Fiction and History', *The Crane Bag*, 7: 2 (1983), pp. 118–24

Brian Friel, *Essays, Diaries, Interviews: 1964–1999*, ed. Christopher Murray (London: Faber and Faber, 1999)

Colin Graham and Richard Kirkland (eds), *Ireland and Cultural Theory: The Mechanics of Authenticity* (Basingstoke: Macmillan, 1999)

Patrick Grant, *Breaking Enmities: Religion, Literature and Culture in Northern Ireland, 1967–97* (Basingstoke and New York: Macmillan and St. Martin's Press, 1999)

—— *Literature, Rhetoric and Violence in Northern Ireland, 1968–98: Hardened to Death* (Basingstoke and New York: Palgrave Macmillan, 2001)

Nicholas Grene, *The Politics of Irish Drama: Plays in Context from Boucicault to Friel* (Cambridge: Cambridge University Press, 1999)

Seamus Heaney, *Preoccupations: Selected Prose, 1968–1978* (London: Faber and Faber, 1980)

—— 'A Field Day for the Irish', *The Times*, 5 December 1988

Stephen Howe, *Ireland and Empire: Colonial Legacies in Irish History and Culture* (Oxford: Oxford University Press, 2000)

Eamonn Hughes, '"To Define Your Dissent": The Plays and Polemics of the Field Day Theatre Company', *Theatre Research International*, 15: 1 (Spring 1990), pp. 67–77

Richard Kearney, *Transitions: Narratives in Modern Irish Culture* (Dublin: Wolfhound Press, 1988)

Declan Kiberd, *Inventing Ireland* (London: Jonathan Cape, 1995)

—— *Irish Classics* (London: Granta Books, 2000)

Thomas Kinsella, *The Dual Tradition: An Essay on Poetry and Politics in Ireland* (Manchester: Carcanet, 1995)

Richard Kirkland, *Literature and Culture in Northern Ireland Since 1965: Moments of Danger* (London and New York: Longman, 1996)

Joseph J. Lee, *Ireland, 1912–1985: Politics and Society* (Cambridge: Cambridge University Press, 1989)

David Lloyd, *Nationalism and Minor Literature: James Clarence Mangan and the Emergence of Irish Cultural Nationalism* (Berkeley, Los Angeles and London: University of California Press, 1987)

Edna Longley, *The Living Stream: Literature and Revisionism in Ireland* (Newcastle: Bloodaxe Books, 1994)

—— 'Poetry and Politics in Northern Ireland', *The Crane Bag*, 9: 1 (1985), pp. 26–40

Jean Lundy and Aodan MacPóilin (eds), *Styles of Belonging: The Cultural Identities of Ulster* (Belfast: Lagan Press, 1992)

Eamonn McCann, *War and an Irish Town*, 3rd edn. (London and Boulder: Pluto Press, 1993)

Conor McCarthy, *Modernisation, Crisis and Culture in Ireland, 1969–1992* (Dublin and Portland: Four Courts Press, 2000)

W. J. McCormack, *The Battle of the Books: Two Decades of Irish Cultural Debate* (Mullingar: The Lilliput Press, 1986)

F. C. McGrath, 'Introducing Ireland's Field Day', *Éire-Ireland*, 23: 4 (1988), pp. 145–55

—— *Brian Friel's (Post)Colonial Drama: Language, Illusion, and Politics* (Syracuse: Syracuse University Press, 1999)

Susan McKay, *Northern Protestants: An Unsettled People* (Belfast: The Blackstaff Press, 2000)

Christopher Morash, *A History of Irish Theatre, 1601–2000* (Cambridge: Cambridge University Press, 2002)
Christopher Murray, *Twentieth Century Irish Drama: Mirror up to a Nation* (Manchester and New York: Manchester University Press, 1997)
Conor Cruise O'Brien, *States of Ireland* (London: Panther, 1974)
Michael Parker, *Northern Irish Literature, 1975–2006*. Vol. II: *The Imprint of History* (Basingstoke and New York: Palgrave Macmillan, 2007)
Tom Paulin, *Ireland and the England Crisis* (Newcastle: Bloodaxe Books, 1984)
—— *Minotaur: Poetry and the Nation State* (London and Boston: Faber and Faber, 1992)
Alan J. Peacock (ed.), *The Achievement of Brian Friel* (Gerrards Cross: Colin Smythe, 1993)
Lionel Pilkington, *Theatre and the State in Twentieth-Century Ireland: Cultivating the People* (London and New York: Routledge, 2001)
Richard Pine, *The Diviner: The Art of Brian Friel* (Dublin: University College Dublin Press, 1999)
Andy Pollock (ed.), *A Citizen's Inquiry: The Opsahl Report on Northern Ireland* (Dublin: The Lilliput Press, 1993)
Norman Porter, *Rethinking Unionism: An Alternative Vision for Northern Ireland* (Belfast: The Blackstaff Press, 1998)
Stephen Regan, 'Ireland's Field Day', *History Workshop Journal*, 33 (1992), pp. 25–37
Shaun Richards, 'To Bind the Northern to the Southern Stars: Field Day in Derry and Dublin', *The Irish Review*, 4 (Spring 1988), pp. 52–8
Marilynn J. Richtarik, *Acting Between the Lines: The Field Day Theatre Company and Irish Cultural Politics, 1980–1984* (Oxford: Clarendon Press, 1994)
Anthony Roche, *Contemporary Irish Drama: From Beckett to McGuinness* (Dublin: Gill and Macmillan, 1994)
A. T. Q. Stewart, *The Narrow Ground: The Roots of Conflict in Ulster* (Aldershot: Gregg Revivals, 1993)
Carmen Szabó, *'Clearing the Ground': The Field Day Theatre Company and the Construction of Irish Identities* (Newcastle: Cambridge Scholars Publishing, 2007)
Norman Vance, *Irish Literature: A Social History. Tradition, Identity and Difference* (Oxford and Cambridge, MA: Basil Blackwell, 1990)
Clair Wills, *That Neutral Island: A Cultural History of Ireland During the Second World War* (London: Faber and Faber, 2007)

Theoretical approaches

diacritics, 34: 3–4 (Fall–Winter 2004) (Special issue on 'Literary into Cultural Translation')
Emily Apter, *The Translation Zone: A New Comparative Literature* (Princeton: Princeton University Press, 2006)
Antonin Artaud, *The Theatre and Its Double*, trans. Victor Corti (London: Calder and Boyars, 1970)
Mona Baker (ed.), *Routledge Encyclopedia of Translation Studies* (London and New York: Routledge, 1998)
Roland Barthes, *Mythologies*, trans. Annette Lavers (New York: Hill and Wang, 1972)
Susan Bassnett, *Translation Studies*, 3rd edn. (London and New York: Routledge, 2002)

Susan Bassnett and Harish Trivedi (eds), *Post-Colonial Translation: Theory and Practice* (London and New York: Routledge, 1999)
Walter Benjamin, 'The Task of the Translator', in *Selected Writings, Vol. 1, 1913–1926*, ed. Marcus Bullock and Michael W. Jennings (Cambridge, MA and London: The Belknap Press of Harvard University Press, 2004), pp. 253–63
—— 'On the Concept of History', in *Selected Writings, Vol. 4, 1938–1940*, ed. H. Eiland and M. W. Jennings, trans. E. Jephcott et al. (Cambridge, MA and London: The Belknap Press of Harvard University Press, 2003), pp. 389–400
Antoine Berman, *The Experience of the Foreign: Culture and Translation in Romantic Germany*, trans. S. Heyvaert (New York: State University of New York Press, 1992)
Sandra Bermann and Michael Wood (eds), *Nation, Language and the Ethics of Translation* (Princeton: Princeton University Press, 2005)
Homi K. Bhabha, *The Location of Culture* (London and New York: Routledge, 1994)
Cathy Caruth, *Unclaimed Experience: Trauma, Narrative, and History* (Baltimore and London: Johns Hopkins University Press, 1996)
Pascale Casanova, *The World Republic of Letters*, trans. M. B. DeBevoise (Cambridge, MA: Harvard University Press, 2004)
Dipesh Chakrabarty, *Provincializing Europe: Postcolonial Thought and Historical Difference* (Princeton: Princeton University Press 2000)
Eric Cheyfitz, *The Poetics of Imperialism: Translation and Colonisation from* The Tempest *to* Tarzan (Oxford: Oxford University Press, 1991)
Michael Cronin, *Translating Ireland: Translation, Languages, Cultures* (Cork: Cork University Press, 1996)
Paul de Man, 'Conclusions: Walter Benjamin's "The Task of the Translator"', in *The Resistance to Theory* (Manchester: Manchester University Press, 1986), pp. 73–105
Jacques Derrida, *Positions*, trans. Alan Bass (London: Athlone Press, 1981)
—— 'Des Tours de Babel', in J. F. Graham (ed. and trans.), *Difference in Translation* (Ithaca and London: Cornell University Press, 1985), pp. 165–207
—— *The Ear of the Other: Otobiography, Transference, Translation. Texts and Discussions with Jacques Derrida*, trans. Avital Ronell (Lincoln, NE and London: University of Nebraska Press and Schocken Books, 1985)
—— 'Foreword: *Fors*: The Anglish Words of Nicolas Abraham and Maria Torok', in Nicolas Abraham and Maria Torok, *The Wolf Man's Magic Word: A Cryptonymy*, trans. Barbara Johnson (Minneapolis: University of Minnesota Press, 1986), pp. xi–xlviii
—— *Monolingualism of the Other; or, The Prosthesis of Origin*, trans. Patrick Menash (Stanford: Stanford University Press, 1998)
—— 'What is a "Relevant" Translation?', trans. Lawrence Venuti, *Critical Inquiry*, 27 (Winter 2001), pp. 174–200
Jacques Derrida and Anne Dufourmantelle, *Of Hospitality*, trans. Rachel Bowlby (Stanford: Stanford University Press, 2000)
Frantz Fanon, *Black Skin, White Masks*, trans. Charles Lam Markmann (New York: Grove Press, 1967)
—— *The Wretched of the Earth*, trans. C. Farrington (London: Penguin Books, 2001)
Hans-Georg Gadamer, 'The Problem of Historical Consciousness', in Paul Rabinow and William M. Sullivan (eds), *Interpretive Social Science: A Second Look*, trans. Jeff L. Close and Hans Fantel (Berkeley, Los Angeles and London: University of California Press, 1987), pp. 82–140
Edwin Gentzler, *Contemporary Translation Theories* (London and New York: Routledge, 1993)

René Girard, *Violence and the Sacred*, trans. Patrick Gregory (Baltimore: Johns Hopkins University Press, 1977)

J. F. Graham (ed.), *Difference in Translation* (Ithaca and London: Cornell University Press, 1985)

Ranajit Guha and Gayatri Chakravorty Spivak (eds), *Selected Subaltern Studies* (Oxford: Oxford University Press, 1988)

Peter Hallward, *Absolutely Postcolonial: Writing Between the Singular and the Specific* (Manchester: Manchester University Press, 2001)

Martin Heidegger, *Poetry, Language, Thought*, trans. and intro. Albert Hofstadter (New York: Harper & Row, 1975)

Richard Kearney (ed.), *Paul Ricoeur: The Hermeneutics of Action* (London: Sage, 1996)

—— *Strangers, Gods and Monsters: Interpreting Otherness* (London and New York: Routledge, 2003)

Richard Kearney and Mark Dooley (eds), *Questioning Ethics: Contemporary Debates in Philosophy* (London and New York: Routledge, 1999)

Thomas Keenan, *Fables of Responsibility: Aberrations and Predicaments in Ethics and Politics* (Stanford: Stanford University Press, 1997)

Reinhart Koselleck, *The Practice of Conceptual History: Timing History, Spacing Concepts*, trans. Todd Presner (Stanford: Stanford University Press, 2002)

Dominick LaCapra, *Writing History, Writing Trauma* (Baltimore and London: Johns Hopkins University Press, 2001)

Tejaswini Niranjana, *Siting Translation: History, Post-Structuralism, and the Colonial Context* (Berkeley and Los Angeles: University of California Press, 1992)

Mary Louise Pratt, *Imperial Eyes: Travel Writing and Transculturation* (London and New York: Routledge, 1992)

Vicente L. Rafael, *Contracting Colonialism: Translation and Christian Conversion in Tagalog Society Under Early Spanish Rule* (Durham: Duke University Press, 1993).

Paul Ricoeur, *Hermeneutics and the Human Sciences: Essays on Language, Action and Interpretation*, ed. and trans. John B. Thompson (Cambridge and Paris: Cambridge University Press and Editions de la Maison des Sciences de l'Homme, 1981)

Michael Riffaterre, 'Prosopopeia', *Yale French Studies*, 69 (1985), pp. 107–23

Douglas Robinson, *Translation and Empire: Postcolonial Theories Explained* (Manchester: St. Jerome Publishing, 1997)

—— (ed.), *Western Translation Theory from Herodotus to Nietzsche* (Manchester: St. Jerome Publishing, 1997)

Freddie Rokem, *Performing History: Theatrical Representations of the Past in Contemporary Theatre* (Iowa City: University of Iowa Press, 2000)

Edward W. Said, *The World, The Text, and The Critic* (Cambridge, MA: Harvard University Press, 1983)

—— *Culture and Imperialism* (New York: Vintage Books, 1993)

—— *Orientalism* (London and New York: Penguin Books, 2003)

Michel Serres, *The Parasite*, trans. Lawrence R. Schehr (Baltimore and London: Johns Hopkins University Press, 1982)

Gayatri Chakravorty Spivak, 'Can the Subaltern Speak?', in Cary Nelson and Lawrence Grossberg (eds), *Marxism and the Interpretation of Culture* (Urbana and Chicago: University of Illinois Press, 1988), pp. 271–313

—— *Death of a Discipline* (Calcutta and New Delhi: Seagull Books, 2004)

George Steiner, *Antigones* (Oxford: Oxford University Press, 1984)

—— *After Babel: Aspects of Language and Translation*, 2nd edn (Oxford and New York: Oxford University Press, 1992)

Lawrence Venuti (ed.), *Rethinking Translation: Discourse, Subjectivity, Ideology* (London and New York: Routledge, 1992)

—— *The Translator's Invisibility: A History of Translation* (London and New York: Routledge, 1995)

—— *The Scandals of Translation: Towards an Ethics of Difference* (London and New York: Routledge, 1998)

—— (ed.) *The Translation Studies Reader* (London and New York: Routledge, 2000)

Rosanna Warren (ed.), *The Art of Translation: Voices from the Field* (Boston: Northeastern University Press, 1989)

Shane Weller, 'In Other Words: On the Ethics of Translation', *Angelaki: Journal of the Theoretical Humanities*, 10: 3 (December 2005), pp. 171–87

Hayden White, *Metahistory: The Historical Imagination in Nineteenth-Century Europe* (Baltimore and London: Johns Hopkins University Press, 1973)

Index

References to Field Day play titles in sub-entries below appear in abbreviated form. Please see p. viii for full details.

Course To Success:
Everything You Need To Succeed Beyond School
2nd Edition.

Published by Granicus Publishing.
http://www.GranicusGroup.com
Printed by Quebecor.
Distributed by Bookworld.
Cover design by Twist Image.

e-book: ISBN 0-9736941-3-0
paperback: ISBN 0-9736941-2-2

http://www.CourseToSuccess.com

INTRODUCTION: THE BIG PICTURE

The Four Pillars of Success

This book will help you get what you want out of life. It will not necessarily offer a cookie-cutter, one-size-fits-all approach to setting your goals, let alone achieving them. Rather, it will provide you with the framework to help you search within yourself to determine what you are passionate about, what you excel at, what you want out of life, and how you can obtain these without alienating those around you.

The problem is that you are your own worst enemy. After all, when it comes to thinking, human beings are generally rational creatures that have limited needs with unlimited wants. In other words, we need to eat, sleep and breathe but we want so much more. This scarcity problem affects society as a whole in one way, individuals in another. For you, this problem creates a challenge: how do you put yourself in a position to acquire what you want without turning others off?

When it comes to acting, we are driven by emotions. This creates an imbalance that explains why some people do not realize their potential.

By examining the scarcity problem in this context, we quickly realize that the goal in life is not to put yourself in a given position overnight, but rather, to take a series of actions that will help you get closer to your original end goal. Ideally, as we take one step after another and improve our positioning in life, we need to reassess our original goals and continue to look beyond our initial objectives.

But to do this, we encounter another dilemma: how to both position yourself as a desirable potential employee without biting the hand that feeds you. Success is absolute in the sense that it boils down to four variables: vision, ambition, execution and, of course, luck. However, in our culture, success is also relative; we measure ourselves against our peers.

If you are reading this, then to one extent or another, you have realized that you need to be in charge of your own destiny even if you cannot fulfill it alone. Fulfilling your destiny will entail the right mix of experience, exposure and education.

Ah yes, education…

The Importance Of Looking Outward: Business Students, The General Student Population And The Common Trap

All disciplines serve to prepare you for life, but none comes close to Business. An amalgamation of every function, Business School provides a framework to not only identify the problem at hand, but solve it as well. Business School teaches accountability and an understanding that everything has a cost: unless you can sell a product or offer a service at a profit, then your business will undeniably be forced to shut down. But this is only half the story.

Business students today suffer from a superiority complex. Much like previous generations saw mothers flaunt lawyers and doctors, in recent years many parents wished their offspring to become stockbrokers, marketing executives and all-star salesmen. If you wish to be rich after all, politics, law and medicine play second fiddle. For this reason the best and brightest have been lured by business' promise of riches. The dot com meltdown notwithstanding, business remains the single best way to strike it rich. And strike it rich many have, so is it any surprise that business students sometimes suffer from a holier than thou complex? Of course not, but does such an outlook help one in life?

Few business students would rarely hear anything if the largest tree in the world would fall just outside of the business campus, let the tree huggers worry about trees. But if the smallest bag of cash would linger around within our walls, one of us would call shotgun ASAP. To many business students, the world ends beyond the business school building. Arts, Sciences, Engineering, Liberal and Fine Arts students happen to share some classrooms and equipment, but the world belongs to the eventual rulers of the capitalist world. Let those Political Scientists run the Free World, who wants that? We want to rule the almighty buck.

But a closer look would suggest that shunning the interests, concerns and input of non-business students could prove to be a mortal blow to your career before it even takes off. After all, while business students will become your colleagues, peers and competitors, most clients and employees – two important shareholders in the grand scheme of business – consist of non-business students.

A stakeholder is an individual or group that has a share or an interest in an enterprise. Examples include employees, clients, shareholders, community, analysts, media, academia, lobby groups and the industry as a whole. Management thinkers have forever argued over who the most important stakeholder is. Former Southwest Airlines CEO Herb Kelleher, one of the best airline executives in the history of the transportation industry put it best when he said that employees are first, followed by clients and only then, shareholders. Incidentally, he handed off a pristine firm to his successor right before the 9/11 attacks. He went on to earn the Morningstar CEO of the Year in 2001 with incumbent Jim Parker after Southwest Airlines ended the year with record numbers.

While it is true that most executives will have some business background on their academic CV, the bulk of their front line soldiers may have little or no business school experience. So when you become CEO, your holier than thou attitude will do little to motivate them.

The next stakeholder in line is the client. Your analysts may understand why your goods' prices have risen, but your clients care little about market share and margins. It is in your interest both today and tomorrow to be able to listen, understand and communicate with non-business students as they will compose two of your most important stakeholders when you are, well, ruler of the capitalist world of course. Herb Kelleher may have been a chain smoker, but he did not smoke anything funny. He knew what he was talking about and his outlook should serve as basis for yours.

Kelleher notwithstanding, others would argue that shareholders come near the top of any list. While many would view a shareholder as a broker, trader or a fund manager, most are individuals who work hard all their lives and place blind trust on others through their Pension Funds. They may one day rely on your judgment and risk analysis, so having a "my way or the highway" view will prove to be detrimental as you will alienate them. Having an open-minded perspective instead of a close-minded one is even more important when someone else's nest egg is at stake.

Ironically, while it is true that most CEOs are former business students, some of the best ones were not. The manager of the century, General Electric's former Chairman and CEO Jack Welch, was an engineer by trade. What he understood was that you need to treat your A players like stars. You then must take your B players and motivate them to take it up a notch. Your C players you may have to get rid of. Business is cut throat and definitely not for the faint of heart but disrespecting others is not the way to go. Welch rarely did and neither should you. After all, second-class citizens go on to form new countries or get rid of the despot. How can you get to become king of the mountain? Read on.

Do You Need A Degree?

The World Wide Web has emerged as the greatest library of all time, bringing a wealth of information to the fingertips of people around the world. In the meantime, conventional wisdom stated that education was the gateway to the good life. This reality has prompted academic institutions to scramble and offer their curriculum online. As a result, people are asking whether formal education is a prerequisite to the good life after all.

The Technological Revolution further accelerated what was started by the Industrial Revolution, migrating popular jobs from production and manufacturing to services and other so-called knowledge-based industries. With the advent of the World Wide Web, deregulated markets and the subsequent emergence of the global village, the skills required to excel and dominate in society have evolved. Today, these forces are urging many to ask whether institutionalizing oneself in a classroom and completing a degree is really a prerequisite to the good life. After all, Bill Gates dropped out of school, as did Michael Dell. Ted Turner had a different plan altogether. This book will look at these and other models of success and try to explain why some make it big while others stumble.

Moreover, completing a regular diploma is at times a lengthier process than taking dreams from concept to reality in the business world. Such rags to riches stories were traditionally confined to Americans. But the reality of the global marketplace has allowed entrepreneurs from Beijing to Buenos Aires to control their destiny and strike it rich.

Experience has emerged as one of the most coveted assets. As a result, shelving education and opting to commercialize products so that they may move off shelves seems like an attractive alternative instead. On the other hand, it is *because* the seeds of great business models have been planted in the classroom that the need to attend business school has accelerated. Which side of the argument is right?

Companies have long hired the cream of the crop. Some have looked beyond the traditional MBA and tapped into the fields of Social Sciences and Liberal Arts to fill their ranks in a hot economy. Fueled by investment, excess and conspicuous consumption, the sky seemed to be the limit. But then almost overnight the economy went into a tailspin: the dot com sector lost $1 trillion in market value; the telecommunications industry evaporated $2.5 trillion in market value, laying off half a million employees in the process. Then the attacks of 9/11 shook our foundation. With the economy about to recover, Enron went bankrupt and Arthur Andersen was decimated. Then Worldcom announced that its numbers were cooked. As hundreds of thousands took another hit, it was clear that the rosy picture of the previous decade was replaced with the grim reality of the 21st century. More than ever, survival was uncertain.

<u>The Obsolescence Of Education</u>

Everything you will learn in business school should be taken with a grain of salt. For example, some introductory classes spend considerable time proving a certain fundamental theory only to disprove it later. How lovely: you pay us thousands of dollars for an education so that at the end of the line, you can prove us wrong. Great business model! But aren't academic institutions non-profit enterprises? Of course not, so what are their purposes after all?

Discipline

The first thing that an education teaches you is discipline. You thought that discipline was not eating that extra bar of chocolate or averting that miscellaneous temptation? You were wrong. Discipline is sitting on your tush for hours (bar stool notwithstanding folks) and listening to a professor go on and on. If you can do this on a consistent basis, then surely you have discipline.

Responsibility

The second thing that an education instills is responsibility. This is most ironic because one can miss a class and go undetected. In your career, missing a day will get noticed and often reprimanded. If discipline and responsibility were it however, why go to school when boot camp would do the trick? Surely we are missing the most important factor: learning.

Learning

Some of you may have never completed your education and are working in a particular field. Do you need to return to school? You may need to at some point, but not now, unless of course completing your degree is the ultimate goal.

Others may have completed studies in high technology sectors like engineering or computer science and now believe that a business education is the way to go. Is it? Well, being in school is not the goal in itself; learning is and should always be. So whether you are learning in your industry firsthand or seek to return to school, you are moving in the right direction as long as you are learning.

A third category of students find themselves in a specific field and ask themselves whether or not they have pursued the right path. Truth is that what you are studying is irrelevant if you are pushing the envelope and maximizing your own development. If you have a technical mind, majoring in computer science or engineering will further your career while majoring in liberal arts may help you present your ideas better. If you are creative, majoring in business may be great in marketing and commercializing your work, but it will not help you become a better artist per se.

A degree also pushes you to learn and expand your knowledge base. The top employers care even less about what you think you know. They want knowledgeable students that can learn now and fast. After all, they believe that their training is superior to your education anyway. The more you think you know, the less attractive you are as an employee.

Completing an education requires dedication, a sense of responsibility, hard work, discipline and an intellectual capacity to learn. This is also the DNA of a successful employee.

Like many, you may find that you belong in none of the three categories. What is the right answer for you? Only you can tell, but sit up straight and pay attention as class is about to get started.

PART ONE: COLLEGE LIFE

CHAPTER 1: OFF SEASON PLANNING

Functions: What's What And Who Is It For?

Declaring a major may be one of the hardest things to do. You can go see various professors to see what you should study but when things are said and done, only you will know what is the best course of action for you. After much soul searching (or a coin flip over a beer), many are attracted by the challenge and upside of business school.

Once you are in business school, the choices only multiply. What to pursue between Accounting, Entrepreneurship, Finance, Human Resources, Information Systems, Management or Marketing requires a good understanding of what each field does.

Accounting

The role of the accountant is two-pronged. On the one hand, accountants represent the historian in a firm. They collect data, gather and file it for others to access and analyze. On occasion, they may be asked to provide their opinion, idea and feedback on an initiative, department, firm or industry. The discipline has been tarred by scandals of late, with the implosion of Enron. The largest corporate bankruptcy shocked most stakeholders because auditors at Arthur Andersen repeatedly said that all was well at the Texas-based company.

In terms of career planning, accountants are risk averse and are willing to put up with long hours in exchange for a predictable career path. After you complete your Bachelor's degree, expect to put in overtime if you wish to get serious and earn the title of Chartered Accountant. This and other like designations grant you considerable authority and power, and turn accountants into auditors.

With the mandate to control and monitor operations and finances, comes the expertise to consult companies on better financial and information systems and optimal operational processes. It is this, more lucrative consulting business that has driven the bottom line of many accounting firms. All major firms have spent considerable resources building a consulting division. Arthur Andersen even endured a much-ballyhooed divorce from its consulting business, Andersen Consulting (which renamed itself Accenture). Subsequently, Arthur Andersen's blurred judgment cost employees, shareholders and taxpayers billions.

The problem with these two incestuous revenue streams is that conflict of interest is bound to happen. Accountants often lose sight of the fact that they are there to critique procedures and systems; prepare financials as well as to give them their seal of approval through audit's checks and balances. Accounting firms became nonchalant auditors in the hope of securing more lucrative consulting accounts.

If you wish to go into accounting, keep two things in mind: this is a very conservative and established industry that operates based on Generally Accepted Accounting Principles (GAAP). The tragedy is that GAAP is outdated; it is practically irrelevant in our complex and interconnected global economy. The principle that serves as the foundation of GAAP is the accrual basis (instead of the cash basis), a utopian approach to reporting so flawed that it makes one wonder how shaky the bricks, windows and roof of the Accounting establishment really are. Ironically, the more loopholes you can find in GAAP, the longer your career, as clients will seek your expertise.

Moreover, the industry is largely self-regulated and this leads to monstrous conflicts of interest. The more problems you find with the system though, the more you will be seen as a troublemaker. The worse you do your job, the higher the likelihood that your client gets into trouble.

But wait, it only gets better.

Decision Sciences & Quantitative Analysis

Decision Sciences & Quantitative Analysis is the term given to statistics in business. Essentially, if you aspire to work in insurance (as an actuarial for example), economics (forecasting trends), productions and operations, quality control and procurement, then Decision Sciences & Quantitative Analysis may be for you.

In other words if you prefer to crunch the numbers to find trends, be quiet and let the figures do the talking, you have found your calling. Decision Sciences & Quantitative Analysis are of great importance in all functions of industry. If you wish to pursue this field, expect to spend considerable time pouring over numbers to find trends. This said, if you have the numbers to back up your recommendation, you could expect a steady rise to the top. You still should master your communication and presentation skills; otherwise you will bore your audience fast.

Entrepreneurship

An entrepreneur is a person who organizes, operates, and assumes the risk for a business venture. Entrepreneurship is a broad term that can include any small business to planting the seed for a future mammoth of a firm. It takes an entrepreneurial spirit to succeed in business so many schools now offer this as a specialization. However, for many years, individuals felt the need to get away from school to test their ventures out on the real world. One of the more prominent examples is Michael Dell.

Increasingly, business school has served as a perfect playground for ideas to be tested and tried before they are commercialized. Top search engine Google is one example. Entrepreneurship allows individuals to launch their ventures as much as it helps corporate employees stand out within a firm, department or team.

Finance

The role of the financial manager is to maximize value. Unlike accountants who mostly collect data, finance students are called upon to examine data in order to come up with some form of conclusion and recommendation. This conclusion is often based on accounting principles, but it needs to take into consideration marketing theories, technological feasibility and good old common sense, something often lacking in business circles.

Financiers seek to maximize return while minimizing risk. In general, they are more risk-taking in nature, reflected in turbulent and unpredictable career paths in line with overall stock market performance. Finance forces you to examine the bottom line of everything that goes on in business to ensure that output is optimal given the level of input.

Human Resources, Management & Organizational Behavior

Human Resources managers get a bad reputation from many but the field remains a very interesting one. HR has more to do with the legal and accounting issues of managing human capital, while Management (with a big M) comes down to psychology in business. In some ways, HR is more for the accountant while Management is for the financier.

Any way you dice it, HR people are knowledgeable in the areas that everyone else sweats over: pay, vacation time, compensation and sick days. Management is important because human capital is everything and motivating people (including yourself) will remain the single greatest challenge you will face in business and in life. Those that can identify winners – be it in sports, music, entertainment or business – are usually amongst the most important people on a company's payroll.

Management Of Information Systems

It is hard to understand how Academia can stay on top of the fastest changing segment of business. Information Technology or Management of Information Systems remains one of the most promising fields in business, whose place in school needs to be examined. After all, by the time instructors make their way to the classroom from their office; the power of computing will double with the price cut in half. According to Moore's Law, coined by Gordon Moore, co-founder of chipmaker Intel, this phenomenon takes place every 18 months. Fine, maybe it takes professors less time to make the journey, but by the time they do make it to class, you know someone, somewhere, just came up with the next big thing.

In any case, MIS is critical because it shows students how to set up a solid, reliable and rigid infrastructure upon which a business can be built on. Glorious ideas have failed because it was not feasible from an IT perspective - either technically or financially (usually both). The telecommunications industry is rife with examples of this.

Marketing

Marketing is a term that encompasses quite a bit. It endures a lot of unfair criticism because it is so much fun. On the surface, marketing appears to be all about ideas, feelings, emotions and messages. These are intangibles that are very hard to measure and almost impossible to track. Despite this knock, most love marketing as it explains why we act the way we do as consumers. It also explains why many companies fail despite great products and priceless services.

Beneath the surface though, marketing is a very tough discipline with methodical statistical research behind every move. It is always more interesting when you blend in financial analysis to show exactly how every dollar spent on marketing impacts a company's bottom line. This is when some students develop an aversion to marketing.

Advertising, Public Relations and Sales all fall under the marketing umbrella. Advertising is the glamorous field in marketing to many. It is a very creative, tough and competitive field with a great upside. Advertisers can control the message they are trying to convey because they pay for it. Public relationists are not so lucky. They tend to spin the facts and figures in the hope that the message gets relayed and interpreted in the best light possible. Public relations and advertising are business terms that make up what the military calls propaganda. Sales are almost treated as an orphan by some in business but it is what makes a company viable. The sooner you realize that selling is the gateway to success and longevity, the more fruitful your career. Salesmen are never in short supply. Business and life are all about selling. Keep reading to learn how to succeed at selling without selling out.

Perhaps no one better captured the notion of selling a feeling more than the late Alfred "Freddy" Heineken. Indeed Heineken is a fine beer, but Grolsch and Amstel are equally good beers from the Netherlands. What made Heineken so successful as a marketer was that instead of selling beer; he was "selling warmth, gaiety." The fact that good ol' Freddy was lacing his warmth and gaiety with 5% alcohol sure did help, but the fact of the matter is that it was the image and aura of Heineken that flowed from those bottles – not beer per se.

International Business

International Business is essentially marketing with a global perspective. It combines psychology and sociology with anthropology and political science. In this day and age of globalization, deregulation and a fairly free flow of capital, marketing is global whether you know it or not. So taking

International Business over Marketing is recommended even though more and more Marketing classes focus on global issues anyway.

Electives: Choose Wisely – Who Moved The Cheese?

The flip side to choosing a major is selecting which classes to take to complete your degree. Assuming you do not declare a minor and have some room to work with, you will have the choice to select easy, straightforward classes or opt for classes that challenge you.

Take classes that will complete you as a person. Recognize your weaknesses so you can compensate them. Take classes that interest you so you can be ahead of the curve while others are agonizing over their boredom.

Business classes and the overall curriculum are helpful in breaking up any situation and seeing what the various roads to take are. This being said, many of the world's problems cannot be put in a SWOT (strengths, weaknesses, opportunities and threats) analysis in the hope of coming out with a best alternative. The world is complex and no two people or situations are alike. This said, using a SWOT analysis remains very much in vogue these days.

As companies tackle on new issues in new countries, Anthropology and Geography classes will help you in your development as an entrepreneur or a company person. History classes will help you understand the future. Sociology and Psychology classes will give you a leg up on everyone else when it comes to getting the most out of your employees, earning the trust of colleagues and the respect of your boss – if you choose to have one that is.

Choosing Your Major: Is That Your Final Answer?

An obvious statement is that whatever you decide to do at school will play a large role in your career thereafter. A less than obvious statement is that most of the classes you enroll in will become irrelevant. School is important because it shows you the basics, provides you with a framework to solve problems, tests your discipline and gives you an ability to learn. In general though, it is understandable why so many employers tell you to forget everything that you have learned when you start your first job.

Your major could play a determining role in what industry, company and department your first job will be. Once your career is launched though, expect it to have very little bearing on where you will end up.

It is thus important for you not to pigeonhole yourself as a function of what you major in, which electives you choose and how much (or little) exposure you get to the outside world. This world includes the Social Sciences, Health Sciences and the Creative Arts.

The suggested course of action is to leave as many doors open as possible. Moreover, your major should teach you a "differentiating factor" and not simply consist of lectures on ideas, concepts and theories.

Finance, Accounting, Decision Sciences & Quantitative Analysis and Management of Information Systems are majors that are considered to offer substantial differentiating factors. After all students need to understand Generally Accepted Accounting Principles (GAAP) to master accounting. A financial advisor needs an understanding of security investing and tax concerns in order to give advice on investments. The same applies to MIS students who learn about technical aspects of the business world.

Management, International Business, Entrepreneurship and Marketing are equally important fields of business but offer less in terms of differentiating factors. As examples of fields that are heavier in concepts, theories and ideas. Management's main differentiating factor may integrate some classes in labor laws, employees' legal rights and payroll accounting – but the concepts of management issues, employee motivation and compensation are more accessible to the general student population than concepts in Accounting, Finance and MIS. Marketing students learn quite a bit about marketing research – a rigorous empirical field – but determining what works in advertising and what would backfire in public relations is fairly straightforward for most to grasp and master. A major in these fields ensures that you have the proper terminology down for each field, but it will do little to ensure that you will rise to the top of your profession.

This being said, the last thing Finance, Accounting and MIS students should do is adopt a holier than thou attitude. In fact, many mock Management students, but they are the most sought after employees when questions about severance, pay and vacation time arise.

Professors

The Good, The Bad And The Ugly

Professors set the tone in class. Two sections of the same course could be radically different depending on who gives it. At times, some students vilify professors without taking the time to better understand them. Like all human beings, professors share good, bad and ugly traits. It is up to you to understand professors in order to benefit from the good traits while eluding the bad and ugly ones.

The Good

Good Professors take a student's interest and raise it several notches. They also create an environment where students feel free to question the institutional imperative. The best ones even take apathetic students and nearly transform them into interested parties. Good business professors are those that view students as clients. Good non-business professors view students as patients. Fitting, since most are doctors anyway thanks to their Ph.D. They aim to satisfy, teach and foster growth. Good professors are always in control even when it seems that they have lost it. When this appears to be the case, the professor is letting the students' minds and emotions take over – knowing very well that control can be regained. This professor commands the respect of all students.

The Bad

The bad ones seem to be disinterested. After years of repeating the same shtick, they see their students, lectures and curriculum as a downward spiral that becomes increasingly blurry as the years go by. While they leave a mark on students, it is not because they had a hand in their development. Bad professors view students as a necessary evil. They have control but for the wrong reasons. One reason is that discussions are rare and the lecture is a monotonous soliloquy. Some professors teach for the balance that it provides them. Some like the power, others are risk averse and enjoy the security of tenure, while many like the stature.

The Ugly

The ugly ones are those that after years of repeating the same thing become almost bitter and sarcastic. What role they play has a lot to do with what kind of students they "teach." Confident students tend to see these professors as a challenge, a hurdle to overcome in their quest for growth, success and happiness. Sadly, less confident students tend to see their development damaged by ugly teachers. Sometimes, much needed grades are not earned as a result of a combination of teacher inaptitude and student apathy. Ugly professors view students as burdens. Such professors feel that they could be doing more and take it out on students. This teacher has no control even though he is usually the only one to talk. Students do not feel the need to share their views as this incites cynicism from the professor. Interestingly, the professor lacks respect and control although he feels that he has it. Some of these individuals could do great things and benefit from a turnaround by working in industry, but they do not make the jump: either out of fear or complacency.

Beauty Is In The Eye Of The Beholder

Whether they know it or not, all professors play a part in students' development. Class clown, class loudmouth or class president, seeking respect from teachers sets the tone for a successful career. After all, in an ideal world, a professor only needs 20% of the class time to spur 80% of class discussions from students. This is an extension of the Pareto principle, a widely used metric that states that 20% of causes will create 80% of the results.

Expect The Unexpected

You may have had a wild range of different professors in your past but expect it to get wilder in business school. While a Master's and or Ph.D. are still required for most teaching posts, some business professors are younger, more energetic and at times do not fit in the traditional "profile" of a teacher.

Who's Who And How To Handle Each One

No two professors are alike. In fact, even individual professors seem to go through a Dr. Jekyll and Mr. Hyde persona. But the fact remains that business professors ultimately fall into the following profiles. Some may be a combination of different types.

The Lifelong Academic

The lifelong academic usually means well. This is one that has highly idealistic notions of education, school and what a degree ought to be about. While some may at some point go through periods (at times prolonged) of cynicism and sarcasm, they generally revert back to their high sense of idealism. Expect to be tested. They view a student's success as a reflection of their work and will see to it that you succeed – so long as you play the Game. Examples of failing their test include not doing assigned problems only to learn that the exams consisted solely of these. Most will even go to great lengths to get your name right. Some may even call you by your last name as a throwback to their days in the classroom. These professors are devoted and they believe that teaching is their calling. They view it as a craft or an art form (whether they wish to admit it or not). Since these professors have usually completed a Bachelor's, Master's and Ph.D., they want to see you go through some of the pain that they endured. Call it tough love. A professor once asked a certain running back at the University of Michigan how he felt about Coach Glenn "Bo" Schembechler ripping into him day in and day out. The player simply answered that it made him feel good knowing that the coach cared. Indeed tough words and tough love make for some of the best professors in school.

The Business Minded Academic

This individual is also an academic at heart but they view school as a business. They love school and cherish the purity of the classroom. They like the idea of playing a part in your development. They may have done their thing in business and now wish to give back. Whether they failed or succeeded is not the issue. What is important is that they drill you the way a senior manager would. They view teaching as a job, and your education as a form of military training. Teaching is not so much of a calling for them as it is a mandate. They view the classroom as a department, the class as a project and themself as the division manager. This type is smart enough to know that some of you will either go on to become a prized commodity in their company or excel at another firm. They do not believe in suffocation but emphasize discipline and responsibility. This professor will never make your life easy, because "business and life is not easy." Their tough grading is a reflection of this mantra. The business minded academic makes for the best business teachers of all.

The Seasoned Executive

This is the type that either has or still is experiencing success in their profession. They are teaching because this is an extra mandate that they have accepted. They enjoy it and it adds to their credibility. Some may do it for personal gratification, others because it compliments their work. Some are wealthy, others are a financial mess that could use a lesson in money planning from you, but they have some notches on their belt and you have been blessed to have their wisdom for the next semester. Enjoy it; others pay dearly for it. Well, you are too in a way so listen to every word.

The Burned-Out Executive

This is the person that has accomplished more than you will in a lifetime. They have taken a break – permanent or temporary – from the business world and have found refuge in the classroom. They are in control of matters now and can go at their own pace. This type has tremendous intellect as well as a mid- to high-level of education that accelerated their career at an early age. But this made them burn both ends of the candle at once. The problem is that at some point in their career, they realized that they had seen everything go by and now wish to take the scenic route. You will get valuable lessons from this person, the most important being to enjoy things. If you encounter him early on in your schooling, you will be blown away by their persona. If you meet them later in your academic career when you have narrowed the gap, they may not give you the credit you deserve, as you pose a threat. This is the residual effect of a background in a cutthroat environment. Learn from them and ensure that you do not make their mistakes.

The Young Maverick

If ever there was one individual that was both vilified and venerated, it is the young maverick. This is one person that has achieved a lot in their young life. To add to their list of accomplishments, they have taken up teaching. While some in the faculty take this character less than seriously, it only adds to his desire to stand out from the academic crowd. Students also share different opinions on this professor. Some students view the brash individual as arrogant, others view his accomplishments as a blueprint to follow: the balance between industry and academic success. Usually this type is rare in non-business school. Industry success has allowed him to get in front of the class and win a new battle.

The Manager On The Rebound

If ever there was one character whose wings were clipped, this is that person. It is possible that an inflated ego – either from academic or business success – was deflated because of a major setback. This ego has been deflated but they sometimes show off that swagger that made them crash and burn. For the most part, the swagger is still apparent but confidence is low beneath a veil of excess pride. Some may feel intimidated in class so they tend to crush any dissent. This only adds to the frustration of students.

The Nervous Academic

This character is just that: a character. What the professor has in academic credits, he lacks in everything else. You get the impression that he crossed over into the business world and was sent packing. As a result, he often loses control and some students may even undermine his authority. But if you cross his path, do not undermine him. You will look like a bully and it could come back to haunt you later on in business. Work with him as an ally and you will reap benefits when it comes time to managing others, building morale and inspiring the troops.

The Short-Changed Professor

In business, strong-minded individuals are often not given a fair shake because of language, ethnicity or gender. While subjects like English force a professor to master the language, business school does not. As a result, it offers an opportunity to many brilliant minds who have seen their restricted language skills hold them back in corporate settings. These are some of the best brains on campus but they are also the least consulted minds. In order to get the most of him, you may be better off consulting him during office hours.

Teacher Hours

Some of the best tips, advice and guidance are missed out on as students fail to take advantage of teachers' office hours. All professors are required to offer some office hours on top of their regular class hours. Yet the only times that any one seems to take advantage is right before exams, presentations and when papers are due.

Why this is so is obvious. Who wants to go sit in front of a professor that pontificates over academics? Yet this is the single best time to get inside information on everything from the lectures, to exams, to assignments as well as strike up great relationships that will stand the test of time.

Professors are proud by nature. Teaching is their profession but nurturing is their craft. As trite as this may sound it is true. As a result, they expect and hope to see students go visit them (at least the good ones do – the ugly ones may not even be there during office hours). Their egos are at stake. Imagine if one professor gets 6 students per hour while the one across the hall gets 0. How do you think the latter will feel? Now imagine if neither teacher get any students. This is akin to the girl at the dance that does not get asked to dance. Of course, this may have something to do with the girl, but what if the guys are shy and they let a great girl go? This is what happens in school. Professors are wells of information that often go untapped.

Professors are always on the lookout for protégés. They want to take students under their wing. This may mean more work but you will have more opportunities while in school and plenty once you leave. You will get inside stuff that other students do not get in class. Of course, you should relay this to classmates but chances are that not every one will receive this information. This is to your advantage. Professors tend to drop subconscious hints when they feel that students are straying away from the important material.

Challenging Them Versus Heckling Them

Parents, friends, professors, colleagues and bosses are human like you. Celebrities, whether they are athletes, entertainers, musicians and actors are also human like you. For this reason, being in awe of their accomplishments will always render you a groupie. You have to wake up thinking that you are the rock star, the guitarist, the bassist or the drummer. But if you have a virtuoso in front of you, then you should pace yourself until the stage is set for you.

It is important to respect and give credit but the key is to take what they have accomplished, recognize the secret to their success, extend and build upon it. Your parents have probably done many great things. Your father walked barefoot for 10 miles in the snow to get to school, right? Well, you should walk 20 miles. If you choose to walk 10, do it in half the time.

Living in the shadows of anyone is not the way to go. Learn from them, listen to them, study what they did, but always strive to do better. Your folks did not sacrifice and put you through school for you to be a passer-by. They put you through school so you can become better than they are. Now imagine you were financing your own education. Shouldn't you maximize your investment?

Learning Through Osmosis

It is for this reason that students should see what their teachers have done, what they studied, pick their brains and strive to be greater. If your professor studied at the Sorbonne and does his or her job (which is educating you) properly, then theoretically you do not need to go to the Sorbonne, do you? The analogy may be extreme, but you should be able to draw from the professor's experience and education and bypass those steps. View everyone as an equal but seek to better yourself by building upon what you learn from them. This is not being selfish since others should do the same with you.

For these reasons, students should ask about their professors' education (albeit in private perhaps), their studies, their findings, as well as their views on current affairs. If the teacher and pupil both do their respective job, then your knowledge base should have expanded enough to reflect (although obviously not equate) what the professor has learned in his or her life.

Mentors

Professors also make for good mentors, especially if you plan to pursue a Master's or Ph.D. Overconfident individuals seem to think that they do not need a mentor but even the most successful business leaders have one person or numerous people that they turn to for advice. Corporate Board of Director members often adopt a mentoring role to many CEOs. Berkshire Hathaway's Warren Buffett is one such example. As you can imagine, other directors are thorns in their side – as they should be to varying degrees.

Counselor Or Advisor

Depending on the size of your academic institution, some advisors go on to befriend you and serve as good sources of information over the years. These are especially powerful to have since they can help you with registration, class schedules and even grades.

Your Partner In Crime

At other times, you may be working on a particular project with a professor. It could be research, a book or a study for a company or class. Keep in mind that even though you may end up doing much of the nitty-gritty work, it is still their work. A word of advice: be grateful for the opportunity, but be firm about what you should get in return. Proper credit, fair compensation (if applicable) and references are only fair if you have helped them with their work.

Father Figure

Some of your professors may display a paternal touch. Whether he has accomplished what you seek to do in your future or has something else to offer you, it will be detrimental for you to ignore their input. While it is good to have a nurturing individual by your side, what you need is someone tough.

The Critic

The best mentor is one that is not afraid to let you know you are wrong (even if you are not). Some are too busy patting you on the back as this reflects their own good job. The better ones know that tough love is the most effective tool.

Contacts

A main advantage of having good relationship with professors is that they are connected at your school as well as across the nation. If you ever need to get in touch with someone at a different school, they may be able to help. When it comes to finding a job, professors are a great reference source.

Beware

At some point, when your goals exceed their scope, you will realize that you have outgrown your mentor. Even worse, your goals may clash with theirs. They may want you to stay to work on your Ph.D. and help them with their research, but you may wish to "turn pro."

Basketball player Kobe Bryant best demonstrated this. The standout National Basketball Association (NBA) player skipped college altogether despite the suggestions made by some coaches, pundits, media, players and teachers alike. Others may have recommended he go to college but he wanted to play in the big leagues. Bryant was then drafted by the Charlotte Hornets with the 13th pick in the 1996 entry draft and subsequently traded to the Los Angeles Lakers for Vlade Divac.

Teachers sometimes come in the shape of a father or mother. No one knows this better than Eldrick "Tiger" Woods, whose father Earl basically raised his son to be the greatest golfer of all time. With all due respect to Sam Snead, Jack Nicklaus, Arnold Palmer, Tom Watson and company, Woods went on to win more at a younger age than anyone else. The fact that he won one of the most prestigious titles in golf, dominating the field in the process as a visible minority in a traditional Caucasian field revolutionized the sport. His triumph at the Masters in Augusta, Georgia was just the beginning and came at an age when most golfers struggle on the development tour or in college. But the lesson here is not one of swings and approaches, it is that Tiger had a passion for golf. Earl Woods may have laid the foundation, but it was Tiger Woods that built upon it afterwards.

Academic Contract

Whether you choose to accept it or not, you have entered an academic contract where you will have a give and take relationship with professors. Some may offer you quite a bit and never ask for anything in return. In other instances, you may have to help them with projects for meager short-term returns. In any case, professors may require much from you, they will use you for leg work, a fresh perspective, as well as having an ally amongst their ultimate judges: your fellow students.

In fact, many great teachers are remembered because of their students. For example, most of what we know about Socrates is through the work of his famous student, Plato. The ascension of Socrates as a Great Thinker marked the Golden Age of Greece. When Socrates was forced to drink hemlock and take his own life, Plato left Greece for over a decade and returned to establish The Academy, a school that operated for about 900 years.

In turn, Plato had a student who was none other than the great Aristotle. His views laid the foundation for the Western thought that permeated across the globe until the 17th century.

Subsequently, Aristotle became Alexander The Great's tutor. Alexander's father, Phillip of Macedon, ended the Golden Age of

Greece when he united the Greek City-States to create the Greek Empire.

Stepping Out Of Others' Shadows

At some point though, you will wish to stand on your own merit. You must respect what your mentors have accomplished, but do you think they are investing in you because they like sheep? Of course not – they are getting something out of it. This is applicable not just to teachers but family as well. Many of you will be stepping into business school knowing that parents or older siblings have set the bar high for you. How do you carve out your own niche to step out of their shadows and live up to the expectations?

Stinkin' Feet

Musicians often have influences that they have never met. Ozzy Osbourne was the original front man in one of the most influential bands in rock music: Black Sabbath. Together they released Heavy Metal classics such as *We Sold Our Soul for Rock & Roll* and *Paranoid.* Circumstances led to his departure from the band in the late 1970s. A couple of years later he recruited Randy Rhoads who was viewed as one of the most gifted and original rock guitarists of all time. Rhoads was a classically trained guitarist from the University of California in Los Angeles (UCLA). Rhoads was seen by many as an equal to Eddie Van Halen. Tragically, Rhoads was killed in a plane crash on March 19, 1982. However, before "the day the music died," Rhoads left his mark on two stellar hard rock albums: *Blizzard of Ozz* (1980) and *Diary of a Madman* (1981). Osbourne went on to release over a dozen albums afterwards but it was those two that taught legions of guitarists afterwards.

After a couple of replacements for Rhoads (including Brad Gillis and Jake E. Lee), Osbourne recruited Southerner Zakk Wylde. Wylde was talented beyond belief but anyone in his position would have been intimidated. What was Wylde to do? Play his own style or imitate Rhoads, Gillis and Lee?

Wylde's first release with Osbourne was *No Rest for the Wicked.* The album was typical Ozzy quality but one could not help but feel that Wylde was playing under a microscope. His next release was *No More Tears* on which Wylde was increasingly at ease. He was blossoming into his potential as a guitarist. But it was not until the double live album *Live and Loud*, that Wylde established himself beyond the shadow of his predecessors by shredding through the Ozzy and Black Sabbath track repertoire.

Wylde needed several albums before feeling comfortable in his own skin in order to step out of his predecessors' shadows. You will have your own pace. Make sure that you are comfortable with your development and you too will shred through life.

Some of you will not only have shadows to step out of, but you may even have legendary parents or siblings. Take the case of two racecar drivers, Jacques Villeneuve and Ralf Schumacher.

Daddy's Shoes

Jacques Villeneuve's father, Gilles was the lead driver for Ferrari. Racing in a Formula 1 car is tough, but the stakes are much higher when you have the keys to a *Scuderia*. Gilles had so impressed Enzo Ferrari that his signing led to the departure of Niki Lauda from the team. Gilles first raced for the team in 1977 at the Grand Prix of Great Britain. His first victory came in 1978, in Montreal, Canada. Fitting since Villeneuve was Canadian. In 1982, at the Grand Prix of Belgium, Gilles died during the try-outs.

About 15 years later, a young wunderkind by the name of Jacques Villeneuve was winning over fans by the truckloads. Young Jacques had just won the Indy 500 in the Cart Series and had sealed the year with the Championship. His sights were now set on Formula 1 racing. Jacques made the move to the Williams Renault team, arguably the top team in the circuit. With everyone watching, and most asking about his father, Jacques remained respectful of his legacy and did what he did best. He finished second to Damon Hill in his rookie season, but capped off his second year with the title, surpassing the legendary Michael Schumacher. The younger Villeneuve could have signed with many teams, but he strayed away from Ferrari and focused on building his own reputation.

Big Brother

After winning Championships for the Benetton racing team in the 1990s, Michael Schumacher switched to Ferrari for a cool $25 million. Within a couple of years, Schumacher began to dominate the Formula 1 circuit as never before.

A couple of years later younger brother Ralf made his entry into Formula 1. Being the younger brother of the world's greatest racecar driver was not easy, but unlike Villeneuve who was never comfortable in his father's shadow, Ralf Schumacher never hid his respect, admiration and emulation for his older brother. By the time that Williams BMW lured Ralf, the younger Schumacher had blazed his own trail in Formula 1.

The lesson is that Zakk Wylde, Jacques Villeneuve and Ralf Schumacher all had big, stinkin' shoes to fill. The shoes were big in terms of accomplishments and stinkin' in terms of them being hard to ignore. They all managed to fill them splendidly by doing what felt comfortable. You too may have big stinkin' shoes to fill, so start smelling the roses and just do what feels natural. The rest will fall in place.

A Little Bit Of Patience, A Lot Of Success

It is not always necessary to fly out the gate. Back in 1984, Steve Young was projected to be the number one pick in the National Football League (NFL) entry draft. Young was ambitious and knew that he would never start in his first year. Instead of holding the clipboard on the sideline, he bolted to the United States Football League (USFL), signing with the LA Express. LA was closer to his home in Provo, Utah and he knew that he would have a chance to start. In 1985, the USFL folded and Young returned to the NFL team that had drafted him in 1984, the Tampa Bay Buccaneers. After 2 seasons and 19 games, the San Francisco 49ers acquired the promising quarterback.

But the 49ers had a legend and a future Hall of Famer at the position: Joe Montana. When Montana was injured in 1991 however, Young came in and led the NFL (NFL) in passing efficiency, his first year as a starter.

Despite his ambition, Young has always maintained that the years he spent watching Montana were invaluable. With football minds like Bill Walsh, Mike Shanahan, Mike Holmgren and George Seifert and a legend at wide receiver like Jerry Rice, Young was provided with the best tools to succeed as well.

Of particular relevance is the fact that Young did not waste his time as a backup. He pursued his law degree at Brigham Young University the entire time, completing it in 1994. In addition,

Young also founded the Forever Young Foundation in 1993, committed to helping children in need. Now that is a class act.

Incidentally, as eager as Steve Young was to step out of Joe Montana's shadow, coach George Seifert was just as eager to do the same with regard to Bill Walsh. Perhaps this is what made Young and Seifert such a fruitful partnership.

Seizing The Moment

Professors can make great mentors for those who seek to study after they complete a Bachelor's. Master's students need their support while Doctorate students rely on the mentorship, tutelage and guidance of someone that has walked the path before.

When things are said and done, mentors may point you in the right direction but they will not get the job done for you. As a result, you must remain in control of your destiny and do what feels right for you.

When the time comes to go in your own direction, remain grateful but avoid feeling forever indebted. At some point, we all have to cut the umbilical chord – the key is to remain cordial.

<u>Students</u>

Regardless of whether you are the type to hang out at school during your break or bolt out of class to leave campus, you will come across some great people at college. The first week of the term is when you should make friends and establish contacts. While professors change course outlines, alter dates and make important announcements, confusion reigns in and out of the classroom. As a result, you should approach fellow students and break the ice. Some of the best connections are made this way.

Out With The Old, In With The New

It may be difficult to adjust to so many new faces, but remember that you are no longer in high school. It is imperative to acknowledge your reality and evolve with it.

Business is all about change: changes in environment, dynamics and stimuli. Companies like personnel who move with the times, so the sooner you become an agent of change, the better.

The first front on which you will experience change is with people. You will interact with some who will become friends, others who will not. Imagine you start a new job and choose to stay friends only with colleagues from your old job: is that healthy? Now imagine you get a transfer to a new department: would you snub your new colleagues? Finally, imagine you are an athlete who gets traded to a new team: do you maintain contact with your old teammates? Of course, but you must make connections with your new teammates as well.

Change is a fact of life, so get used to it.

Blurry Lines

Teammate Or Opponent?

Athletes get traded often. They sweat, bleed and cry for their team but must shift their allegiance once they put on that new jersey. While most will remember their old team, fans and teammates with fondness, their goals become consistent with those of their new team. If you want to win the Championship, you may have an old enemy cast as an ally.

Colleague Or Competitor

Students need not work with those they dislike. After all, they can avoid one another and choose their own group mates. But in some instances, professors may lump students into groups based on schedule, majors and other factors.

If you find yourself in such a setting, you may be prone to ask for a proverbial trade. The recommendation is to stick it out and emerge as a diplomat as opposed to a prima donna.

Officially you will be working against the average and not other students. As a result, you will rarely have actual opponents in school even though you will all be fighting for the same pool of jobs down the road.

Friend Or Foe?

Football is a consummate team sport. The offensive line must move the defensive line in one direction and open up holes so that the quarterback can hand off the ball to a running back. The entire team must repeat this over and over until they get those magical 10 yards, only to start from scratch.

Sometimes however, when time is running out and the team is desperate, the offense must get aggressive. The team can change the momentum of the game in one swift play, but they can also lose everything outright. Imagine that the signal caller now decides to go long – very long.

In sports, the weather affects both teams. The home crowd affects the away team adversely but adds pressure to the home team. The only thing that either team has to worry about is the opposition.

In school, you will have to work with students who will one day compete with you for employment. As a result, even members of your own group may have unclear allegiances and shifting loyalty.

Imagine the quarterback asking the offensive line to block so that his wide receivers have enough time to make it downfield to catch the hail marry pass.

Enemy Or Ally?

How would the quarterback feel if his running back sacked him to end the game? Confused, angry and bitter.

This would never stand in football but in business and in business school, it is all too common.

Parts, Pride, Profit, Patriotism And Penalty

All About The Bottom Line

When it comes to crossing the line in business, no situation comes close to the scandal that hit the automobile industry in the late 1990s. The tale of espionage at General Motors and Volkswagen remains the largest ever in corporate history.

The story traces back to a proud Spaniard named Jose Ignacio Lopez de Arriorta. After a decade of effective cost cutting and dramatic showmanship, Lopez was promoted to Vice President of Worldwide Purchasing at GM in 1992.

But a mere 10 months later, Lopez defected to one of GM's cross Atlantic rivals, Germany's Volkswagen. While the defection sparked some concern and attention, Lopez maintained that he did nothing wrong, despite the fact that he had recruited several lieutenants to VW and had made off with several boxes of trade secrets.

Lopez was forced to resign from VW amid a hailstorm of lawsuits. In hindsight, Lopez defected for many reasons, one of them being that he wanted to build a plant in his native Spain. Dubbed Plant X, it was to be the most efficient auto factory in the world. While VW was receptive to the idea, GM was not.

Most corporate espionage cases are settled quietly but this one generated a great deal of attention. In the end, VW agreed to pay GM a $100 million fine and buy $1 billion worth of GM parts.

The lesson is that in life as in business, the lines are drawn clearly in the sand. If you choose to cross them on your terms with total disregard for others, expect there to be warfare. The irony is that the more valuable you are, the greater the stakes and the subsequent headaches. Even though Lopez has maintained his innocence and is revered as a corporate giant in his home country, his reputation abroad is tarnished.

All About Victory

The line between traitor and pawn is sometimes blurry. Spain is also the setting for one of the most high profile soccer defections where a blurred line clouded a situation. At the turn of the millennium, Portugal's Luis Filipe Madeira Caeiro Figo was about to make waves, huge waves. As one of the top 3 players in the world by virtue of having scored 30 goals in 172 games, Figo was transferred from FC Barcelona to Real Madrid for a then record $56 million. To measure the impact of such a change of allegiance, this was the equivalent of the league superstar being traded from the Boston Red Sox to the New York Yankees; the Montreal Canadiens to the Toronto Maple Leafs; the Dallas Cowboys to the Washington Redskins or the Los Angeles Lakers to the Boston Celtics.

Even though FC Barcelona had refused to re-negotiate Figo's contract, the Barcelona faithful had no choice but to view Figo's move as treasonous, tantamount to a slap in the face.

What many failed to realize was that Figo was a pawn in a greater Game. Madrid sought to steal Barcelona's shiniest star: Figo.

Expect to face treasonous students in school, dubious colleagues in business and questionable acquaintances in your personal life. The way to come out on top and remain unscathed is to see the forest through the trees.

Class Is In Session: Lectures

To Go Or Not To Go?

No one really wants to go to school. After all, when that alarm clock goes off and you press snooze a dozen times, there are many places you would rather be than in class. Sparing you the trite reasons why you should go, here are the *real* reasons why school is worth your time:

Avoiding Surprises

Professors are usually on the ball, sticking to the outline and the exam schedule. But sometimes, some teachers could use a lesson in organization. When you skip class, you presume that what was written on the outline is carved in stone. There is no such thing. Dates change, chapters shift and plans go down the tube. Surprises are inevitable so make sure you are prepared to deal with them.

The Inside Scoop

Despite what many think, teachers do not want to see students fail. As a result, they either give away what will be on the exams or drop obvious hints. Even if you daydream of better things, being there at least gives you a fighting chance of getting the inside information.

The Whole Truth…

Fellow students listen as much as you do, so a lot gets lost in the "translation." Competitive students will never give you the whole story: why should you be able to have your cake and eat it too while they toil away in class?

Critical Path: When You Must Go

Your attendance should be mandatory at the following:

The First Class

This is when all the important information is divulged, changes to the outline are specified and important dates announced. This is when you gauge whether a class and professor will provide you with a good experience or sleepless nights (or both). It is important to be there while you still have the time to drop classes. What you do afterwards is your prerogative but being present that first day is your duty.

The First Week

This is when group projects are assigned and groups are composed. These are critical matters that will ultimately determine your grade.

The Lecture Before The Exam

Some students drop this class and opt to study. Whether they are studying beer consumption, anatomy or sleep disorders is secondary to the fact that they are missing class. This is as counter productive as one can get. It would seem obvious that the teacher will get into or discuss some of the material on the exam, but many students seem to think that time is better spent cramming.

During Others' Presentations

Many students seem to think nothing of skipping class while other groups are presenting. First of all, this is rude as the other students will feel that they are being snubbed. Second, some professors take material from presentations and test students on it. Not being there will make you miss precious material. Third, you miss how the professor reacts and what questions he asks. These are critical for your own presentation. Finally, students often have fresh ideas that the professor does not bring up. Not being there could mean missing the boat on a potential genius.

What To Do When You Miss A Class

Wishful thinking aside, you will miss many classes over your school years. While some missed classes may be worse than others, this is never the end of the world. But here are some things that you should do to even out the playing field:

Ask Around

Even if some students will not tell you everything or cannot remember all that was said and discussed, asking a couple of students gives you an idea of what went down. Think of it this way: each version of the story is a piece in the puzzle – get enough pieces and the puzzle starts to represent a clear image.

Contact Teacher

Whether you decide to email, call or visit your professor, it never hurts to send a message that your absenteeism was not a disregard for the course or a sign of disrespect. Simply explain why you were not there without being defensive; it is after all your prerogative to go to class or not. This shows that you are serious and will make up for any missing information.

Class Discussions

Size Matters

Business school is optimized in small classes where discussions, debates and exchanges can take place like they would in the real world. But with academic institutions forced to do more with less, some classes will be crowded. Should you behave the same way in both settings? Of course not, as each has a unique set of dynamics.

Although class projects, assignments and exams account for the main part of your final grade, professors also grade your class participation.

A Taste Of Things To Come

This is consistent with company performance evaluations. Both small and large firms have them. While small firms tend to be less formal, large ones have no choice but to have established, standardized and uniform evaluations to see who is doing what.

Younger firms may have informal evaluations as they are too new to have thought of evaluating employees. Nonetheless, senior managers and founders tend to judge you on the fly no matter how young the firm. After all, all companies seek to maximize shareholder value; producing the most using the least amount of resources.

At small companies, this input versus output metric is identified immediately. At larger firms, top performers also get noticed but they are not rewarded fairly. Outliers (those that are in the top percentage and bottom percentage of performers) get noticed in both small and large firms. The difference is that in small companies, the worst ones last a day or two while the good ones rise to the top fast. In large firms, the bad ones can stay in their cocoon for years while the top performers hit the proverbial glass ceiling.

At small firms, everyone is too busy working to stop and congratulate you. You did a good job? Make sure it is great next time.

At large firms, excellence is seen as a threat. Mission statements emphasize it but in reality, top performers are seen as troublemakers. They cannot be evaluated in the same way as others and this is a headache for Management.

In small companies, the more creative, productive and valuable employees stand out like a basketball player among jockeys. Such individuals "write the book" and help their firms emerge as leaders.

Large Versus Small

To a certain degree, such differentiation also applies in business school. If you are in a class of 100 students or more, the professor is usually rambling and students are itching to get out. Whether or not you understand is only a concern to you. Sure, you can ask a question, but who will hear when 99 others are doing their own thing and want the class to move as fast as possible? This is the equivalent of attending a conference where it would be rude to cut off the speaker in the middle of his speech and ask a question. Timing is key when in a large lecture.

In small classes, it is your duty to ask questions and participate. After all, this is not a conference. And since the class is smaller, your professor is likely taking notes on who participates and who does not. This will reflect your grades and can make quite a difference down the road.

"Hello? Hello? Anybody Home?..... Think McFly, Think."

In any class, it is easy to sit there and pretend to understand when you flirt between listening and dosing off.

Others feel that writing every word is the answer. Even after taking notes, we sometimes look back at them and ask, "what on earth is all this supposed to mean?" Active listening requires you to do more than acknowledge a message.

Don't Just Do Something, Sit There!

Yet sometimes the best action is to just sit there. Yes, you read that right.

Students should on occasion put their pens down, sit and listen carefully to what the professor has to say. When something seems more important than the rest, go ahead and take notes. It is crucial to differentiate between mundane information, important information and critical information, in order to avoid information overload. Does anyone look at his or her notes upon getting home?

Another thing that could be done is to follow in the textbook and see what topics come up both in the class lecture and the textbook and highlight those.

Hear, Understand, Remember, Interpret, Evaluate, Respond

Sometimes having a clear measure between what is taught and what is being processed allows students to determine their own strengths and weaknesses.

The HURIER model is an acronym for Hear, Understand, Remember, Interpret, Evaluate and Respond. How does this apply to business school and your curriculum?

When you listen to the professor, ask yourself:

How Does It Relate To Business In General?

Do you know of any business cases where this has been tried and tested? Did it work or did it fail? In your experience, have you gone through the same thing? If the answer to any of these is yes, it is worth noting as it will make studying and remembering easier.

Apply To Other Classes In Your Major.

It is normal for a common concept to appear in a wide number of classes within your major. The impact of currency fluctuation is discussed in international finance, portfolio management and security analysis for example.

Apply To Other Classes In Business.

Often the same theories resurface in various classes. The concept of a product life cycle in Marketing comes up in Finance. Making this link is key to understanding in business school and excelling in the corporate world.

Apply To Other Majors.

Many times, business theories are simple extensions of theories from other fields. Management is an extension of Psychology and Sociology as applied in business. Most of us have taken Psychology and Sociology classes, so if a concept seems familiar, make a note of it and mention it in an exam for extra credit.

Debates

A typical scenario in business is a young employee trying to convince a more experienced one of his point of view. Seldom does the young employee even get a chance to speak. On rare occasions, the senior manager gives the rookie a shot. However, his advice usually falls on deaf ears. Why?

It comes down to a question of power. Everyone wants power. If you wish to be the student who wins the debate or the rookie who captures the attention of senior management, you will have to understand the sources and origins of power and how you can obtain it.

Perception plays a large role in who comes out on top in a debate between equals. For the time being, you will be debating with people who outrank you, so here are ways to level the playing ground.

I've Got The Power

Essentially, there are two kinds of power: power based on personal traits and power based on the position one holds.

Position

Needless to say, some people are in positions of authority. The President of the United States is perceived to be the leader of the free world and can do what he pleases. However, if he steps out of line, like some have, he is subject to impeachment.

Legitimate

Since your address is not 1600 Pennsylvania Avenue, let's come back to reality and use professors as an example. They are the ones that determine the class outline, formulate exams and hand out grades. Like it or not, they are in the driver's seat and you are lucky to be in the trunk.

Reward

The second type of power based on position is reward power. If you happen to have a job at current time, you can attest to the fact that your manager has reward power. He or she can give you a day off, a raise or a promotion. True, the reward is based on productivity and personality, but the buck stops with them.

Coercive

Surely you have heard the expression "do as I say, not as I do." Well, if you ever failed to listen to the adage, you probably got some form of punishment. This is an example of coercive power.

Information

A fourth form of position power is information power. Technical information is a precious resource these days. When a computer site gets hacked, every one wants to get in touch with the technical people to see what happened. After all, they have or are perceived to have the precious information to solve the problem.

Personal

If you are sweating bricks because you have few if any of these types of power, fear not. Most develop these over time. You may be in luck. You see, there is a second form of power relevant to someone in your situation.

Rational

Traditionally, philosophers were seen as wise men that held answers to life's many questions. These were men with extensive educational backgrounds. So while it can be argued that they held rational power, their power stemmed in fact from position. It was their position that allowed them to become students of the greatest thinkers at the time. For example, Alexander The Great was Aristotle's student, who in turn was Plato's student. Plato himself was Socrates' apprentice.

An example of rational power is probably typified every day in your classes. Surely you have met, come across or worked with that one student that not only has off the wall ideas, but who can present them well and manage to convince the students and professor of their feasibility. If you have not yet, give it time. Who knows, maybe you fit the profile. Regardless, this is one form of rational power. These individuals tend to make good lawyers and even better businesspeople.

Expert

After the 9/11 attacks, CNN regularly featured terrorist expert Peter Bergen. His information and familiarity with the suspected terrorist ranked him as an expert in the matter. But it was not the information alone; after all, anyone could search online and find information on Osama bin Laden. Bergen was an expert because he was one of the few men (still alive that is) who had interviewed the leader of Al-Qaeda.

Referent

Some people are liked more than others. They are either better human beings, make others laugh or share mutual interests. These people have significant power over others.

Charisma

The most important source of power, when everything is weighed in, is charisma. Bill Clinton was a Rhodes scholar from Oxford, but what made him victorious in 1992 and allowed him to emerge as one of the most popular elected officials in the history of politics was his charisma. Statesmen, businesspeople and entertainers will all tell you that this is the single most determining factor in their success.

Setting An Example: Stand Up And Shout

Presence Versus Attendance

Attendance refers to roll call. In French, attendance is translated as presence. This is ironic because Presence is a trait required to succeed in business. What do politicians and business leaders have in common? Most have an explicit aura around them that is second to none.

What do you think athletes and musicians have in common? Usually, an implicit and quiet presence highlighted when they perform. Athletes are often shy or reserved; it is for this reason that leaders are priceless in sports. Many teams trade elite scorers for leaders to take their performance to the next level. Musicians and entertainers are rarely bombastic. They can tear down a stadium but are more reserved offstage. Their work speaks volumes and gives them a Presence.

In business, some technical geniuses are like this. Shy and reserved on the surface but quite strong-willed beneath it. Microsoft's Bill Gates is just one example. His partner, Steve Ballmer has always been the more vocal leader while Gates has been the one to lead by example. Gates is more analytical and a technocrat, while Ballmer is the salesman with the booming personality. Gates will identify a mistake in millions of lines of code, but Ballmer will tell you whether or not the code will make a product that will sell. Great teams are complementary and no team is a better example of that than Gates and Ballmer. One can plant the seed for a beautiful concept while the other will see it to reality. It was Ballmer that championed the Internet while Gates was more focused on the Computer. Fitting since the Computer is largely closed off while the Web is wide open – similar to their personalities.

Business minded people have no choice but to have Presence because everyone wants a piece of the pie. Those that can legitimately claim a piece are those that advance a career or business. Silence may not rock the boat but seldom does it add value.

Here Is Your Future Boss

What is key to remember with your professors and classmates is that each may resurface as a boss. Some teachers are personable while others are tough. Bosses will be the same. Some instill fear while others show you compassion. For better or worse, you should view each one like a rite of passage, a step in your development.

Never make a professor look bad. They have accomplished too much in their lives to let a young punk derail it. A veteran of many battles, they have found in school a fiefdom to call their own. You will often find similar characters in business. While you feel that you will one day achieve greater things, you should come back to reality and acknowledge that you live in their shadow at present time.

Chain Of Command

In school, business or the military, respecting the chain of command makes the difference between moving up and being shoved aside. People like to be in control. Those who are in charge want undisputed authority and tend to dislike loose cannons. Rebels are seen as threats and are contained because they endanger the mission.

Talented individuals are often given leeway to be themselves but when push comes to shove, being too independent could land you in hot water, or worse, in shark infested waters.

President / Professor

There is an interesting case that needs to be examined even if it does not fit the traditional professor-student model. This case is relevant because it involves a President-Professor model that received a lot of press because of the actors involved.

Dr. Cornell West is a prominent professor at Harvard University in African-American Studies and Philosophy of Religion. Dr. West can be credited with making Harvard's African-American Studies Department the envy of all other schools in the world. For various reasons, he released a rap album in 2001.

What made more noise than the music on the vinyl was that Dr. West went against the wishes of his superior. Harvard President

Dr. Larry Summers publicly stated that Dr. West's action was an example of "conduct unbecoming of a Harvard professor."

To put the matter into context, it is interesting to describe the three cycles of rap music as coined by Grandmaster Flash, the father of DJing in rap music. Grandmaster Flash argued that rappers, or MCs, sang about three topics: things, people and places. Many would argue that rap's golden age was marked by the era when Public Enemy and Arrested Development sang about people and places. When Dr.West released his album, rappers were singing about material things. In some ways, Dr. West sought to claim the throne as the next Chuck D, founder of the influential rap group Public Enemy. Chuck D had once labeled rap music the "Black CNN" and as you can imagine, CNN's claim to fame stems from the stories it reports on people and places, and not necessarily the stories on material things.

Whether or not Dr.West's music backed up his website's claim that he ranked as "one of the preeminent minds of our time" remains to be seen. The bottom line is that Dr. West did nothing other than extend his thoughts and public relations skills that have made him to be one of the leading African-American intellectuals in the US.

Nonetheless, Dr. West should have considered Harvard's culture first. Asking for permission would have violated what he stood for, but it was understandable for the President of Harvard to feel slighted by the unusual act of a Professor releasing a rap record. The content was not the issue. It was how Dr. West ignored the *context* that he was in.

The irony is that President Summers' appointment to Harvard was intended to address the institutional imperative at Harvard. It is a school steeped in tradition yet critical of mainstream thoughts throughout American history. The bottom line in this case is that both individuals remain unscathed because of their track record and their popularity. Until you have a track record to back you up when you choose to go against the grain, protect your neck and understand the culture you live and work in.

Vice-President / Coach

In the military, mutiny can lead to a discharge. In sports, having a coach lobby for the General Manager position to the team President can lead to dismissal. Butch Carter did this as coach of the National Basketball Association's (NBA) Toronto Raptors when he tried to usurp GM Glen Grunwald's position. He was subsequently fired and Grunwald went on to build upon Carter's foundation. The stigma that was attached to Carter did not leave.

True, some professional athletes and coaches may have their own set of laws, but you are an amateur. This is why it is important to think twice about whom you choose as mentors and whom you view as an ally. If someone takes you under their wing, they may expect you to follow their lead and not side with their rivals – or try to steal their thunder.

Be careful when you ask for advice. You should think twice before going against what is suggested to you. You are free to do as you wish, but doing so may burn bridges and turn the advisor off. Debate is positive, dissent rarely is. It leads to mutiny or a discharge. Neither looks good on your résumé. Lose strong and powerful allies and doors will slam shut before you can ask "how high should I jump?"

President / Coach / Player

One of the most shocking and bizarre incidents where the chain of command was disrupted involved one of the most esteemed professional sports franchises, the National Hockey League's (NHL) Montreal Canadiens. The date was December 2, 1995, the Detroit Red Wings were in town and the "Habs" were being humiliated by legendary coach Scotty Bowman's troops. Detroit had jumped out to a 9-1 lead with eight minutes left in the second period before goaltender Patrick Roy got pulled. Like switching pitchers after they give up considerable hits or runs, it is customary in hockey to pull a goalie during a weak performance. With the Habs trailing the Wings, Roy felt that he should have been replaced well before the ninth goal. A decade earlier, coach Mario Tremblay had been a roommate of Roy in his last season as a player with Montreal in 1985-86; the same year that the rookie Roy led the team to a record 23rd Stanley Cup. In any case, Tremblay did not pull Roy as early as he should. Some have argued that Roy may have been insulted to be replaced midway in a period. After all, even though Roy had played poorly, so did the entire team. Who knows, maybe Tremblay wanted to make things easier for Roy and pull him between periods. Whatever the case, the two had a history. Roy had been quoted as saying that he could not help but take a cold shower when he found that that Tremblay, his former road roommate and overall average player, was to coach him. At the time, Roy had been a three-time first-team All-Star, a two-time playoff MVP and a one-man franchise, having led Montreal to two Cups.

In that infamous game against the Wings, Roy made a seemingly easy save and received a sarcastic ovation from the crowd. Roy reacted by taunting the fans and was yanked. What took place then was odd. Roy got to the bench, took off his mask, sat down, wiped the sweat off his forehead with a towel, got up, walked passed Coach Tremblay and exchanged a death stare with his former roommate. As Roy walked passed Tremblay, he stepped up to seat 105 in the Red section, looked at team President Ronald Corey and stated that this was to be his "last game for Montreal." Roy then walked back to the bench.

What happened was that the authority figure – President Ronald Corey – had to side with the rookie coach over the respected 2-time Stanley Cup Champion to prove that the chain of command should not be broken.

The outcome was ironic. Going against the chain of command may hurt you in the short term, but Roy was "punished" by being traded to his first preference, the Colorado Avalanche franchise that had just relocated from Quebec City (then called the Nordiques). That first season with Colorado, Roy led the Avs to the Stanley Cup Championship. The Montreal Canadiens subsequently missed the playoffs for consecutive years.

The lesson was important though: when your superiors lose confidence in your abilities or judgment, you are living on borrowed time. So avoid getting in this situation in the first place. Remember that loose links pose a threat and will be cut, sooner or later. Unless you are the greatest at your position – like Roy was – your chances of coming back are slim. But if you know that you have been short-changed, then go ahead and give the "thumb in the air" salute and stand your ground. Risky and bold perhaps, but who said life was safe?

Finding Equilibrium In Exchanges

People Precede Profit

In Chinese philosophy, two principles interact and produce the world in which we live in. The two polar elements do not necessarily represent any two matters, but their balance is sought in order to foster a healthy life. An imbalance between the two will lead to disease.

The yin and the yang represent all the opposite and conflicting principles found in the universe. The creation of the yin from the yang and vice versa explains why in the long run, balance prevails and no one force will dominate the other. If there is an imbalance, then things get unhealthy. In other words: you're gonna die.

Seeking a balance in your learning process will create an environment in which you will lead a healthy business career.

In business, you will often see a contrast between two forces: reactions and feelings (people) clash with the hard facts such as numbers, margins and percentages (profit)

Consider these two forces as a yin-yang that have to strike a perfect balance over the course of your professional development and your personal life. These present a spectrum, with one focusing on profit while the other emphasizes people.

Now extend this circle-like shape to represent the proverbial pie in business. This could be the same pie that represents market share in an industry or the size of a firm within an industry. The goal for business people is to grow the size of the pie, or yin-yang. Entrepreneurs often say that one needs to give up a piece of the pie in order to increase it.

However, as you will learn in business school, the corporate world and in life, trying to push profit by neglecting people will harm your business. Conversely, before getting too sentimental, remember that numbers are equally important. In *Fortune*'s 2002 list of the 100 Best Companies to work for, Kinko's failed to make the list for the first time in years. The firm's CEO and President, Gary Kusin emailed employees telling them that he shared their disappointment, but reminded them that in previous years, Kinko's had made the list despite disappointing financial results. Kusin was right in reminding his employees the importance of strong operating results over good public relations.

At either extreme, understand both the fundamentals of managing people in order to maximize profit. Recognize trends that impact both and keep an eye on the bottom line in order to grow.

These two forces, represented by black and white, must work in tandem for the yin-yang to remain healthy and that the pie grows. This applies to business and life in general. Jerry Seinfeld was right about the black and white cookie... perfect harmony is what it's all about.

The Ugly Truth

Perception Is Everything

Everyone wants to win, be it in debates, campaigns or sports. But who comes out on top has more to do with perception than with right or wrong.

This is where psychology comes into play. The numbers may be on your side, but unless you can appeal to people's emotions, you will lose the Game. This has happened over and over again, so ensure that you do not become another statistic.

Confrontation And Intimidation

Exchanges reveal conflicting points of view with no way to quantify that one is more right than the other. In order to win, you must let the other point of view emerge. Let the other person have their angle. When it is your turn, isolate some key points and take them apart. Attacking an entire position is being negative for the sake of being negative. Take a couple of the more obvious statements and show that you agree with those. This will make others perceive you as fair. After that, when you make your case, it is like catching an opponent off-guard with an upper cut.

Slippery When Wet

The one thing one should never do is to get on the Slippery Slope. The slippery slope is a vicious downward spiral that two sides get into when they wish to win an argument at any cost. The debate may begin on one matter, but then everything and anything under the sun is brought up. Once it gets personal, then you have skidded off the slope into the danger zone.

CHAPTER 3: SCHOOL POLITRICKS & RED TAPE

Schedules

It is rather simplistic to believe that you will be able to get into the best classes, with the best professors and at the times that you prefer. The fact of the matter is that booking classes can be a frustrating and time-consuming task. The powers that be will have no choice but to restrict some of the choices. To make it easier, here are some tips:

The Classes

Research The Class

Titles and descriptions seldom do any justice – if you want a good idea of what some of the electives cover, either ask the teacher or an advisor. Find out if any of your classmates have taken the class and make your decision an informed one.

Familiarize Yourself With The Professor

More than anyone else, your professor will be master of your destiny, so ask around and see what to expect. Is the teacher tough or easy? Will the grading be fair or not? Whatever you do, take the input of students with a grain of salt. After all, one student may have never shown up to class and may opt to slag the teacher to explain his poor grade.

Credit

Make sure you understand what it will take to earn your diploma first. Find out whether your classes meet the required credits towards your degree. If you only need three electives and you have taken three already, taking a fourth elective may mean that you will have to take an additional class to meet the requirements of your major.

Fit In The Schedule

It is noble to want to take a class out of interest, but keep in mind that it should fit in your overall schedule. There is no sense in having a huge break because of one class, unless of course the class is really worth it. Just don't complain about it afterwards.

Pick What You Like

You should seek to pick classes that you find interesting. After all, it is one thing to book a class, it is another to have to sit through it and do the required work.

<u>The Booking</u>

Use Your Head, Call Ahead

Most of you will probably have to book by phone. Many will also only be allowed to start registering by a certain time, say 9am. But what happens is that everyone gets on the phone at 9am and this makes it hard to get through. By the time you get through, it is 4:20pm in the afternoon and you have to choose between "The Life and Times of Martha Stewart" or "Learning to Sew 101." What you should do is call at 8:45am, check out some other features like your payment status and only choose the register option at 9am. Since you are already in the system, you will have secured a free line.

Substitutes

Even in the best-case scenario, you will have some of your first picks filled to capacity. Make sure to have back ups for each choice based on a) what you need to complete your degree; b) time preference; c) level of interest and d) preferred professor.

Go Over It

Once you are done, it does not hurt to ask to hear all of your selections before you hang up.

<u>Administration</u>

Bureaucracy is omnipresent in most institutions and business school is no different. There are opportunities to take advantage of and drawbacks to consider.

Deadlines

Mark important deadlines for everything in your calendars and agendas. This is fairly straightforward advice but it is often neglected.

The Paper Trail

There are some things that are in your control, like not missing deadlines. There is no reason why you should get sloppy and stress for nothing. When you do make payments, keep a record because the burden of proof falls on you. Failure to make payment means no credit for your classes. Even if you can fix this oversight with the administration, you will have to waste precious resources (sleep and time for example) that would be better spent studying or partying.

Financial Aid

In terms of opportunities, there are always great scholarships, grants and bursaries available to students. Business students are blessed with an influx of corporate gifts and parachutes. Check corporate websites, school websites as well as non-profit organizations to know all of the opportunities that are available.

Advisors

Advisors are there to help. But because of their limited resources, they often find themselves shorthanded and tend to give varying degrees of help to different students.

Is this fair? Of course not, but advisors are human beings and it is normal that they may lobby more for some than others.

Recruiters and headhunters on the lookout for talent often contact advisors. But whom these advisors recommend has little to do with the best fit. Usually it is a matter of giving preferential treatment to a protégé. While such behavior is indicative of poor human resources management, it presents a perfect challenge for the aspiring business student. In other words not everyone will be your fan, fewer yet will root for you, but it will not be the last time. So prove everyone wrong by overcoming the odds.

The scenario is akin to when the flashy Internet executive (advisor protégé) was a darling of the stock market when performance did not really count (in school). Yet the same executive was seen as irrelevant when profits mattered (in business). In this analogy, advisor support replaces the venture capitalist's financing.

When the protégé graduates, the tide turns as he is left to work on his own. Only then will everyone – including employers – see that they are swimming naked.

A greater problem arises when advisors serve as stumbling blocks for less fortunate students. You may have some wishes and aspirations, but some advisors may send you to interviews to fill their slots even if the position is not the best one for you. After all, they are doing you a favor by sending you to interview with Procter & Gamble, but is it really a favor if you wanted to go interview at Kimberly Clark? Throughout history, Procter & Gamble has been a favorite of students. This begs the following tip: make sure you know why you wish to work for a firm. Just because the masses want something does not mean that you should too. Warren Buffett referred to this as the "institutional imperative." Everyone else calls it being a sheep.

Faculty

Like advisors, professors have their preferred students. Professors are coaches in many ways: they are responsible for your development. However, the difference is that athletes volunteer to take part in athletics; students do not have that luxury. Understand the rules in order to win their Game.

Fellow Classmates

Indeed some of the best friends you make will be those you meet in business school. One of the most powerful brands and products to have emerged in the past few years is the popular search engine service Google. Google was co-founded by Ph.D. students Larry Page and Sergey Brin, two Stanford Doctorate students that are now making sense of the Web and sorting the overwhelming information found on it better than anyone else. This theme must have been contagious in Palo Alto because Yahoo! was created by two Stanford students as well.

David Filo and Jerry Yang were two Ph.D. candidates in Electrical Engineering at Stanford University. As the Web was starting to spiral out of control, the two started a guide out of their campus trailer in February 1994. Their goal was simple but ambitious: to keep track of their personal interests on the Net. As the Web grew, so did their guide and the time they spent on it – at the expense of their Doctoral dissertations. With time, the endless links morphed into directories, and with it the seed of Yahoo! was planted.

Originally called "Jerry's Guide to the World Wide Web," the site eventually took on the name of Yahoo!, an acronym for "Yet Another Hierarchical Officious Oracle." The founders maintain that the name stuck because of the connotation of something "rude, unsophisticated, uncouth." Whatever it was, in the fall of 1994, Yahoo! was serving over 100 thousand unique visitors.

Like Yahoo!, Google was a project that turned into a business. Ironically, in 2001, the industrious Google search engine began to power the millions of daily searches on Yahoo! A match made in heaven? Well, Stanford anyway.

Projects lead to teamwork and gamesmanship. But teamwork and competition engenders opposition, jealousy and envy.

It will not take very long before you see why business is a rat race. Before your first term is over, any idealistic notion you have of school being fair will go down the drain. It is often in business school that the seed of ambition, success and teamwork is planted and nurtured. But it is also at this juncture when envy, jealousy and backstabbing infect you.

In any student's development, a time comes when that greed muscle starts pumping and paranoia sets in. Intel co-founder Andy Grove said it best in his book *Only the Paranoid Survive*: if you want to get ahead, you should learn to trust others but always sleep with an eye open.

Some will be happy to see you succeed. But envy and jealousy are too innate in human beings for most to be genuinely happy for you.

The Game: Avoiding Enemies

At the risk of using trite analogies, business school is a Game. Whether you wish to call it a rat race or snakes and ladders, you will encounter more challenges and injustice than you would like to. But as you can imagine, life is unfair so get used to it.

What school does is teach you different tools that you can use once you graduate. It is thus in your interest to pick up complementary tools. These tools come in the shape of classes, theories and contacts. Most of these may become obsolete by graduation day, so make sure that you can detach yourself from them.

Friends may become foes, enemies may turn into allies, the key is being able to judge characters and personality traits. The same way that you must be able to detach yourself from redundant tools, you just may have to detach yourself from the people you befriend.

Sad, but true.

PART TWO: PRACTICE MAKES PERFECT

CHAPTER 4: BATTING PRACTICE - IN THE STUDY

Exams

Study Tricks

Class Notes

While some of these tips may seem obvious, students fail to use them. Not all of them may work for you but some may prove to be new, more streamlined ways of studying. You never know unless you try.

Take Notes In Textbook

Many students take notes but this makes for too much material to review when studying. An alternative would be to open up your textbook and follow along with the lecture. You can take notes on some things, but many students duplicate material found in the textbook.

Highlight What Is Repeated

What is the professor emphasizing that is also in the textbook? Is there a key central theme in the chapter that your professor is repeating over and over? If yes, then chances are that it will appear in the exam.

Be Concise

If you do plan on taking traditional notes, be concise. Do not write everything down word for word, take down the key information and listen to what the professor is saying.

Add Notes From Other Classes

Chances are that in any given class, material from another class will come up. Such cross-referencing will make topics clearer and problems simpler to solve.

Add Examples From Industry

If any theory strikes a chord with something a company did, jot this down. Obviously it helps to read business magazines, watch industry reports and the like.

Revise Notes

It is also a good idea to go over your class notes that same day or at least that same week. Sometimes notes are incorrect and incomplete and it is good to detect that immediately. The best thing is to add comments afterwards and ask questions the next week. These are steps that go the extra mile.

Problems

Sometimes you may have no choice but to go that extra mile and do the problems you have been assigned. Many do not even crack a textbook though. What you must ask yourself is how much of the exam will be made up of exercises. If the percentage is high, then get cracking.

In each function, there is one thing that you can do to help yourself understand the material, score the best marks and even use some of the stuff down the road. Sadly, there are no "silver bullets."

Accounting

The first thing to tackle is what is Generally Accepted Accounting Principles (GAAP), why are the rules there and what are they? Do they work? How do you interpret and apply them to business?

It would also help dramatically to do the problems. You can understand the lingo but unless you can apply the theory, you will sit there looking at your exam sheet with a blank stare.

Follow up on all the accounting scandals. Between Waste Management, Sunbeam and Enron, you have enough case studies to last you a lifetime. Incidentally, all three firms were Arthur Andersen clients, one of the most respected Big Five accounting firms, at least until this "trifecta of deception" made the news. If that were not enough, Worldcom joined the party with its own scandal at a time when investors were already hung over and fed up.

Find out why accountants lost focus and how this impacted stakeholders. Was anything done to be remedied or did the rules stay the same?

It would also help you to understand Generally Accepted Accounting Principles (GAAP), how it came about, when it is irrelevant and why it is inconsistent. The sooner you develop your critical thinking, the sooner you will emerge ahead of the pack. Just because everyone swears by the accrual basis does not make it right.

Finally, understand the intricacy and interdependence of the various financial statements. Understand what each financial statement reports. The income statement shows how much the firm earned over a specified period. The balance sheet represents a firm's financial health at a given time, providing a glimpse into whether or not it is solvent. The statement that shows changes in financial position (or cash flow statement) is very important as it explains why some firms hit the wall with liquidity problems. How does each transaction get reported, in which statement would it get reported. Break up the income statement between financial and operational matters to see the full impact of any deal or initiative. How will it impact the firm from an operational and financial perspective? Everything above Earnings Before Interest and Taxes impact operations, everything below has an effect on finances.

Decision Sciences & Quantitative Analsyis

An initial obvious tip is to actually do the assigned problems. These exams are often filled with number crunching or formula churning questions. The second tip is to master the software that you will be called upon to use, and yes, this means more than Excel. Third, read up on the General Electric-popularized Six Sigma initiative, find out how quality control applies to the automobile and hardware industry, as well as retail businesses like clothing and appliances. The more you immerse yourself into industry applications of classroom theories, the better off you will be. Dell's built-to-order model is a perfect manifestation of Just-In-Time theory.

Marketing

The trick is to view industry as sport, a company as a team and individuals as athletes. At this point, it becomes easy to follow and understand. The most important thing for marketing students to do in the studying process is to stay on top of what is happening in the business world. Which companies are leaders in market share? Which ones convert market leadership into high sales? What about profitability? Is a company delivering superior products to new markets or is it slashing costs to gain market share? Did the company launch a new advertising campaign that is working?

The idea here is that while textbooks will teach you the lingo and theory behind marketing objectives and goals, they do very little to explain the business world. Business magazines help to bridge the gap. Pick up any magazine and see how much of the terms you understand. When you read about strategies that worked or failed, is it clear to you why? The more you read, watch business news programs, the more you will be on top of matters and stand out in class and in exams. After all, if every student is writing on definitions and theories and you come out and explain why a certain company beat a competitor, you are now head and shoulders above the class. Popular examples are the Cola Wars between Coca-Cola and Pepsi Co., the computer industry with Compaq, Hewlett-Packard, IBM and Dell and for some reason, everything surrounding Procter & Gamble.

Management, Human Resources And Organizational Behavior

A recommendation is to comprehend why, when and how management shakeups take place, how they impact the company from top to bottom and when. Procter & Gamble, Citigroup, Gateway, AOL Time Warner are just some of the examples of companies that have recently gone through some interesting management changes.

It is also helpful to be on top of labor laws, how they came about and why they change.

For example, understanding why Procter & Gamble chose John Pepper to succeed the reputed Prince of Darkness himself, Ed Artzt, only to replace the likeable Pepper with militiaman Durk Jager later on is important. This will help you understand how the balance between people and profit is often manipulated.

Assessing why the mega-merger between Citicorp and Travelers Group led to the departure of heir apparent Jamie Dimon will better explain corporate turf battles. Dimon was widely seen as the successor for both Travelers Group's Sandy Weill and Citicorp's John Reed. He left and eventually wound up at Bank One. But if people are everything why would Reed and Weill let the young Dimon go? These are the lessons that once understood, will make you into a well-rounded businessperson. Merely knowing what the definitions are will not.

Finance

Follow the market. See which stocks go up and which ones go down. For every company that you study in class, request a financial statement from their head office. Read it, look at it and study it. Read the notes in the back, this is where you find all of the best information. Visit their websites and get an idea of how they present themselves and how they view their online efforts.

Find out how each term relates to one another and why stock prices move. Apply the theories to your own personal financial situation, no matter how simple it may be.

The single most important variable in finance is understanding what the cost of capital is and how it relates to a company's value. Since a company's market value is the present value of all future earning power, students can solve this number by dividing the purest form of income by a certain cost of capital. Research the study on Economic Value Added by the brain trust at Stern Stewart & Co. and you will see what makes companies winners and investors rich.

Management Of Information Systems

One of the fields where resources from outside of the classroom are important is MIS. After all, by the time the textbook is researched, written, printed, shipped and purchased – technology has probably improved at a reduced cost. Most IT classes teach you frameworks to understand how to analyze, assess and choose what technology to go with.

A good way to be leaps and bounds ahead of others is to read IT magazines. Visit corporate websites of non-tech companies. A web site says a lot about a firm's IT outlook. Then visit IT companies' websites and dig deeper to find out what they do, what their solutions are and how they make money. This is very important as some have poured billions to do some great things but lost money in the process. If you are really hard-core about IT, then it may be good to visit message boards where techies discuss details of the industry, new developments and debate future trends.

You should remain a silent witness and absorb information like a sponge until you feel confident enough about what you would say.

Projects

Choosing The Perfect Topic: Individual Or In Group

Pick An Interesting Topic

In some classes, professors hand out topics and give students no choice. Others prefer to list some choices and let students choose topics. In other cases, students are told to choose a topic and run it by the professor.

In the latter case, here are some ways to help you make a good choice. Do your homework to make sure that you know what you are talking about. The following is very subjective, but when it comes to choosing potential topics, ask yourself whether the professor and class will care.

The following to do list before you start a project can be applied in school and in business. Going over these simple steps will ensure that you will not waste your time and that your efforts bear fruit and not frustration.

Research

Before the advent of the WWW, researching was an arduous task involving considerable time spent at the library, going through archives of newspapers and dusty books to find the occasional needle in the haystack.

That was then, this is now. Today, you have no reason not to sit in front of a computer for 5 to 15 minutes to conduct a query with a good search engine. Immediately you get a feel for what is out there in terms of existing data, previous research and potential leads. This first step alone can save you time and headaches.

Pick A Topic Where Research And Facts Exist

For example, the Black Market is a very interesting topic for a marketing project, but where and how do you plan on identifying, gathering, analyzing and presenting the facts. If you do too good of a job, you just may find yourself face down in a back alley.

Inquire: Do You Smell What I'm Cooking?

Once you have an idea, do not be shy about visiting your professor. Pitch the idea and use them as a soundboard. After all, they will be in full control when it comes time to grading, so you want them to have a taste of what to expect.

Some professors will tell you straight up that the topic is impossible, or whether you are getting too specific or too general. Whatever the case may be they are a great beacon throughout the project so make sure that their input is considered.

Most professors have an element of nurturing in them, while some are downright father figures, so consulting them from Day 1 provides them with a sense of ownership. They may be easier on you when grading is conducted.

It is also possible that professor cynicism over a project will challenge you more. If this is so, good for you, go for it. But understand that you are taking a risk.

Get In The Driver's Seat

The first thing you should do when you step out of the office is relay all of the information to group mates. While some may think that the ideal would be for all group members to be present at the first teacher meeting, this is not so. Too many cooks in a small, crammed office spoils the broth. What you should ensure is that the group leader is there with another person.

Reassess

These two people (or at least the leader) must then present the feedback in order for the group to reassess the project status.

Identify

What are the topics to be addressed and which functions deserve more emphasis? The SWOT analysis comes up often, but is that all there is to it?

Plan

Whether you decide to proceed or re-evaluate the project, at this point, you have to plan. Planning in business comes down to determining how much and which resources must be allocated; to whom these resources must be allocated to; and how the work must be measured and analyzed.

While it would be great for all members to work in unison on all facets of the project, this is time consuming, short sighted and will insight control freaks down the road. It would be great to work together step by step as the project takes shape, but this would entail that everyone oversee the arcane, mundane and trivial details.

The better, albeit riskier alternative is to break up the work amongst the group based on individual interest level and area of expertise.

Breaking Down A Problem

Macro (Industry) Versus Micro (Firm)

The top down approach involves looking at matters from an economy level, then industry and company. The alternative bottom up approach is also good. The challenge and main determinant of success is knowing which one to use.

Operations Versus Finances (Income Statement Approach)

The second way to look at a problem is to see how it involves the operations and finances of the firm. Break up the problem by looking at the company income statement.

Income Statement Versus Balance Sheet

Finally it is important to evaluate how the problem will impact the firm's balance sheet and income statement. Only then has the analysis examined both sides of the equation.

Write In A Consistent Style

So many groups end up dividing the project and having one person cut and paste all of the work into one master document. This is efficient in terms of time allocation but it usually makes for a mess of a document with different styles, messages and sentence structures.

What you should do is either single out one person that will handle everything from presentation to editing or ensure that the group leader also handles these tasks. It is undoubtedly more work, but it will make the difference between a B and an A, or in some cases, an F and a C.

Follow Up

Once the project is underway, pay the professor another visit. Hopefully the teacher will see an improvement and factor that into your grade. If the project is still off the mark, the group will have to assess whether it is worth going back to square one.

Set Deadlines

The problem is that sometimes you may feel that you are ahead and get complacent. This is when the deadline will sneak up and bite you. So set deadlines like you would in any project and meet them. Setting and respecting deadlines is perhaps the single most underrated skill in business.

Individual Projects

Move!

Business is all about speed and business school is no different. The second you enter class on that first day, the sand begins to filter through the hourglass. So before you find yourself behind the curve, initiate a strategy.

Consult The Professor

The sooner you visit the professor, the better. Asking what is expected, how the grading will be done and what is not desired will help you considerably when others hit the wall. This will help you choose topics that will meet the professor's expectations.

Research

Like anything else, research your topic to make sure it makes sense, is interesting and is feasible to attack.

Outline

Plan an outline and make sure that the key points are addressed. Determine whether the necessary time is spent on each topic.

Guiding Light

The professor should set some parameters as to what they want, so do not be shy to keep them in the loop and throw ideas off them. Surely some will be more helpful than others. Welcome to school.

Approve

Run the outline by the professor. Watch their eyes when they first see it. This is more psychology than business but it helps with getting to the bottom line: will he like the topic?

Draft

Begin to draft by having the main ideas serve as the foundation before you fill in the blanks.

Edit

Edit often as you progress. Leaving it to the end will create an unbearable challenge and lead to a sloppy paper or presentation.

Touch Up

Make sure that the paper is clear, clean and fun to read. Sadly, some teachers do judge a book by its cover.

Group Projects

The Draft – Choosing The Team: Checklist Required

Unlike Major League Baseball (MLB) you will have to assemble players to make up your batting order without the opportunity to trade them.

It is often in the first classes when groups are made up. Perhaps there is no need to draw enemy lines in the sand, but your group mates will form the battalion that will take on the class. Students commit prolific mistakes when it comes time to choosing groupmates, so here is a basic checklist to consider.

Schedule Conflict

On the one hand, business students are taught how to be effective with resources such as time. On the other hand, finding a common time to meet your group is a challenge and can be a waste of time. After all, in business, you all work to find a common meeting time. In school, schedules are rarely in sync and this causes a problem.

Students should make groups according to schedules. Do you prefer to meet on weekends or weekdays? Are you a night person? If so, make sure your group is too.

Topic

In a perfect world, your group should be composed primarily as a function of the topic. This is utopian. Reality is that the topic is secondary to getting the job done and it should come as no surprise that time is the single most important variable in all schoolwork. This said, the topic makes or breaks a project so make sure you all agree on one and are confident in getting the most out of everyone.

Friend Or Foe

Conventional wisdom would dictate that you get together with friends. After all, you know one another and can avoid any surprises. But one goal of group projects is to meet new people who become valuable contacts in your professional career. So maximize the opportunity. Group work is not glamorous, it is grueling, time consuming and tiresome. As a result, finger pointing could arise. Do you wish to lose a friend over something as trivial as this?

Recruit From Different Majors

One factor to consider is that any class project – regardless of the particular class – requires input from all functions. In Finance, you may be asked to develop a new security but a complete analysis would also entail the marketing of the security to the appropriate investor market.

A marketing project may entail the development of a new product to target an emerging market, but without a rigorous break-even analysis, the entire context is irrelevant. Remember the car that Homer Simpson designed at his long-lost brother's request in *The Simpsons*?

Each student brings a different expertise and will bring their class lessons and textbook know-how to the project. This completes the analysis and gives a more thorough, real life end product.
This also instills responsibility in other members and challenges them with a mandate, making each member a department head or VP.

This may pose a leadership conundrum, but the reality is that being leader is far more than a figurehead position.

Leadership: Assume A Role

It would be blasphemous to discuss anything on leadership without first providing a definition in the words of Dr. Warren Bennis, Distinguished Professor of Business Administration at the University of Southern California and the founding chairman of the school's Leadership Institute. Dr. Bennis has written 18 books including *On Becoming a Leader*, *Why Leaders Can't Lead*, and *The Unreality Industry* (with Ian Mitroff). Dr. Bennis' definition of leadership comes down to: "the capacity to create a compelling vision and translate it into action and sustain it."

These words do little justice to a lifetime of work. Nonetheless, in any team, individuals are called upon to play a role to achieve the common goal. A prime example is in the movie *Gladiator*. When Russell Crowe was first thrown into the role of a gladiator, he did not stand up and proclaim himself to be the best. What he did was check his previous accomplishments at the door and proceed to fight, defend and kill the others to emerge as the uncontested victor.

His actions paved the way for his ultimate role as leader. What do you think would have happened had he said that he was the best fighter? Someone who viewed him as a threat would have killed him. What happened instead was that he earned the respect of his peers.

Green Bay Packer coach Vince Lombardi put it best: "Leadership is based on a spiritual quality; the power to inspire, the power to inspire others to follow. Having the capacity to lead is not enough. The leader must be willing to use it. Leaders are made they are not born. They are made by hard effort, which is the price which all of us must pay to achieve any goal that is worthwhile."

International Flavor

Many students tend to shun the foreigner with the accent because they fear that their oral presentation may suffer and that cultural differences may cause problems. This is a mistake because international students bring a global perspective. Intel's Craig Barrett said it best when he said: "The US represents 4% of the world, I will take the other 96%."

Students from your neck of the woods may also overlook emerging markets while those from abroad seldom do. Jack Welch's book, *Straight From The Gut* emphasizes that all businesses should develop a thorough strategy for China. While China's enormous size makes it an obvious pick as an interesting market, it took Welch many trips and business deals to come to this conclusion.

Whether the international student is on an exchange program from INSEAD in France, the London School of Economics or CEIBS in China, their addition can make the difference in a group project.

Gender Mix

If possible, try to have an even composition of males and females. The business world has become a diverse arena with women in positions of authority. They have emerged as formidable partners and competitors.

Hewlett-Packard's Carly Fiorina and Lucent's Patricia Russo are two of the more prominent women in CEO positions, but you need not be a CEO to be in charge, as many females you encounter will be. Meg Whitman is running one of the best firms to have emerged from the dot com economy at eBay.

In fact, MBA programs are going out of their way to extend an olive branch to females. Not only will this improve class dynamic, but it should also improve the corporate talent pool

On a lighter note, men should be careful about choosing a group based on the women. This strategy can backfire – but that is for another book.

Earn The Votes

The reason to include international students and women is Machiavellian but obvious. Half of the world is made up of women and the majority of people hail from outside your native country. Business school is no different so strive to cull support from the collective. The fact that women and international students have to work very hard does not hurt either. If your group and future company are a reflection of the real world, you will gain the advantage.

Gameplan

Winning

The venerable "Conductor of Madness" Ozzy Osbourne once boasted "it's not how you play the game, it's if you win or lose." Ozzy may not have been the clearest thinking fellow but those words should stick with you if popular convention states the opposite.

Competition

School is about learning while business school is about winning. This is why competitive people migrate to corporate life. While learning is part of the process, the goal is to come out on top. If mediocrity and complacency were the norm, we would all become government employees.

On Your Marks...

Victory can mean a pass for one student and a high grade for another. What is key to remember is that on day one, all group members are on the same level. Some are better students than others, some are faster, some may not care about school and a few will not even be there when the group meets. Each member will go through a transition between day one and presentation day.

Understanding Each Member

The sooner you realize the challenge inherent in managing people, the easier group work will be. If you rush to split up the work and do not take into consideration strengths and weaknesses, you are setting yourself up for failure. The difficult but rewarding part about group work is getting more out of students that a) thought they had all the answers; b) did not care; c) were lazy or d) all of the above.

The implicit leader of the group must play the role of arbitrator and coach. The last thing he or she should do is let the ego run amok. This is when personal issues arise.

Chicago Bulls and Los Angeles Lakers coach Phil Jackson was a master at managing the greatest player of all time, Michael Jordan. But he also had to manage him within a system that included stars Scottie Pippen and Dennis Rodman. He subsequently presided over a dominant Lakers team with stars Shaquille O'Neal and Kobe Bryant.

Few managers in professional sports understand the dynamics of teamwork better than Joe Torre of the New York Yankees. Torre has led the Bronx Bombers to World Series Championships by understanding how to manage a millionaire's ego. At one point or another, Torre has managed Roger Clemens, David Wells, Derek Jeter, Darryl Strawberry, Paul O'Neil, Orlando Hernandez, Andy Pettitte, John Wetteland, Mariano Rivera, Chuck Knoblauch, Tino Martinez, David Cone as well as numerous other all-stars.

Torre took a page out of Vince Lombardi's play book. The great Green Bay Packer' legend said that: "Individual commitment to a group effort – that is what makes a team work, a company work, a society work, a civilization work. Teamwork is what the Green Bay Packers were all about. They didn't do it for individual glory. They did it because they loved one another... The achievements of an organization are the results of the combined effort of each individual. People who work together will win, whether it be against complex football defenses, or the problems of modern society. "

Level Playing Ground

Students cannot rely on their track records when groups are formed. You must go into a group project willing to work and prove yourself for the first time. This is the challenge.

Basketball's best player ever, Michael Jordan, did everything imaginable as a star guard for the National Basketball Association's (NBA) Chicago Bulls. After his "final" retirement, he did the unimaginable and returned to play for the team he co-owned, the Washington Wizards. His partner in Washington, Ted Leonsis was once asked what the dynamic was like when he had a conflicting view with his "Airness." Leonsis went on about their exchanges, but Jordan had the final word: "At some point he just puts up his two hands, showing off the 6 rings." Enough said.

You can put up either hands or one finger; chances are that you will have no championship rings to show off. So be diplomatic and let your performance do the talking.

Motivation

The next stumbling block is ensuring that each group member ends up tackling a task that interests him or her. Or at the very least, that they can handle. Group work has both sexy and mundane aspects. The former should be assigned to the most bored students. From then on, you can determine who the more eager, ambitious and driven students are, and then determine what gets their juices flowing.

Cheerlead

Whatever you do, be positive and encourage others. There is nothing more valuable than a pat on the back when the going is tough. Make people feel like stars when they are doing well, but also make them feel confident when they are having some trouble.

Teamwork

Chances are that you will end up with a lame topic to address, but hey, at least everyone is happy and the division of labor is done. Maybe you would have liked to go out as the star quarterback but find yourself blocking for the running back, but this is trivial in hindsight.

By being able to motivate others, you end up honing your skills and become a complete manager, in terms of business acumen and people management. It is unreasonable to expect the least motivated students to give a hoot if you give them the boring parts and you demand to take on the more interesting topics. By volunteering for the mundane subjects, you emerge as unselfish and will end up having more clout when you make recommendations.

If your position is based on what benefits you most, you will emerge as nothing but a selfish dictator. Business colleagues and business school classmates want a leader who leads for the greater good. Everyone despises a despot.

This illicits a crucial point. In school, the spoils are often split. In business, they are not and this impacts motivation and teamwork substantially. General Electric's Welch was legendary for compensating his lieutenants with widely disparaging remuneration.

Individuality Versus Teamwork

Business school is rife with group projects. While Arts and Sciences students emphasize independent work, business students must learn and understand from the start that corporate success is a function of group work.

The *Fortune 500* consists of huge enterprises with thousands – sometimes hundreds of thousands – of employees. Even smaller firms have dozens of employees who interact with one another to attain a common objective.

Before tackling the issue of teamwork, it is important to note that one function within business stands out. Finance is perhaps the one area where students work alone more so than in teams. The end goal is to beat the average and make money. This is how fund managers and brokers are measured.

Decision Sciences & Quantitative Analysis employees spend considerable time on the field interviewing employees, studying operations and analyzing processes. While much of their analysis may be on an individual basis, a bulk of the work will be with fellow employees that they may have never met. This once again shows the importance of tact and diplomacy in communication.

Marketing is all about developing an abstract idea that may be intangible and hard to grasp. It takes a group to nurture and work on it in order to make it into a reality. Only then will it be seen, heard, felt, touched and ultimately (and hopefully) bought. Effective marketing starts from data collection to planning so that at the end of the chain, someone can pitch an idea. If a single person fails, the sale does not get completed. While it is true that sales is about individual accomplishment with quotas to meet and commissions to earn, you need a group to develop a product first so you can sell it. Even when the product is sold or service rendered, customer service and technical support are key to ensure a repeat customer.

Accountants may spend much time by themselves, but since all of their clients will be other people within the company (departments, shareholders or board members), they too are team players who must abide by Generally Accepted Accounting Principles (GAAP). There is little individuality in that endeavor. In fact, while it is true that some of the most sought after accountants stick out of the masses with their ingenuity, creativity and interpretation of GAAP, accountants remain a uniform body. After all, uniformity, consistency and ubiquity seldom lead to individuality.

Management is all about the recruiting, management and promotion of human capital. Managers work best when the team and company are placed ahead of themselves. This relates to school advisors being unselfish and fair with all students.

MIS may encourage technologically inclined students to be independent in their research and work, but unless their ideas, solutions and applications answer the needs of the greater good, they are useless. Moreover, technical minds need to communicate their findings and applications to the greater society.

We All Make It Or Nobody...

While you may wish that your final grade be a reflection of your understanding and productivity solely, fact of the matter is that your work will only represent on average, half of your final grade. Your final grade, the one on which your job opportunities will rely on (admittedly unfairly) will be equally determined by how well you interact, manage and work in a group setting. Ah yes, the group setting.

Team Sports Versus Individual Sports

Whether you prefer team sports such as basketball or individual competition like track and field, you want to win. For better or worse, in team sports, you can be the best darn player on a field and walk away a loser. Conversely, you can be a role player that gets no recognition yet walk away a winner.

We Win Together, We Lose Together

Imagine being in Roger Clemens' shoes as a member of the Boston Red Sox 1986 team that was an out away from winning the World Series Championship and breaking the Curse of the Bambino. How would you feel to see a ball slipping through Bill Buckner's feet alter the course of history and hand the championship to a New York City franchise? Well, you should realize that you only really got there thanks to the team and that you learn through failure. Ironically, Clemens won a couple of World Series with that other baseball team from the Big Apple.

Give and Take

Wayne Gretzky retired as the greatest hockey player of all time. His 4 Stanley Cups were surpassed by many other players but his 61 National Hockey League (NHL) records were second to none. Gretzky was never the fastest or biggest player on the ice, he was the best for two reasons: First, he anticipated the play better than anyone else. He always said that he knew where the puck was going and that is what explained his success. Second, Gretzky set scoring records but it was his ability to make his teammates into better players that was his trademark. He often earned more assists than others notched total points. He drew defenders and opened up the ice for linemates. This was what elevated his career to the next level. This is also why sponsors rushed to associate themselves with #99 – his Presence drew the crowd – both on the ice and in the corporate landscape.

If you wish to be the greatest, learn from Gretzky and understand that it is how you make others look around you that matters most. In other words, you have to pass the puck if you want others to pass it to you.

Some athletes opt for individual sports where they control all the variables. Sprinters are the quintessential examples. They block everything out and win and lose on their own merit. The weather and track may have something to do with their performance, but if they win, it is because they are the best. But even sprinters need to give and take, be it with coaches or with trainers.

The problem is that this is an all or nothing course of action. An injury to the sprinter means the end. In team sports, any one player can go down – even Wayne Gretzky – and the team still has a shot at victory.

The Sum Is Greater Than The Individual Parts

The greatest lesson that sports teaches you is that Gestalt psychology truly does make sense. What is Gestalt psychology?

Gestalt psychology is based on the belief that the whole is greater than the sum of its parts (Gestalt is German for "form" or "shape").

A good example is the St-Louis Rams of the National Football League (NFL). Here is a team filled with talent. Quarterback Kurt Warner is a two-time MVP. Running back Marshall Faulk was the best Offensive Player of the Year two years in a row. Receivers Isaac Bruce is a Pro Bowler. While every single one of these players has a legitimate reason to ask for the ball, none of them do. They all trust that the quarterback will do what is best for the team in order to win.

The players are not focused on getting the ball more often; they want the offense – as a unit (of the Greatest Show on Earth as some dub it) – to put up the numbers necessary to win games.

Of course, it is much easier to say that you are a team player than it is to be one.

Think of the 1993 #1 draft pick by the New England Patriots, quarterback Drew Bledsoe. Four years after being drafted with the top pick out of Washington State, Bledose led his team to Super Bowl XXXI, but lost to the Green Bay Packers 35-21. In 2000, Bledsoe was awarded a 10-year, $103 million contract. In the 2001 season, he was injured early in the season. Backup Tom Brady emerged as an unlikely star and Bledsoe ended up being the backup. Surely his ego was bruised and his pride shaken, but on the surface Bledsoe remained upbeat because the Pats were winning. And for better or worse, sports is about winning.

Bledsoe could have whined, complained and acted like the prima donna we have come to expect of professional athletes. Did he? Not at all. Bledsoe handled himself with such class that you could not help but feel happy for him when he had to come in to the AFC Championship game against the Pittsburgh Steelers. Cool, calm and collected, he studied at practices and paid attention in order to help the Pats win.

And win games the Pats did, making it to Super Bowl XXVI and meeting those same St-Louis Rams. Despite Bledsoe's efforts in the AFC Championship game, Brady was chosen to start in the Super Bowl. In an upset for the ages, Brady emerged as the MVP and the Pats rode to victory thanks to a last second field goal by place kicker Adam Vinatieri.

Despite the loss, the dominant Rams team remained intact and resolute. The unselfish core of star players wanted to be the best at their position. This is a great lesson for all. At various points in group projects, business settings and personal predicaments, you will be cast in a different role. Sometimes, you may be coach. Other times you may be asked to be the waterboy.

What is important for you is to be the best darn waterboy in the history of H2O. After all, playing *The Waterboy* worked wonders for *Saturday Night Live* alumni Adam Sandler. If you do that, you will help your team win and be entrusted with the ball on Super Bowl Sunday (or at least be the waterboy of the Super Bowl Champions).

John Madden, the former Oakland Raiders coach and current football announcer, has a tendency of referring to practically every player he describes as "the best at his position." Is every single player he refers to really the best at his position?

Probably not, but the lesson to be learned is that you should strive to be the best at *your* position. No one is the best at everything, but they may be the best at one specific thing. If you set your sights on being everything to everyone, someone with more focus and attention to detail will emerge and steal your starting position.

The Rams example is a simple extension of the central argument in Plato's Principle of Specialization, as described in *The Republic*.

Plato argued that for society to work in harmony, people should do what they are best at so that the entire society can accomplish more. The St-Louis Rams players were arguably the best at their position. None of them wanted to do the job of one another, they showed up to work, did their thing, may have even wanted the ball more but never asked for it. Why? Because calling the plays was the coach's job, the players wanted to execute them.

Whose Team Are You On Anyways?

There are not too many waterboys in your class or your group. Rather, you will have aspiring Kurt Warner's waiting to toss the ball on Super Bowl Sunday. The concept of team player will come up, so be careful about how it is used *against* you.

Yes, yes, it is good to be fair, humble and modest and put others ahead of yourself. But others will place themselves ahead of you as well, so someone may as well think of you, and that may as well be you! Huh? Read on.

Next time someone questions your commitment to the team, ask them which team they are referring to. For one, if they do not like that you are excelling and helping the team, it may be that they crave the spotlight. This does not make them a team player, so they should not be able to question *your* commitment.

A Necessary Evil

Group projects are a necessary evil, perhaps the single sharpest double-edged sword in business school. Business is all about team play, competition, planning and execution. What better way to teach this than lump students into groups?

The End Goal

Class projects usually consist of a paper and a presentation. This is consistent with the business world where a formal recommendation, proposal or pitch may be submitted as well. Depending on how many people are responsible for the verdict (as in, will all students grade your work or only the professor?), each component plays a role. Students make their decision based on the oral presentation while the professor relies on the paper as well. Hence why both are important.

In school, while only the professor will read the paper, students see the presentation. While a professor can have a short-term impact on your career, it is some of your fellow students who will play a greater role down the road.

Some professors make up their mind about the paper as a result of the presentation. To what extent this is true remains a function of the particular professor, but you can imagine why presentations exist in the first place in school: call it *Cliffs Notes* for Professors who do not wish to read a bunch of papers.

Celebrate

Whether you end up just passing or acing the project, there needs to be downtime afterwards. If for nothing else, you want to revert back from a stressful relationship to one that is cordial. After all, you never know whether members of the group will resurface as group mates, classmates (some of whom may eventually grade you), colleagues and even bosses. Head out to the school or local bar for a drink or two. This is just another example of balancing business with pleasure.

The Teacher, Coach, Quarterback And The Waterboy

Every group has an end goal. Some are attainable while others are not. Some face little obstacles, others face monster ones.

The way to approach any group setting is to use your skills as a judge of character and recognize talent in each person for the group to achieve great things.

An example of a team success despite considerable obstacles is the University of Miami Hurricanes football program. In 1995, the National Collegiate Athletic Association (NCAA) imposed severe sanctions on the school that would have killed other top football programs in the nation. Soon thereafter, the Hurricanes named Butch Davis as their coach. Miami Director Paul Dee knew that he had the man who could perhaps restore some of the program's lost luster.

Dee spoke highly of Davis: "Butch exemplifies the qualities we were seeking in our search for the complete coach. He demonstrates outstanding leadership ability, integrity, commitment to academics and is among the best recruiters in the country."

As confident as Dee was in Davis, one got the impression that Davis was equally confident in himself. Davis had been part of one of the greatest turnarounds in sports history by helping the Dallas Cowboys go from 1-15 in 1989 to winning back-to-back Super Bowl Championships in 1992 and 1993 as a coach on Jimmy Johnson's staff. But while Johnson and staff were blessed with young talent in quarterback Troy Aikman and a steady flow of draft picks from the Hershel Walker trade to the Minnesota Vikings, Davis was not to be so fortunate in Miami.

Miami had been ravaged by the NCAA sanctions. Davis saw the number of scholarships he could hand out to players drop by 31 over the first three years of his tenure. Scholarships offered to promising high school players allow teams to recruit the best talent to win at the Division I College Level.

Entering the season, Davis' goal was to return the program to the level of unprecedented success that it enjoyed in the 1980s and early 1990s. During that era, Miami spawned Championship teams, legendary college quarterbacks in Vinny Testaverde, Gino Torreta and future National Football League (NFL) gunslingers Jim Kelly and Bernie Kosar.

Incidentally, Davis was the defensive line coach of the 1987 Miami team that won the Championship. At the time, he coached an All-American roster of defensive linemen in Russell Maryland (#1 pick overall by the Dallas Cowboys), Bill Hawkins, Cortez Kennedy, Daniel Stubbs and the late Jerome Brown. Miami also produced linebacker Ken Norton and free safety Thomas Everett. While various coaches could have done well given the talent at Miami, Davis was the best one suited for the job. First, Davis thrived when challenged; a true test of a champion. He was a leader to his team who stayed upbeat even when things look bleak. He was also a visionary on and off the field.

He knew the importance of laying down the foundation for a solid football program. He convinced the Athletic Director to invest over $2 million dollars to bolster the Greentree Practice Field. He also understood the importance of doing things right from step one. Corporate heavyweight General Electric mastered this "basic" concept with the Six Sigma quality program. The process that Davis wanted to perfect was one of human development; the same focus that former General Electric Chairman Jack Welch emphasized throughout his tenure, which started in 1981.

Davis understood better than anyone else that building a Championship team on the field was only half the equation: Davis wanted to mold his recruits into all-star human beings off the field as well.

He fostered an environment where players volunteered in the community. He once said: "I want us to be successful, but I want kids to leave here with the sense that there's more important things to individual success than winning football games."

Davis saw the bigger picture and prided himself in building Miami back to the powerhouse level it once was. Sure the flattery was nice, but the results were what mattered to Butch. He made the players see that if they cast aside their egos, commit themselves to success both on and off the field, act responsibly and work as one, then anything would be possible.

This was the focus that Director Dee was hoping for when he recruited Davis. From there, he let Davis take over and the results spoke for themselves.

Miami won the Bowl Championship Series (BCS) in 2002 with Larry Coker as coach. But to some extent, it was an extension of what Davis had started. By then, Davis had been recruited as the National Football League's (NFL) Cleveland Browns coach.

Davis must have taken a cue from one of the greatest coaches in professional sports, Vince Lombardi, who said: "They call it coaching but it is teaching. You do not just tell them...you show them the reasons." Davis showed his players and his superiors just that.

CHAPTER 5 - STAYING AHEAD OF THE CURVE: GOING THE EXTRA MILE

Business Magazines: Breeze On Through To The Other Side

Strategy classes often look at popular firms at a crossroads and ask students to formulate a game plan. The problem is that in most instances, these cases are outdated. Imagine a group in the year 2000 being asked to examine Pepsi Co.'s strategy in the late 1990s! At the time, the company was dabbling in capital-intensive restaurants (KFC) and snack products (Frito-Lay) while trying to catch up to Coca-Cola. After years of trailing the most popular brand in the world, by the mid-1990s Pepsi Co. was contemplating what to do next. Now imagine if in 2000, a business class is asked to come up with a plan for Pepsi Co. from 1995 onwards and the choice came down to the status quo or shedding the capital-intensive divisions to focus on beverages. Some students would argue that the firm should keep the chicken, tacos and chips. Pepsi did this and its stock continued to fizzle and trail Coke. Years later, Pepsi finally shed those assets and focused on beverages. The stock recovered and analysts jumped on Pepsi Co. at the expense of Coca-Cola.

While the specifics of Pepsi Co.'s strategy are irrelevant, what is important is that anyone that would have been aware of what was happening in the industry in 2000 would have known what the right and wrong strategy was.

The lesson is that while schoolbooks are great and irreplaceable, the proof of concept can be found in business magazines. The hallowed halls of Academia may provide valuable lessons, but they are secondary to the hands-on experience you get on Wall Street, Madison Avenue and Main Street. Business magazines can teach you in the short and mid term – textbooks in the long term. But magazines and textbooks are like a recipe book; you can read it as often as you like but unless you try to cook yourself (and accidentally poison a couple of people along the way), the book is almost useless.

But beware: some textbooks and magazines like to permeate myths. Separating fact from fiction and reality from myth is necessary in order to come up with conclusions and make informed decisions.

Moreover, while business is not rocket science, some would like to make it seem like it is in order to justify stratospheric salaries. Care for an example? The Black-Scholes options pricing model was developed in 1973 by Fischer Black and Myron Scholes. Options are financial instruments called derivatives because their value is a function of another security, usually a stock. The duo's breakthrough led to the following put-call parity to help students make sense of options pricing, as well as anything else in life. After all, parity is all about balance. Let's see how it applies in finance and then relate it to sports business and how VPs and General Managers go about deciding when to trade a player.

Put-call parity states that:

$$\text{Call} + \text{Cash} / (1+\text{Rf})^t = \text{Put} + \text{Stock}$$

A call is an option giving the holder the right, but not the obligation to buy the underlying security at a given time at a predetermined price (in the case of European options, with American options, the investor can buy anytime before maturity).

A put is an option giving the holder the right, but not the obligation to sell the underlying security at a given time at a predetermined price (in the case of European options, with American options, the investor can sell anytime before maturity).

The denominator of $(1+\text{Rf})^t$ refers to discounting the present value to find what is needed today to eventually reach the strike price. The strike price (or exercise price) is the predetermined price.

Maturity is how long until the right expires.

Finally, the stock is a share of the company.

The Black Scholes pricing model is fairly sophisticated but put-call parity is fairly straightforward: imagine two investors, one is an optimist (Bull) and the other is a pessimist (Bear).

The Bull decides to buy the stock and ride the wave. But having taken a couple of finance classes, he realizes that he should probably hedge himself and buy the right to sell, locking in some protection in case something goes wrong.

The Bear figures that cash is king and decides to wait on the sidelines with good old fashion cash. But not wanting to miss out on the next Microsoft, he decides to invest in a call, giving himself the right to buy the security.

As you can see, these two investors have managed to take mirror positions. The Bull has the stock but can get rid of it. The Bear is hoarding cash but has given himself the right to buy. This much we agree, but why are these positions equal?

The value of the cash is obvious. The value of both the put and call are a function of the underlying security: the stock. If the stock rises, so will the right to buy (call). If the stock plunges, so will the call's value. Conversely, if the stock falls, the put's value rises but the call's value will sink. This mechanism must strike equilibrium; otherwise, there is the potential for a riskless profit, or arbitrage. However, there is no such thing as a riskless profit, the risk may be small, but it will still be present.

The final question is why is the cash discounted at the risk free rate, especially if there is no such thing as a riskless profit? Risk is synonymous with uncertainty. There is little uncertainty in cash, so that is one reason. Second, all of the uncertainty in the values of the put and the call are a direct function of the underlying security. This elimination of added risk means that a "common denominator" has been attained and all excess risk avoided. For this reason, the risk free rate can be used.

Now, how is this applied to the sports world? Well, consider the stock as an athlete. The call is the team's right to renew the athlete's contract, the put is their right to trade him away. The cash is the amount they will need to sign him when his contract expires (European option) or at anytime before (American option). They only need the present value of his signing amount since the cash will compound interest and reach the athlete's demanded salary (strike price). VPs and GMs use this formula implicitly and to a large extent, we all do too on a daily basis.

Title Descriptions: This Job's For You

Firms have specific commercial goals and they strive to accomplish these given their capital and human resources. Although specific titles vary, the firm's President or Chief Executive Officer (CEO) comes up with the goals and policy that will help the firm attain these. The following are some general descriptions to help better understand each person's role.

Chief Executive Officer (CEO)

A company's Chief Executive Officer plans, organizes, directs, controls and coordinates the operations, finances, infrastructure and administration of an entire organization as well as the interaction of its major departments, initiatives and programs. The CEO interacts with other senior executives to make sure that the firm's actions reinforce their quest to meet their goals. Depending on the size of the firm, the duties of the CEO range from managing day-to-day operations to overseeing the general interaction of a firm's various departments. The CEO is ultimately held accountable for the firm's financials, actions and results. The CEO reports to the Board of Directors.

In large firms, the CEO may handle the strategic planning and corporate development of the firm but he will recruit a lieutenant to execute the business plan of the firm.

Chairman of the Board

The Chairman of the Board is the main liaison between management and shareholders. Depending on ownership, power and the individuals, the CEO can at times also serve as the Chairman of the Board. In publicly traded firms, it is the Board of Directors that is ultimately responsible to shareholders, the community and all other stakeholders.

Chief Financial Officer (CFO)

The Chief Financial Officer sets the firm's financial goals and targets. The CFO controls operations and monitors results so that the company steers away form trouble and gains market leadership. CFOs work with marketers, researchers, developers and salespeople but the buck stops at their office. CFOs also oversee all of the firm's financing efforts to be used for the expansion of the firm and the execution of the firm's business plan. They also measure the performance of its investments and allocate capital throughout the firm. Most consider various merger and acquisition opportunities. Unless the firm has a separate general counsel, CFOs are responsible for most government affairs, ranging from financial reporting, audits, taxes as well as respecting industry standards.

Chief Information Officer (CIO) Or Chief Technology Officer (CTO)

Chief Information Officers or Chief Technology Officers oversee the technical issues of the firm. As technology plays a greater part in business and society, so does the role played by CIOs and CTOs. As a result, they must work closely with all to ensure that a firm's strategic plan is feasible from a technological perspective. They oversee operations issues such as procedures, budgets, hiring, procurement and all matters pertaining to administration, such as financial reporting and payroll. It is for this reason that the lines between technology head and CEO will become increasingly blurry in the future. If you are confident in your technical skills, you may wish to major in MIS and minor in Marketing or Finance for a fast pace rise to the top. CIOs work with CFOs and oversee equipment purchases, technology upgrades and new capital expenditures. As one can imagine, CIOs must be well aware of emerging trends and technological innovation to ensure that a company does not slip in the marketplace.

Chief Marketing Officer (CMO)

Not all firms have CMOs because there are two sides to Marketing at firms. The first one involves developing the marketing strategy in the marketplace vis-à-vis the brand and positioning. This entails coordinating advertising, public relations and corporate communications. The other one involves generating sales for the firm. Depending on the firm, there could be one person overseeing all of these duties. Often however, the areas are split between different lieutenants.

Exams

Essays

Read Question

To most this should sound obvious, but the first thing to do is to read the entire question. Make sure that you grasp the question being asked.

Highlight The Important

Words in a sentence can be put in a different sequence to yield an entirely different question. The key here is to highlight with your pen or pencil the key question in the problem, what the unknown variable is and if applicable, how you are asked to solve it. In case you need to move on to the next question, having highlighted the important things will facilitate your task when you get back to it (if you have time).

Outline

Once you are ready to get cracking, remain patient and visualize your response. Rambling on aimlessly will only frustrate the grader. Take a couple of minutes to break up your essay into a simple, powerful and concise outline that will be easy to develop. Add specific chapters, company examples and class lecture references as you write the outline. It may be the single best investment in your exam.

Blitz

Once you start to put down your ideas on paper, do not get cute. Get the job done, but work smart.

Reread

Wishful thinking perhaps, but it is always important to reread your answer.

Multiple-Choice

Time Per Question

At the expense of breaking into panic mode, divide the time you are allotted by the number of questions you will have to answer to get a feel of how much time you should spend on each.

Move On

A mistake many students make is to spend too much time on one question. Depending on how many questions you must answer in the time frame, keep in mind that partial marks are not up for grabs in multiple-choice exams. Thus time is critical. Spend a minute reading the question, highlight the important stuff as you go along and try to answer promptly. If this takes more than a couple of minutes, move on to the next one. Every second counts.

Plug It In, Plug It In

Imagine you are given a numerical problem and asked to solve it. Many can compute the problem but a shortcut is to simply plug in the choices. Only one alternative would yield an equation. If the numbers add up, you have found your answer.

The Ultimate Sales Pitch: Tips For A Stellar Presentation

Finalize The Pitch

Before presenting the project before a class, always practice numerous times to make sure that the presentation is smooth. Have other classmates or at least group mates ask questions in order to make sure that everyone is on the ball.

Time yourself. Some professors will think nothing of cutting you off at the precise time limit. Since conclusions are very important to make your case, make sure that you are well within the time restriction.

Understand Your Audience

Unlike professional meetings and industry conferences, class audiences often have a limited knowledge base of your topic. So do not get too smart and lose your listeners.

Content Is King!

What Is The Big Idea?

Every presentation should have a Big Idea. Everyone presenting should have it serve as the guiding light in the back of their mind. Every point that is discussed should unequivocally emphasize and reinforce this one principle.

Frame The Message

Do not beat around the bush. Once you have a big idea, you should ensure that the rhetoric does not fly all over the place. James Carville, the popular political advisor emphasized in his book that rhetoric was a main reason why Ronald Reagan won in 1980. He closed his debate by asking Americans whether they were in a better position in 1980 than in 1976, the year that Jimmy Carter took office. Americans felt that they were not. He framed his message masterfully and served two terms as one of the most popular leaders in American history.

Give Industry Examples

Another way to frame the message and hit the nail on the head is to offer industry examples and relate to previous references in lectures.

Set A Tone

If there is one thing that makes for bad presentations, it is when the desired tone is lost on the audience. Are you trying to be informative, funny, assertive or controversial?

Whatever you try to do, it should be clear in the tone of the pitch. Imagine you are going up there and questioning a fundamental theory in business: would you go about it the same way as presenting an ad campaign strategy? Of course not.

The Mechanics

Make The Person That Is Most Nervous Go First - Sometimes.

This could be a risky maneuver but it beats having the nervous student sit idle as the group members speak one by one. The better the performance the higher the stakes. The higher the stakes, the more nervous they will be. Besides, introducing a topic is far easier than driving home the main points or concluding a pitch. However, if you need a solid intro, then make the nervous person go second.

Being The First Group To Present?

Being the first group to present is the way to go when you are confident about everything and are about to do something that no one else has. This is the perfect way to set the bar so high that all following groups will offer seemingly diluted presentations with obvious conclusions. Having to follow up great presentations also throws the other groups off. Sure this is slightly Machiavellian, but so is business. Get used to it. The other reason to go first is when all projects more or less cover the same topic. You can avoid anyone accusing you of stealing ideas by going first. The flip side to going first is that others will have more time to prepare and can even get inspiration from your work. Usually, professors are well aware of this and grades and feedback take this into consideration.

Being the last to present means you have had more time to prepare and have been able to gather notes from the feedback that the first groups have gotten. Whether or not this is implied, it means that you must deliver a finished product that is almost twice as better as anyone else's. It also means that your findings will have to be deeper. Similarly to presenting first, the great thing about this is that if you have something up your sleeve, no one else can copy you thereafter since you are the last to go. The flip side is that someone may voluntarily or involuntarily steal your thunder if they pull the same stunt before you do.

What this question comes down to is your level of confidence and how well you will have prepared.

Presentation

PowerPoint Or Slides

PowerPoint is a great tool and sometimes a prerequisite, but slides are safer and provide students with far more control. Sometimes professors demand PowerPoint to see how well you manage the software. Conversely, some professors want little downtime between presentations so they opt for traditional overhead slides.

In individual presentations, PowerPoint is fine because control is in your hands and not other classmates. In group presentations, you need extra preparation and practice to time the PowerPoint presentation. The advantage is consistency in everything from the background to the font as well as the content layout.

Ask the teacher in private what they prefer. Giving them what they want is better anyway. In business, companies that ask for submissions, pitches and proposals usually provide specifications of what they want. Start early by finding out what that is (in marketing, potential clients ask for Request For Proposals, aka RFPs, that you submit in the hope of getting their business).

Whatever you choose to do, make sure that you get enough practice with PowerPoint as you will be expected to be able to use it in any job or setting once you get out of school.

Achtung Baby!

Information Overload

Unlike conference attendees, most students do not have the faintest idea what your presentation will discuss, so do not get cute and get in over your head. Keep it simple and stick to the facts.

Read My Lips: Engage The Audience

The key to successful oral presentations is to engage your audience as opposed to drubbing them with a monotonous tone that will put them to sleep.

The best professors are the ones that pull you into their lecture as opposed to those that push their rhetoric on you. You should do the same thing when the mike is in your hand. Talk to your audience as opposed to speaking over them. Communicate ideas, ask questions, use common examples, read their eyes and adjust your presentation on the fly. This may be tough for first time presenters, but with time, you will be able to adjust the tone, speed and even the message of your presentation to win over the audience. This is key because in business, you may be able to present key material but if you are going too fast or not fast enough, your message will get lost and your audience will lose interest.

Take Your Time

Unlike professional meetings and industry conferences, class presentations have audiences that are not very familiar with your topic. Rushing through your presentation will only do one thing: make your audience sit there and say "come again?"

In all likelihood they may even dose off as you ramble on. You must strike a perfect balance between addressing your audience as a panel of experts or a group of inexperienced amateurs. Professors often lose control of their classes because some students are downright lost while others feel that the professor is going too slowly. The speaker must act like a conductor of a symphony. Adjust the rhythm and pace to make sure that everything flows and keeps people attentive.

Yet when it comes time to answering questions, students feel like they are in the hot seat and often fudge answers. And speaking of hot seat, TV recently set adrift from shows playing music videos or sports to focus on reality programming and game shows. The reality was that the cost of playing music or sports was too steep. Reality programming such as *Survivor* or *Temptation Island* were primal or hedonistic. Game shows emerged as well. Two of them were *Who Wants To Be A Millionaire* and *The Weakest Link.* To show how taking your time makes a difference, it is interesting to note that on the former, host Regis Philbin would give the contestant as much time as possible and more often that not, the contestant would come to the right answer. On the latter, host Ann Robinson would either insult or ridicule contestants on the spot. As a result, many contestants would fudge seemingly easy answers. The lesson is to bide your time and answer at your own pace to avoid foolish mistakes.

Come Again? Repeat The Question Please

When you field questions from classmates and professors, the only rule is to make sure you know exactly what the question is. The number of times that students rush to answer and open a can of worms in the process is staggering. Questions are usually far more straightforward than one can imagine.

<u>The Final Product: A Killer Document</u>

Many of the criteria about selecting a topic, researching it thoroughly and getting feedback from professors apply here.

Presentation

A Cover To Judge A Book By

Presentation is very important in every paper that you hand in. Throughout your life, many have reminded you that not all books should be judged by their covers, yet teachers often forget this.

Your cover page must stand out. Either have the company logo or product image on it. You can even go the extra mile and present the paper as the final product in the appropriate packaging (in marketing classes especially). Make Management papers resemble censuses. Make IT assignments look like white papers. Something as simple as a different shade of color for the paper will make a big difference. If the paper is for a Finance class, make it look like an actual annual report, analyst recommendation or government proposal.

Presentation comes to many other factors though. In group projects, names should be alphabetized. Some rank them in order of input, others by student number. The latter makes sense but the former does not. If your professor wants a sense of who contributed the most, he will ask.

Consistency

Presentation also means consistency. Page numbers, headings and font must be consistent.

Eye Grabber

Whatever you do, ask yourself what would make the teacher notice it from the pile and give it a read while he browses the others. A solid presentation with out-of-the-box thinking could lead the professor to be curious about the contents of the paper.

Vicious Spiral

Some teachers may prefer that you not bind your paper. Many will ask that you staple it in order to keep it thin. Imagine a poor professor that must carry dozens of thick papers back and forth between the classroom, office and home. Ask your teacher about their preference.

Table Of Contents

The table of contents should be attractive. Be descriptive and add some interesting headlines.

Try to ask yourself what would jump out and grab your attention as a reader. Try to use terms that come up in classes and business such as "Flight To Quality" or "Conspicuous Consumption." Better yet, take something conventional and add a personal twist to it. The Federal Reserve Chairman Alan Greenspan once criticized the American investor for irrational exuberance. If your position is that stock prices are not irrational (Greenspan was suggesting that they were), use a title like "Rational Exuberance" and readers will know that you follow the market yet have a different take. Chances are that the grader will pay more attention to the rest of your paper.

Bibliography

The bibliography is one of the most important parts of a paper. While some disciplines like Psychology, History and Sociology rely on bibliographies more than Business, never underestimate the importance of backing up your claims in consistent, methodical fashion.

Business is for ambitious and bold individuals who think out-of-the-box but this does not mean that all of your ideas will be accepted at face vale. So back every idea, quote and fact with an appropriate source.

Index

It may also be a good idea to include an index of people, topics and events to help the teacher "frame" your presentation better. This shows that you laid out a game plan and then executed it. It also demonstrates that your ideas and findings are not just based on personal pontification.

Success Versus Failure

Perception

In professional sports, even though the final score is indisputable, there can be debate about whether or not the winner deserved to win. Even in victory, there may be naysayers who argue that a team benefited from poor officiating, the most famous example of this was in the 1986 World Cup of Soccer. Argentina defeated England on the controversial "Hand of God" goal by legend Diego Maradona.

In athletic competition, when that last whistle is blown, it is obvious who has won and who has lost. Business is not as black and white. Some firms may lead in market share, market value or sales, but are they "The Best?" If they were, why so many conflicting lists on most admired firms and best places to work. Because business runs over cycles, it is never easy to quantify success. For this reason, perception is everything.

It is sometimes more effective to make it seem like you have won than to actually have won. Whatever the prize may be.

Perception, hype, public relations, noise is everything in business and in life.

Define Success

Business is often about setting, communicating and managing expectations. Whether it is a company's results or their product development, mismanagement of expectations has cost top CEOs their jobs and senior executives promotions.

Defining success is of utmost importance in business and school as well. Whether you are in a group setting or working alone, ask yourself what is required to conquer the task at hand. Do you wish to score an A? Do you wish to pass? Or are you content to make it out of bed and make it to class?

There are two types of people: those who meet a stated goal and those who surpass them.

Baby Steps: "Measures Of Merit"

Thought it may be difficult to know what to expect from classes and professors, get in the habit of surpassing each and every goal that you set for yourself. But how should you go about setting goals?

Wesley Clark, the former NATO Supreme Commander who led forces during the conflict in Kosovo speaks often about the concept of "Measures of Merit." NATO then broke up the task at hand in numerous baby steps, each serving as a measure of success.

"Measures of Merit" is not too different than the notion of "Baby Steps." The concept spoofed in the film *What About Bob* with Bill Murray and Richard Dreyfuss is a staple of modern psychology. Countless people find success to be far more attainable when they break it up into baby steps. By doing so, the long journey toward an end goal is far easier to manage.

From Concept To Reality

Nobody said that creating something was easy. And even once the creation is complete, critical acclaim is no sure thing. At the age of 35, Vincent Van Gogh painted *The Night Café*, *Artist's Bedroom at Arles*, *Pink Peach Trees*, *Café Terrace at Night*, *Woman Reading a Novel*, *Garden at Arles* and many more in what is considered to be the greatest output in any one year by an artist. A poor man, he took his own life shortly thereafter, having sold only one painting.

There is a very fine line between success and failure. What separates the two?

Envision A Dream But Visualize The Outcome

To be successful in life you need vision and passion. You can build a better mousetrap but imagine a solution to capture mice and send them to Disneyland without having to harm them. Now that would get you some press.

Decide On A Course, Focus On The Goal, Formulate The Strategy, Establish Steps And Act

The idea is that to leave your mark in life, you have to think bold, think creative and do something that no one else has done. This is the golden rule in everything from the arts to product development. Once you have an idea in mind, you must then act.

One thing that will get you in trouble is a lack of focus. You have to choose one goal and set your sights on success. When you have done so you can try to execute on several fronts. Set a time line, some rough objectives and make the right decisions to attain your original goal.

Design, Implement, Create And Execute

After this, it is a matter of taking your concept from dream to reality. This is tricky as many ideas are beautiful in your head but impossible or quite ugly in reality.

The notion of taking an idea from concept to reality will come up in your business career but has been integral in your life as well. Some of you have had the "pleasure" of having worn braces. For most, this was a lengthy process. For one moment, forget how you felt and imagine what the orthodontist visualized as he sought to shape your teeth and jaw over time.

But the pace and specific trajectory of an idea is often hard to forecast. In the initial period, the greater picture is more important than individual steps. But as the date the braces were supposed to come off approach, specific points begin to emerge. If you thought the expression "pulling teeth" referred to a challenge, think again. "Moving teeth" is the greater challenge. This analogy is as key in school as it will be in business.

In school, all you are concerned about in your first term is to complete your degree. How long it will take and what stops you will make on the journey are secondary to your diploma. But as you get closer to finishing your degree, the specifics do matter, and they will even play a large role in your career.

In business, whether you pursue an entrepreneurial path or a corporate route, you will have to visualize an end product at first and gather the resources available. As you get closer to the end zone, you will have to make adjustments on the fly.

Setting Goals

Most parents push their children to do better in school, set goals and exceed them. If we promised a C but received a B-, they would be happier than if we promised an A but delivered a B+.

Do you wish to play it safe, or do you wish to get serious and blow people away?

Motivation

Motivation is an exercise in psychology where you must understand where people come from, what makes them tick and what will fire them up. Everyone has something that will arouse their passion - find out what that is and you hold the key to motivate them.

Respect

Respect is an important factor in life and business. Some work to pay the bills, others because they like to work. It comes down to respect: both from others and for oneself. If you want others to respect you, start by respecting others.

Appreciation

Thank everyone even when you do not need to. Showing appreciation will get more out of people in the long term. The waiter and hairdresser may be doing her job, but not saying thank you and leaving a tip will tempt them to spit in your salad and butcher your coiffe.

Not saying thank you gives others the impression that you take them for granted. You lose talent, lose accounts and ultimately lose your business. During the economic boom when firms had to beg to recruit talent, eTrade CEO Christos Cotsakis used to convince candidates about why eTrade was a great place to work. This kept him humble and down to earth. It also showed potential employees that top brass would appreciate them.

Over Achievers

Setting a goal for oneself and trying to beat it is a trait of a champion and leader. An overachiever is someone who surpasses expectations.

Imagine finding out that the class clown is mayor but the class president is now a municipal councilor. What would this mean to you and what would it mean to them?

Need To Have Positive Reinforcement

Some individuals need a pat on the back to motivate them. This is often the case with people who have a superior skill set (either creative or technical), but who lack confidence.

One advantage of outside approval is that it acts as a control mechanism. If you are working on a new product and management agrees that it will meet a need and sell at a profit, the positive feedback could serve as wind in your sail. If people are critical, then you may get dejected.

While this type of overachiever may accomplish great things, they will not usually break new ground. In other words, they may improve an application or compose great music within an existing genre, but they will not revolutionize life as we know it. A moderate risk taker by nature, they engage in careful and methodical planning.

Nice To Have Positive Reinforcement But No Need For It

The second type of overachiever is someone that appreciates the positive feedback but does not need it to execute their vision. These are people of great intellectual capacity that have confidence and charisma. While some may consider them to bc arrogant, such criticism is disregarded.

These people defy the conventional paradigm and think outside of the box. They may create little in terms of tangible goods or products but their outlook is different. As you can imagine, anyone with a bold version is a target for controversy.

Two men worth looking at in the computing industry are Steve Jobs and Michael Dell. Steve Jobs was the founder of Apple and developed both the Macintosh computer and the laser printer. In 1983, Jobs recruited Pepsi Co.'s John Sculley. A power struggle ensued and Jobs departed from the firm. He founded Pixar Studios and went on to produce the blockbuster films *Toy Story*, *Toy Story 2*, *A Bug's Life* and *Monsters Inc.* Not too bad for someone that started Apple out of his garage. When Jobs was not at Apple, the company was saddled with two inept chief executives, an ineffective Board of Directors, and a weak product line. It racked up $1.5 billion in losses in 1996 alone. Apple was suffering from a lack of focus when Jobs came back, giving himself a $1 salary and basing his compensation on company stock, which was at an all-time low.

He struck a deal with supposed enemy Bill Gates and Microsoft. While critics gave Jobs little "positive reinforcement," he knew what he was doing. Microsoft would invest $150 million in Apple and the two firms agreed to share some technological knowledge.

Apple slashed the product line and focused on 4 segments: the iMac (a huge success), the iBook laptop, the Cube (this was not successful) and a new iMac. This innovative form of thinking has earned Apple ardent supporters around the world. Such creative and original thoughts were also in two marketing campaigns: the famous 1984 "Big Brother" ad during the Super Bowl as well as the more recent "Think Different" campaign.

Michael Dell has often been beyond reproach. His firm has emerged as the standard in computer related services. The value of a Dell stock has appreciated roughly 50,000% over the 1990s. In 1999, the *Wall Street Journal* named Dell tops amongst total return to shareholders over the previous three, five and ten years. Both Dell and Jobs have proven to have tremendous intellectual, technical and creative capabilities. While Jobs thought outside of the box in terms of product development, Dell thought outside of the box in terms of the process and the business. But for the sake of this analysis, he has never jumped on bandwagons nor created something radically different (his computers were originally the hardware that stored Windows software and Intel chips). But it was this simplicity that made his company the leader in its field.

This nuance is also at the forefront of one of the great questions in Rock 'n' Roll music: who was better, Led Zeppelin or The Beatles (although many would also consider Rolling Stones and The Beatles as a good toss-up). Liverpool's Fab Four put out gems like *Revolver*, *Sgt. Pepper's Lonely Hearts Club Band* and dominated the charts in the 1960s. Led Zeppelin's self-titled first album was released in 1969 and contained blinders like "Dazed and Confused" and "Communication Breakdown". The album set the stage for what ensued in rock music, classic rock and heavy metal.

While many view The Beatles as the greatest band ever, some give that credit to Led Zeppelin, if for no other reason that their *genre* of music was out-of-the-box. This is not to say that The Beatles lacked creativity.

So, who was the better band? You can answer that if you can say who between Steve Jobs or Michael Dell is the better mind.

Praise

DJs are fascinating people. Not the Rick Dees or Howard Stern type but the Paul Oakenfold, Tiesto, Max Graham and John Digweed breed.

A popular DJ was once asked whether he preferred composing and producing his own music so that others could listen and play it or selecting and playing music for others to enjoy. The response was that he preferred the former.

A deeper consideration of his answer reveals how right it was. After all, repeating what others do is simple. So what if you are paying respect to the original producer by playing their music. What have you added to the industry?

What you should do is be the trendsetter. Compose the work that others will select, play and pay homage to.

Do You Want It?

Your mission, if you choose to accept it, is to play fair and to win. Someone once said that the "world was a tuxedo, but they felt like brown shoes." You should strive to be the bow tie or better yet, the tux. If you wake up feeling like the sock, you will end up stinking.

Be front and center. Even if you are shy and reserved, let your craft do the work. Be bold and crave success because it rarely lands in your lap. And even if it does, it may slip between your grasp like a fumble in the end zone.

Double Edged Sword: Risk Versus Return

When you enter your first job, you can choose to rise slowly or look for the elevator and take the express route. But if you do plan on going that way, make sure that you understand your audience and remember that those you report to will have become jaded.

Not All Risks Taken Lead To Great Results

The greater your wish to triumph, the higher the likelihood of defeat. Success does not come without a cost. Ask Alex Zanardi. Zanardi was a high-flying, speed-loving and thrill-seeking racecar driver. His initial stint in the Cart Series was very successful. He then crossed lines and moved to Formula 1 where he experienced moderate success. When he came back to Cart, he sought to get back everything that he once had.

In the 2001 campaign, everything changed. Fellow driver Alex Tagliani of the Players Forsythe racing team hit Zanardi's car. In a tragic accident, Zanardi lost both of his legs.

The more you want out of life, the more you must be willing to lose. Business is often similar. In the wake of the Enron debacle, J. Clifford Baxter, one of the loudest critics of Enron management during the swindle ended up taking his life in January 2002 after word of his involvement was publicized to the media.

Overcoming Setbacks

While we would like to think that everything we touch will turn to gold, it may all turn to dust. Former Coca-Cola CEO Doug Ivester was a brilliant CFO under the legendary Roberto C. Goizueta. When Ivester took over for the late Goizueta, everyone was expecting the dealmaker and numbers guy to continue in his footsteps. What happened was that Ivester spent 26 months at the helm and was replaced by Australian Doug Daft.

One reason why Ivester was replaced was that many things went wrong at once. Some say that he was the victim of a perfect storm: food poisoning case in Belgium and competition from Pepsi. But Ivester never called a time out after his first setback. He simply grew more arrogant.

Whenever you suffer a setback, check your pride and call timeout. Self-reflection will help you bounce back.

The industrious cyclist Lance Armstrong is one example of an athlete who has become synonymous with resilience and determination. Armstrong came back from testicular cancer and won several consecutive Tour de France races.

But if you thought his comeback was impressive, check this out. On December 31, 1984, Rick Allen, drummer for British band Def Leppard, lost his left arm. Allen got into a race on a slippery road and his car flew off the track, shredding his arm clean off. Despite the shock, pain and trauma, Allen learned to drum one-handed and he and Def Leppard came back with a vengeance with *Hysteria*, the

1987 release that sold over 10 million records in the US alone and over 15 million worldwide.

The National Hockey League's (NHL) star Mario Lemieux also came back from adversity. He was diagnosed with Hodgkin's Disease, a form of cancer, in 1993 and missed 62 games. When he returned, he set a blistering scoring pace and despite the lay-off, led the league in points for the season. Lemieux had to say goodbye to the game he loved without knowing if he would ever skate again. But when he came back, he seized the opportunity.

Another example of an unbelievable comeback belongs to National Football League (NFL) running back Garrison Hearst. Hearst was named the Associated Press NFL Comeback Player of the Year for the second time in his career following the 2002 season. This is not the kind of trophy one wants to win once let alone twice. But Hearst has persevered and his case is worth examining. After a devastating injury, Hearst went through two years of painful rehabilitation and kept the faith when no one thought he would play again. Students can learn many things from Hearst. While the setback was a physical one, very few people could overcome the psychological obstacle inherent in such an uphill battle.

Vince Lombardi stated that "it's not whether you get knocked down, it's whether you get up. Once you learn to quit, it becomes a habit. It's easy to have faith in yourself and have discipline when you're a winner, when you're number one. What you've got to have is faith and discipline when you're not yet a winner. Winning is not a sometime thing: it's an all the time thing. You don't win once in a while; you don't do the right thing once in a while; you do them right all the time. Winning is a habit. Unfortunately, so is losing."

Next time you think something is impossible think of the National Hockey League's (NHL) Saku Koivu did. After learning that he had cancer the NHL star was sidelined for most of the 2001-02 season, but came back in time for the playoffs and led his Montreal Canadiens like a champion.

In business, setbacks are not of a physical nature as they are in sports. They are a combination of creative and technological changes, competitor actions and overall market forces.

Business comebacks are feasible thanks to the following assessment.

Analyze The Process

What happened? Where did the process go wrong? If there was one thing, such as poor research, then the company, project or person can bounce back. After all, all you need to do is research better next time or make sure that the methodology is improved. However, if everything else was a failure, then it will take more to remedy the situation and may require drastic changes.

Assess The People

Be Tough

Who did what? Who were the players involved? How did each contribute to the results? This is not finger pointing but acknowledging that because people work differently and bring varied expertise, the output fluctuates. General Electric's Jack Welch was famous for remunerating people with different pay and equity packages. He also knew how and when to imply tough love. Jeffrey Immelt, who succeeded Welch in 2001, was ripped apart by Welch in private for missing numbers at their annual Boca Raton management meeting. Immelt recovered because he knew that he would be measured based on his work and not his relationship with Welch. The rest is history.

Be Fair

Be careful how you single out an individual for reproach. After a couple of years in Indianapolis, the National Football League's (NFL) Indianapolis Colts' coach Jim Mora was fired after a lackluster 6-10 record. Mora failed to build on the early success of star quarterback Peyton Manning and the rest of his team. Despite Manning, strong running back Edgerin James, wide receiver Marvin Harrison and a solid support cast, the Colts underachieved. Mora stated that he wanted to come back after the last game of the season. But he had made one big mistake during the year. After a horrendous game by the Colts, he pointed the finger at Manning, whose poor performance was matched by the entire team. This may not have been the main or single reason why Mora was dismissed, but it played some part. The other, larger reason was that Mora failed to assess his people. Owner Abe Pollin felt that his defensive coordinator was not the man for the job. Mora stood by his coach and both were asked to leave.

Temporary Or Permanent Setback?

In school, not all setbacks are identical. Those at the end of the term tend to be of greater importance. Mistakes made at the end of your degree are even more damaging, though not fatal. Imagine if you learn during the second lecture that a student in your group dropped out. Is this anything worth fussing over? Of course not, but imagine if on presentation day, the student with the disk for your presentation fails to show.

Some setbacks are temporary. Imagine working on a marketing paper and hoping for your preliminary research to demonstrate the viability of a new product. What do you do when results show that consumers would not pay for it? Do you panic? No, perhaps people were not informed of the benefits. In such a case, the product label and packaging can be altered to entice consumers to buy. A temporary setback such as this is a good thing in the long term.

Other setbacks are more permanent but in the long-term fade in importance. The airline industry suffered a mortal blow as a result of the 9/11 attacks. Imagine if a class was assigned to devise a marketing approach for an airline company to increase sales. Before 9/11, the group may have looked at increasing sales generated by business travel. This would have been a feasible approach but after 9/11 the reality was that the business flyer was the first one to cut back on airfare. Corporate spenders sought to shun spending on airfare and focused on teleconferencing. A group in business school would have had to take this into consideration and alter their recommendation if they were studying the case in the fall of 2001.

Other times, setbacks are permanent. A firm losing a top executive has to reshuffle the cards and hit the drawing board again. The best ones though do not miss a beat and walk to the beat of their own different drummer.

Was This The War Or Just A Battle?

Another question you should ask yourself is whether a setback represented a lost battle or a defeat in war? There is a big difference and how well you size up the situation will determine what remedy you prescribe.

Remain Positive

What Goes Up What Must Come Down

Yahoo!, the popular site that began as a directory and is now quite more is an enigma in business. After all, when was the last time you paid money to Yahoo? Some people have but they are few. For this reason, after the stock market began to emphasize revenue growth and sustained profitability, many questioned how long Yahoo! could survive as a going concern and whether it would have to use its cash hoard to make an acquisition or sell to a larger media firm.

As these questions lingered, CEO Tim Koogle retired and room was made was for former Hollywood studio chief Terry Semel. Independent of criticism, Semel maintained his cool demeanor. Many experienced Internet executives – treated as saints during the golden era but ridiculed when they could not show a profit – would have been great for Yahoo! But Semel was one of the few people to keep his composure when the market was falling beneath his feet.

What Is Down Must Come Up

If Yahoo! was a high flyer that came crashing down, Trent Dilfer was a high flyer who crashed, rose, then crashed again, only to make an ascent once more. Dilfer was the Super Bowl XXXV MVP in Tampa Bay who led the Baltimore Ravens to victory over the New York Giants. What made the victory and MVP title even sweeter was that the Tampa Bay Buccaneers - host city of the XXXV Super Bowl - drafted Dilfer.

The season before the Super Bowl triumph, Dilfer had been cast aside. His previous team, the Bucs, had favored the younger and more athletic Shaun King. When Dilfer was released, he signed as a backup to Tony Banks in Baltimore. By default, he was given a shot to run the offense and he rewarded Ravens management with a victory in the Super Bowl. Weeks later, the Ravens disposed of the MVP and signed Elvis Grbac. The more things change, the more they stay the same.

Afterwards, Dilfer was asked what his greatest asset was, and he answered that he had endured many failures but had overcome adversity. It is true what they say, you need to hit rock bottom to get back up.

Marathon Or Race

The shortest distance between two points is a straight line, but it is not always the most fun and interesting route. If the journey is a race, take the straight line but be advised that in business you should take the scenic route once in a while.

This applies both to business school and the corporate world. In school, some of you have a 1-year, 2-year, 3-year or even 4-year degree to complete. Does this mean that you have to complete it in the given timeframe? Not necessarily, but technically this is an endeavor with a predetermined budget and time deadline, so respecting them is key.

That being said, finishing on time and within budget does not guarantee success. Think of James Cameron's *Titanic*: his movie surpassed its budget and was completed later than anticipated yet it went on to set records at the box office. So good things do come to those who wait.

The Empire Strikes Back

If ever there were a case of an industry looking like it was down and out for the count, it would be the recording industry. The Recording Industry Association of America (RIAA) is essentially a cartel that represents the major labels, their interests and their investments.

At the turn of the millennium, Napster began to gain steam in the online music marketplace through a superior yet simple service that answered a clear need. Word of mouth advertising spread the music like wildfire and industry insiders found themselves on the outside looking in.

But Napster's viral effect was almost too good for its own good. It had grown from a couple hundred users to over 50 million worldwide by allowing people to share music files with one another – for free. The majority of you may have used this or another file sharing service and seen that it works wonders. Dr. Dre and Metallica may have even warned some of you to stop. Thanks to these artists and government intervention, the RIAA managed to shut down Napster. Even though alternative services managed to pop up, the RIAA gained momentum.

Despite owning the rights to the music catalogues, none of the major five labels has been able to do much in distributing these online. After all technology was not the barrier as Napster and other services like it proved. What was the barrier? Greed, arrogance, misunderstanding and a lack of customer service to start with. Ultimately it was because the labels did not go on the offensive. When Napster was around, the labels considered such rogue applications as the villain. They were willing to cut deals with one another and make concessions. But with Napster out of the way and a legal precedent to take out new services like Scour, Morpheus, Kazaa and WinMx, they began to think about seizing the opportunity. Even then, they were too slow to make a difference.

The key term in this debate is "consumer" and not "user." After all, people were accustomed to downloading tunes for free but were not going to the record labels to do so.

The labels had to encourage users to pay and become consumers. This entailed offering a superior service. Jupiter Media Metrix forecasted sales of online music (in the form of download services and CD purchases) to hit $5.5 billion by 2006. The stakes were high but the labels were bleeding from the initial shock of Napster. According to one record label, the growth of piracy contributed to a decline in overall music sales of between 5 and 10% in 2001. What did the RIAA do?

Instead of taking action, the association hedged its bets. Labels joined forces and developed their own services. But this strategy did not win over consumers. The problem was that the RIAA took a defensive strategy against Napster and with one another. Yes, Vivendi Universal and Sony backed Pressplay while Warner Music, BMG, and EMI backed MusicNet. But having segmented services was counteractive to enticing consumers. They failed to recognize that the war was two pronged, with a short-term conflict against piracy and a long-term conflict against other music labels. Incidentally, Bertelsmann bought out Napster in May 2002 for a mere $8 million even though its BMG Entertainment subsidiary was suing it. This demonstrated that Bertelsmann had two distinct strategies for the short and long-term.

When Sony came out with the Walkman, they may have encouraged competitors to produce their own version as well. Only once demand picks up do you go after your competitors. This is the key difference in building primary demand versus selective demand. Palm Computing licensed its operating system to archenemy Handspring because in the words of their management: "you have to give up a piece of the pie to increase its size."

While Palm and Sony were open-minded and saw the greater picture, the RIAA were close-minded. Palm passed knowledge to Handspring because it knew that it was the best way to benefit the entire industry. In business, winning takes offense. Defense makes you a great takeover target.

Grades

To most students grades represent the end goal. To be fair, they are often what goals are based on: we compute class averages to see where we stand against our peers. As we consider potential employers, we often have our dreams shattered when we realize that some have a minimum GPA requirement.

While all students should keep an eye on their grades, they should also understand that GPA means very little to most employers and will diminish in importance as you embark on your career. Goldman Sachs and McKinsey may only look at students with a certain minimum GPA but many top-notch employers do not. The reason behind the Goldman and McKinsey requirement is that they want students that can absorb knowledge like sponges and learn on the fly. They assume that a good student will make for a good employee. But it is this kind of "inside-the-box" thinking that makes such employers horrible for young professionals who wish to control their destiny and think out-of-the-box.

A successful career in business does exist beyond the perceived elite companies. The question comes down to whether you wish to be in a comfortable company having to sit silently and pay your dues or whether you prefer being seen as an equal who can contribute with the best of them.

Bottom of the barrel employees do nothing but flatten the beer. Those at the top give the beer the much needed crispness and element of freshness. Anything can make a beer go stale but very few things will make it taste good. If you believe that you are at the top of the mug then choose to be in control of your destiny.

Evaluation at work is one thing. Grades in school are another. There is a certain level of hypocrisy in grades and group work specifically. Weak students will rely on strong group mates to increase sagging individual exam grades. Better students will see their stellar individual exam grades be pulled down due to weaker group performances. Unfair? Well, so is life.

If you get excited about the challenge of getting that group performance to improve so it match your personal output, that says a lot about which path you might take and where you will end up. If the idea of seeing your results tarnished by others angers you, then another path may be best for you.

Vince Lombardi said it best: "You never win a game unless you beat the guy in front of you. The score on the board doesn't mean a thing. That's for the fans."

In school, grades are for professors, parents, potential employers, and yes, to some extent, for you. But what is really important for you is the performance and the effort. That is what everyone sees and remembers.

PART THREE: TEMPTATIONS, DISTRACTIONS AND THE OUTSIDE WORLD

CHAPTER 7: MUCH ADO ABOUT EVERYTHING

Balancing Work With School While Studying

We would all love to coast through our college years without worrying about paying the bills. Having money to enjoy life is not as easy as we would like it though. Money does not grow on trees after all and even if it did, trees do not grow in malls, bars and schools. So the reality for most is that college life is either partially or fully financed by concurrent employment.

Get Your Butt Out Of Bed

While it is not easy for all students to benefit from the best of both worlds, it is possible. Sure, sleeping a bit more is always good and coming straight home after class is less stressful, but juggling work and school is the single best tool in your overall development.

Waves Today, Smooth Sailing Tomorrow

For one, being able to pull everything off at such a young age will make your schedule after graduation seem simpler. While colleagues at work will be struggling and breaking a sweat to put in their workweek, you will be hitting cruise control once school is no longer on your mind.

Synergy Baby, Syner-What?

The main thing is to find a job that is complementary to your education. If you are a finance student, get a job at a bank, either in customer service or in credit and loans. The advantage of working in customer service is that it will give you a great look into areas like public relations – something that marketing students read about while you are mastering it. Working in credit and loans is an obvious plus for the aspiring financier as you familiarize yourself with risk analysis, portfolio management and bad loans. Credit shows the dark side of business: some clients will have trouble making ends meet. This is a valuable lesson in balancing the quest for profit with respecting people and treating everyone with courtesy. Credit or debt is often overlooked but in business, understanding a company's obligations goes a long way to explaining its health.

Marketing students can also benefit from jobs at banks but they may wish to get in the field and work as part of independent public relations firms or as a member of a corporate public relations department. Taking part in grassroots programs and hitting the streets provides students with healthy exercise and priceless experience. Going up to others and filling out surveys may seem tedious and pushy, but marketing research and consumer behavior are key areas in marketing so this will give you a taste of what you are getting yourself into. And if you are running out of job ideas, do not forget where you spent some time in your earlier years; places like malls and stores provide great retail experience in everything from inventory management, customer service and shelf space management.

Working in a bar is a great way to acquire street smarts. First, it gives you an opportunity to earn money while others are burning theirs. Second, it gives you free time during the week when you would be working at some other job during regular hours. Third, you get to work with liquor companies and see how they operate. Sponsorships, inventory management and deal making with these companies will prove valuable in various career paths.

All of these tips will not only help pay the bills, but create synergy with your future employment. They will also provide you with great reference material in class and exams as well. You may sometimes forget a specific theory but dropping a real life example will make the grader see that you know your stuff.

Textbooks are outdated for the most part. Their cases often outline what specific firms did years ago. Having experience helps put your finger on the pulse of the economy, specific sectors and individual companies as you breeze through school.

Doughboy In The House

Textbooks are sometimes out of date and provide cookie cutter theories or approaches to business. Your experience with the retail store, bar sponsorship or bank error will give you insight into real business, as opposed to what is being taught in classes. Work experience will stand out in an exam when others regurgitate what the teacher said or what the book preached. This is the kind of thinking that will serve you well in the workforce.

All In The Family

While some of you work to pay for school, others may work for pocket money. Some of you may actually be working to support family – either now or later on upon graduation.

A large group of you may be in school because your parents "convinced" you. Another group may have something to prove to your family. Whatever the case may be, it is always important to balance your education with your parents and siblings. This is as important as balancing your career with your significant other and your children.

Two athletes – one far more famous than the other – provide interesting perspectives for students.

When Barry Bonds broke Mark McGwire's record of 70 home runs in a season, his children were by his side. Incidentally, McGwire's own son was a major focus when he was traded from the Oakland Athletics to the St-Louis Cardinals. When Bonds was negotiating his contract after his record-breaking season, he was firm about what he wanted: stability for his family and fair market value so he could provide for them. One would think that the new home run king would state a World Series Championship, but he did not.

This is by no means a knock against the much-vilified outfielder and son of former San Francisco Giants great Bobby Bonds. It is a tribute to a man who realized that his family was first and foremost.

A much less known story is that of National Football League (NFL) running back Antowain Smith. Smith was a very skilled athlete who only began playing football in his late high school years. Smith earned all-district honors at Stanhope Elmore High in Alabama. A natural athlete who favored basketball early on, Smith received but turned down a scholarship offer from Auburn. Not only would few high school graduates turn down a Division I scholarship, even less would say no to the famed football program at Auburn that produced Bo Jackson. Smith took a job as a cloth dyer, earning $3.65 an hour and working the 3-to-11 shift. He did his best to help his cancer-stricken grandfather and grandmother, who needed dialysis due to kidney problems.

When his grandparents passed away, his high school coach referred him to the coach at East Mississippi Junior College. For a man who was tendered a scholarship to Auburn, his willingness to even consider this "lesser" opportunity spoke volumes about his maturity and perspective.

After a 40-yard dash and a vertical leap, Smith was offered a scholarship to East Mississippi Junior College. With that, his journey resumed on the football field and Smith – who sports a "In memory of John and Clara Smith" tattoo – fulfilled a promise he made to his grandmother to go to University.

Smith was drafted in the National Football League (NFL) and played four years with the Buffalo Bills. From there, Smith signed on with the rival New England Patriots and he walked away a winner in Super Bowl XXXVI.

Money

If ever there were a time when we wished that money would grow on trees, it would be in school. Because cash is short, banks target students with credit cards.

Credit Cards

Credit cards are everywhere on campus. They are also in malls, stadiums and pretty much anywhere else "you want to be." Even if we anticipate getting rejected, we seem to fill applications out to get that free hat, pen or mug. But despite thinking that we would not get approved, lo and behold, a credit card is granted through the mail shortly thereafter.

Credit cards are great as they allow some students to make ends meet: buy groceries and purchase much-needed books as exams get close. But credit cards represent a double-edged sword. Here are some questions any student (or adult for that matter) should ask before filling out that application.

Do You Need One?

As basic as this may sound, some students apply for cards when they do not need one. If you have a job or your parents send you money, a credit card may be good to manage your cash flow. But if your goal is to collect points, this is not a good strategy. If you like the insurance offered by most credit cards, keep in mind that regular bank debit cards offer the same thing. So use debit cards for most purchases and you will be covered.

A good reason to use credit cards is to keep track of your expenses. Do this if you have the discipline to budget your finances.

Another good reason to get credit cards is that one is required to make purchases online. The Web offers great discounts, comparative shopping and a wider variety than offline outlets. But transactions are made using a credit card, not a debit card. Unless you plan to call mom and dad so they can make basic online purchases for you, you should look into getting one.

Buyer Beware

You should keep in mind that despite never having signed for a purchase, the combination of your credit card number and the expiry date constitutes an authorization.

Why Do You Need One?

Managing cash flow is one of the main reasons why you need a card in college. Some of the benefits may be attractive as well. Unless you are disciplined and will not run up debt, then you may prefer to stay away. If you have ample cash in your account, then you may not need a card. You may perceive to have your credit limit on top of your cash to play with and this is a recipe for disaster. Nothing in life comes for free after all.

Call And Inquire

You may be offered credit cards from day one at school. The problem is that the people offering you cards are salesmen and not customer service representatives. CSRs are the most informed people on the rules and regulations. Keep in mind that banks send you a document with all the rules and regulations when you get a card and you are expected to know what they are. Signing an application without knowing all of the facts can be dangerous. At least call the bank in question if you are thinking of getting a card, speak to an agent and ask questions.

Get The Agent's Name

You should always get the name of the person you are speaking with. This applies to banks, governments, school and your personal affairs. Without making it sound like you distrust them, ask for their first and last name. If they refuse, at least get a phone extension or file number. This only makes them double check all the information and gives you some proof. Do not forget, when dealing with large institutions, the burden of proof falls on you!

Everything Should Be Documented

Keep all documents and related information on file. Open files for school, bank accounts and company interviews. It is even useful to have a one-page summary of all of the correspondence, paperwork and calls you make in order to keep track of developments and stay on top of your affairs.

What You Do Today Will Impact You Tomorrow

Students often think that whatever they do in school does not follow them to business life. While this may apply to your drunken antics, financial records often stay on your file for up to 7 years. So late payments, missed payments and other black marks may hurt you get a loan, mortgage or even employment down the road. This is why credit cards can be so vital yet dangerous.

What To Use Them For

Things like groceries and textbooks are fine – so long as you intend and are able to repay the amount sooner rather than later. Clothes are fine so long as you need them. You should pass on the excessive spending, so no lobster and Cristal for now. This said, you do live once, so treat yourself once in a while.

What Not To Use Them For

Who does not like to party? Everyone does but using plastic on booze is not a good idea. It is dangerous and will get out of control very quickly – for both your drinking and credit card balance. If you are seeing someone, you can buy the odd flowers or even pay for entire date, but do not go overboard and think that you can live like royalty. For one, you will be in debt fast. Second, you will give the wrong impression to your date.

Pay Everything Off

As a general rule, it is always good to pay off your debt. Warren Buffett once said that you never know who is swimming naked until the tide goes down. So leave your credit cards for underwear and cover up.

Pay At Least The Minimum Amount

You must always pay the minimum amount even if you cannot pay the entire balance. Missing the minimum payment is a signal that you are in financial distress. It goes on your permanent financial record. One missed payment is not the end of the world but a second should be avoided at all costs.

Banks often force credit card holders to pay either 3% or 5% of the amount owing. Some banks try to lure clients by saying that the minimum monthly payment is only 3%, instead of 5% of the amount owed. This seems attractive but all it means is that you can pay less now, but much more later on and incur far more interest.

Pay Something Every Month

The more you pay each month the better off you are. Get into the habit so that if you have a mortgage in the future, you will pay that off sooner rather than later as well.

Have The Right Card

Make sure that you are using the right card for you. Some may have flashy points programs but they charge you more interest. If you do not carry a balance and do not take out cash advances, you do not pay interest so this is irrelevant. If you do carry a balance, no point program will make up for the interest expense; so choose a low interest rate card.

Inquire about student cards that charge no annual fees, offer good discounts on relevant purchases and are more flexible.

Cash Advance – Step Away From The ATM

Cash advances are loans and thus begin to incur interest immediately. Banks also have complex interest charge formulas that apply until all charges are paid off, thus prolonging the gut wrenching interest paying period.

Are You Interested?

If you do not carry a balance, then you pay interest only if you take out cash advances. Once you start to pay off your loan, the interest accrues until you pay it all off.

Banks also reserve the right to change interest rates. They do have to advise you by mail though, so read everything you get.

Check Your Statement

Banks make mistakes and everything that happens to your account should be reflected on your bank statement. The first thing you should do is cross-reference your purchases and make sure that they are correct and indeed yours. If anything out of the ordinary appears, contact your bank immediately and ask them to investigate. Some bank employees may ask you to investigate on your own behalf with the merchant. This is wrong as it is a basic service in your package. Other times, you may see duplicate charges for the same charge. To their "credit," banks often resolve these matters to the client's satisfaction.

Don't Run With Scissors... Walk

A time will come when you have no choice but to realize that cards are not for you, at least not while in school. You may then wish to cut the card and let your institution know that you are not using it.

One More Thing

One thing that business students should keep in mind is that having a card means that you are building your credit history and rating. The better you nurture it, the easier it will be for you to get a business loan, credit line and mortgage down the road. So even if you cut the card, it may be prudent to leave the account open and pay off the balance over time.

Communicate With Bank

If you decide to close the account, make sure that you specify that you wish to close it on your terms. There is a big difference between closing an account at your request and having the bank close it for you. Get the agent's name and ask that they send you a letter confirming that it was closed at your request. Usually you can reopen accounts that have been closed by simply giving the bank a call. You can only do this within a short time period: usually three months. Accounts that are closed by the bank need to get approved again, so this means a new application and it will take longer. It could also mean that you get turned down.

Stress

Credit card debt is one of many things that stress students and adults alike. Stress is a fact of life; the challenge is to know how to respond when it hits.

Identify The Triggers

What makes you tick? What makes you lose your cool? Isolate the trigger by detaching yourself from the source.

Modify Your Reactions

If the first thing you do when feeling stressed is having a cigarette, then try to eat. If you usually eat, go for a walk. If you drink liquor, make yourself some green tea or grab a cup of coffee if you wish.

Body And Mind

Meditate, biofeedback, massage, exercise, aromatherapy, yoga, breathing techniques, aerobic exercise, working out – some will render you nauseous, others will not. Find out what works for you and just do them.

Support Network

Whether you need to vent, complain or scream, have a group of people that can lend a shoulder or spare a moment to listen. Do not be shy to let it all out. When water boils, steam comes out of the kettle: why should you keep it in?

Fraternities & Sororities

One definite support mechanism to help you cope with stress is fraternities and sororities. These will help you make considerable friends, meet interesting people and expand your network. It will also make you regret some exploits and pickle your liver.

Contacts And Connections

Fraternities and sororities are also helpful when it comes to finding employment once you graduate. Interviewers and recruiters that hail from a common "house" may give you a break and contact you.

You are after all a brother or sister. The problem is that you still have to back up the interview with your own merit.

For better or worse, the factors that determine whether students get the best jobs often have little to do with grades, vision, skill or execution. Connections work in business as they do in school. Landing that first job is important and fraternities and sororities can definitely help.

Come Here Brotha!

As students try to juggle work, time for friends, family and school becomes scarce. The result is fraternities and sororities have lost some of their cachet. For this reason, some go to extreme lengths to attract potential pledges.

Sisters Come Correct...

As in relationships, never try too hard to join a fraternity and sorority. Show some interest but make them pitch to you. Some fraternities and sororities have wonderful reputations and serve as a gateway to fun in college, but fun can be had anywhere at school. Furthermore, remaining a free agent allows you to socialize with everyone at your own leisure instead of being forced to go along with others.

What Have You Done For Me Lately?

How long have they been around and what is their mission? What do others say about them? What are their academic, social and athletic records?

Fraternities and sororities should offer you something to satisfy all of school's benefits. They should be involved in the community. Also, find out which fraternities are involved with which sororities to get an idea of who you will be crossing paths with.

Hazing

Have you heard about hazing? Surely you have no intention of eating manure, sleeping in pickle juice and doing other crazy things. So do some research: if the hassles do not justify the benefits, chances are that you will fare better making friends from all walks of life and not just confining yourself to one fraternity or sorority.

Where Are They Now?

The main thing is finding out where some previous members have ended up today. Will you have sisters in high places? Have any previous brothers become famous or successful? If former members are now serving on death row, perhaps you should look elsewhere.

Fraternities and sororities help students come out of school with solid leads for jobs, great contacts for interviews and even potential partners in business. This is especially important if you decide to relocate to a new city and need contacts to get a new job.

Size Matters

Schools in large cities have so much to offer that joining a fraternity or sorority may not make sense. Imagine wanting to visit many places in such a short time but having to follow a schedule imposed on you by others.

In some small towns, there is not all that much to do for students. In these instances, joining a fraternity or sorority is great as it provides for a group to interact with.

One final consideration: how shy and reserved are you? If you are very shy and live in a shell, you will have to open up at some point. Fraternities and sororities are perfect for this. If you have too much personality though, no frat house will be large enough to contain it. So if you have no difficulty making friends and having fun, you may be better off as a free agent.

Sports & Athletics

Get involved in athletics while in school. Not only does it build leadership but it also refines your teamwork skills and brews a desire to win. After all, business is about winning whether we wish to admit it or not.

Play To Win

Psychologists at Case Western Reserve University in Cleveland, Ohio conducted a study that aimed to explain the behavior of sports stars. They found that people who see themselves as superior to others may only perform well in competition if they know there is a good chance of gaining glory for themselves. In other words, the findings suggest that in situations where such people have the opportunity to outshine others and obtain self-gratification, they respond very well and their performance levels increase. But if there is little or no chance of winning, they perform poorly.

"Narcissists may outperform others when there is a good opportunity to garner the admiration of others," said researcher Dr. Roy Baumeister. "In contrast, when little is at stake, narcissists may withdraw effort and perform worse than others."

Suffice to say, this is the opposite of what you should do. In life, if you only strive to win and excel if there is a material gain at stake, you become complacent, greedy and apathetic. In fact, you become a donkey who is searching for the proverbial carrot to move his butt.

Former Florida Gators coach Steve Spurrier said it best. When he announced his retirement from the University of Florida Gators in 2002, he left on his own terms and offered the gathering of newspaper reporters the usual mix of tall tales, quick-witted one-liners and the usual Spurrier swagger.

He conceded that his critics were probably expecting him to "cry because that is all those Florida State University Seminoles say" he does. But Spurrier was anything but a crybaby by most standards. He took a team that was going nowhere and made it an instant winner after coaching Duke for three years. The Heisman Trophy winner in 1966, Spurrier was considered arrogant and cocky. He was known for running up the score and speaking his outspoken mind when he should not have.

Hate To Lose

But speak his mind he did upon his resignation: "You can call me arrogant, cocky, crybaby, whiner or whatever names you like, as long as you do not call me a 'loser'." This was one of his more popular sayings, and it perfectly explained how the former Heisman Trophy winner turned things around.

Think Original

It also showed that winning means you have to be original and unconventional. In sports, they say that defense wins Championships. In business, it takes offense. Spurrier was all about unconventional offense schemes. He perfected electrifying double reverses, the questionable punting on 3rd downs and the crazy Emory & Henry formation in his "Fun 'N' Gun" offense.

It definitely was not traditional play calling but Spurrier wrote his own textbook, as should you. Spurrier made the Gator play book an exciting one in a world of boring copycats and unoriginal coaching.

If you remember anything Spurrier said, it should be this: "if people like you too much, it's probably because they're beating you."

You're An Athlete And A Scholar

Another reason why athletics should be at the forefront of your mind is that it can help you earn a scholarship to help offset the sky-high cost of education. Athletic programs offer extensive scholarships to promising high schoolers, but even if you missed the boat then, it is never too late. Some of the most successful stories are about "walk-ons," so to speak.

Contacts

Win or lose, you can get a good sense of someone's drive to win, their level of intensity and their work ethic by how they perform in sports. Such trials and tribulations also make for good introductions down the road when you are on either side of the recruiting process.

Military Education

Some of you may be interested in serving your country and getting an education at the same time. This is a great way to earn a solid degree, build your character and do so in a smart financial way. This is not for everyone of course, so make sure you know what you are getting yourself into. Boxing great Joe Louis, baseball legend Ted Williams and Academy Award-winning director Oliver Stone are three examples of men who served in the military. Louis served in the Army during World War II and even fought and defeated German boxer Max Schmelling in a symbolic bout. Williams twice interrupted his career with the Boston Red Sox to serve as a pilot in the United States Marine Corps – serving in both World War II and the Korean War. Finally, Stone served in Vietnam. The experiences formed the foundation for his trilogy of films: *Platoon* (1986), *Born on the Fourth of July* (1989) and *Heaven and Earth* (1993).

World

Find out as fast as you can about student exchange programs. These are usually offered to international business students but even if you minor in languages, you may have some great adventures in your future. It is always advised to combine school with your interests. So many young people take time off from school to travel. In this way, you can kill two birds with one stone.

Another thing you should do is make contact with students abroad. Whether you meet these people in chat rooms or as exchange students, they will serve as great contacts down the road in both your professional and personal development.

Ask your professors about contacts they may know elsewhere. This is great if you wish to pursue your education abroad, if you need information for a project or even if you wish to move abroad for work after you graduate.

World Wide Web

Indeed, the World Wide Web is the greatest and fastest growing library ever. Whatever you thought was impossible, it is now within reach thanks to the computer first, and the WWW second.

Webisize Your Class

All of your classes will entail some interaction with the Web. What you should do from day one is integrate the Web into everything that you do.

If You Build It, They Will Come…

One thing that all students should undertake is to create a personal website while in school. Many of you will either have to or will choose to do so in a class, but not all programs mandate it. And a Geocities or Tripod site does not cut it after you graduate, but it is fine while you are enrolled in school.

Not only will building a website improve your technical skills; it will encourage you to continuously improve your products and services over the years. The best websites are works in progress, as are the best products and services.

Early In The Product Life Cycle

The Internet has moved from infancy to adolescence and is now growing into adulthood.

The Conception

The Web was a vision for an interconnected, global and ever-growing network over which information could flow. In the early days, the Web was not exactly the information superhighway that we have come to embrace but rather a mish mash of independent sites that were largely in the clutches of the Pentagon. After years of tinkering, the World Wide Web began to gain steam in the 1990s and crossed over to mainstream culture. To stay abreast of its development, follow its founder, Tim Berners-Lee at the World Wide Web Consortium, or online at www.w3.org.

The Infant

Many would consider the turning point of the Internet within mainstream culture as the day when Netscape went public. On that day, the company that served as the real "windows" to the world hit the capital markets. When founder Marc Andreesen woke up in the afternoon and noticed that the share price for a stock of Netscape had risen from $15 to $72, he knew that he was rich, but he also knew for good that the world as we knew it would never be the same.

What resulted from the fallout was a race to change the world and become a billionaire as fast as information could travel from New Zealand to New Mexico.

As more and more contestants entered this information race, some made fortunes while others lost some. Initially, no one really cared about what this creation would one day do or look like. Most were simply happy to see that their creation was alive and well.

The Toddler

Once the infant was crawling, more and more people wanted to choose the shoes, pick out the socks, sow the pants and emblazon their logos on the child's shirts. The problem was that all of this special attention was making the toddler to be one spoiled, rotten kid.

The Adolescent

By the time that the toddler had hit adolescence, something was starting to smell fishy. After all, this child felt that there was no such thing as "no." Accountability, responsibility and common sense were not part of the glossary either.

As this adolescent was happy basking in the oasis of pleasure, a shocking reality began to emerge. The parents had raised this child based on lies, borrowed money and hype. In their excitement for this child's potential, they totally disregarded what had made previous children turn out to be model citizens.

What made matters worse, far worse, was that this was not a typical child that was raised by two parents. This was a child that was indeed brought up by a village. A global army of mercenaries: computer programmers, engineers, financiers and marketers had been busy building the World Wide Web. Sadly, they did not fully consider the cost, time and energy that was put in and the added resources required to sustain it.

When the child was awakened to the blunt reality, the house was foreclosed, the car repossessed and the espresso machine auctioned off on eBay. To make matters worse, the child found out by visiting a famous site that was inspired by *Fast Company* magazine.

The Young Adult

Once the original group of people that had served as the parents had disappeared and returned to traditional lines of work, the adolescent had no choice but to turn a page and try to make ends meet. After all the worst was over.

The first step was building the infrastructure over which the Content of the Web would travel; laying the network that would enable Communities to prosper; and establish the systems that could enable Commerce.

The hits and misses that occurred are normal in times of radical discovery. From there, a flight to quality follows and the line is drawn once and for all between the winners and losers.

Where is this young adult going to go from here? Where do you want it to go?

Giving Back While In School

Business students are always reminded of the importance of respecting stakeholders, being socially responsible and giving back to the community. Academic institutions represent one of the most important stakeholders in your life.

Regardless of your personal experience, you should consider making a tangible contribution to the school where you earned your degree. You may have wanted that scholarship, letter of recommendation or grade and did not get it at one point or another, but carrying personal beefs is trivial. Whatever you accomplish in the future has something to do with what school gave you.

While it is good to give back in the future, it is more important to give now. The problem with charitable contributions is that they seldom find their way to those that need it most. For this reason it is good to give back so you see the politics involved and how your donation (be it in time or money) trickles down to who needs it most.

Also as you can imagine, most recruiters like to see that you gave back when your own resources were thin. Anyone can give when they have the money and time to spare, but those who give when they need it most are worth keeping an eye out for.

Giving back can take various forms. Sun co-founder Scott McNealy named his firm Sun Microsystems after the Stanford University Network, a tribute to the school that gave him his MBA.

Bill Gates has always maintained that he intends to give away the vast majority of his enormous wealth to charity. Not to be outdone, his rival at Oracle, Larry Ellison is also a big time contributor to life sciences, especially to the fight against cancer. In May 2002, the Bill & Melinda Gates Foundation announced it had several new stakes in companies, mainly in health related sectors.

Which brings up whether they spoke to Scott A. Jones. You may not know who he is, despite the fact that you have used voice mail or listened to an MP3 file. Scott A. Jones is the holder of many of the patents that will shape our lives in the decades to come. The companies that he has helped shape and grow have all contributed significantly to the technological landscape around the world. In 1986, Jones co-founded and chaired Boston Technology, Inc. until 1992. Employing his patented voice-messaging technology, Boston Technology, Inc. became very successful and later merged with Comverse Technologies, a Nasdaq-listed firm. The offspring of this merger was a multi-billion-dollar company whose products are used by the majority of telephone companies throughout the world. That was the 1980s. More recently, while the music labels are having a miserable time developing a standard for online music distribution, Jones has developed Gracenote, the technology that allows MP3 players to read information and operate in a more user-friendly fashion.

When asked what would follow the telephone, television, computer and World Wide Web, he answered after a brief pause, "life sciences generally and genomics specifically."

Despite all of this past success and vision throughout, Jones has also not forgotten his roots. He is Chairman of the Indiana Technology Partnership, Chairman of the Gazelle Fund, Chairman of GrowIndiana Media Ventures, and a director on the boards of the state-funded 21st Century Fund and the Central Indiana Corporate Partnership. Through such activities, he plans to establish Indiana as a nationally recognized high-technology region.

The lesson is that while it is important to give, you should also spread the wealth to those that did not play a role in your development. Canadian high-tech pioneer Ted Rogers has built quite an impressive media empire. He not only gave $25 million to the University of Toronto in 2000, he then gave an additional $12.5 to rival Ryerson University. Talk about spreading the wealth.

Students Associations

Recruiters also like students who get involved with student associations. But very much like contacts, business cards and grades, being in a student association is only beneficial if you maximize the opportunity. Many students fall in the trap of using student associations solely to their own advantage. They tend to use the contacts they make at companies for their own subsequent job search. Professor relationships are used to secure letters for this and that. Important information about opportunities and events are usually kept private between association members. Do you seek a career in politics or in business? If you chose the latter, share what you know.

Inequality and injustice are prevalent in society. If you wish to make a positive difference instead of dwelling on the bad and the ugly, then how well you use student associations as a vehicle will determine what you will accomplish in your life.

Please The Masses

What you must remember is that you represent all students, not just your friends and those you wish to impress. In business, they advise you to focus on a niche market as opposed to blasting a message to everyone. In politics, you need critical mass to emerge victorious.

Cater To The Niches

Organizing events that benefit a large number of students at once is hard to do. What you should do is segment different groups and find out how you can make each one happy. Organize events by interest, time and number of attendees. Invite those that the event is targeting but keep slots open for outsiders.

Goal

What are you trying to achieve? Make sure that you know what you are going after to ensure that you reach your goal.

Bring Companies

Academia is nice and dandy but your primary function should be to extend an olive branch to industry. If you are fortunate enough to go to a top business school, firms will come to you. But if you are in the majority, you will have to go after the top firms. Reaching firms will make you stand out as a good marketer.

Talent is omnipresent, not just at the top schools. After all, Duke University and the University of North Carolina may be elite basketball programs, but one of the most dominant players in the National Basketball Association (NBA), Shaquille O'Neal, hails from Louisiana State University. Other notable anomalies include Julius Erving, a graduate of the University of Massachusetts and Hakeem Olajuwon of the University of Houston.

Reach Outside

Do not be shy to promote your events to all students even if a certain group will be served. In order to make sure that these are most targeted, book enough variety at different times to make sure that everyone yields the greatest benefit from each event.

In some events, it may make sense to invite non-business students to participate in one way or another. There may be no need to invite everyone to everything, but here are some cases where opening up makes sense:

Invite Psychology students to Marketing events: they will understand consumer behavior very well and tell you when something will not work.

Invite Economics students to Finance events: how does what you say stand in the greater economic environment?

Invite Engineering and Computer Programming students to MIS events: is what you preach easily implemented in the real world?

Professors Available To All

Professors are often invited to noteworthy events but sadly, association members tend to hog their time to get their attention. As association members, you are already making a good impression on them; they will know where to find you, so let the others students find them when they are at events. Your unselfishness will not go unnoticed.

Follow The Demand, Ask Students Whom They Wish To Hear And Go Get Them

Student association members are either elected officials of academic institutions or self-appointed representatives. For these reasons, some students get disillusioned with the self-serving nature of some association members. What they should do is pass by association offices and inquire about what is planned for them and follow up. Association members have a mandate to serve students.

Finally, very much like fraternities and sororities, associations should meet student demands on academic, personal, social and community matters to fulfill their mandates.

Tutoring And Teacher's Assistants

Follow The Example

Whether you are having difficulty with a particular class or simply need to fine-tune some things going into an exam, tutors and teachers assistants are always present and available on campus. The larger the classroom, the greater the need to tap into the brains and experience of a tutor and see what advice, tips and guidance these have to offer.

In private, find out about their professors, which classes they recommend that you take and which ones you should avoid. How were their exams and how did they perform? How did they get into tutoring? What do they think of the faculty and the administration? Ask them how your school fares against the competition.

Lead By Example

Students that excel in their studies should also consider becoming tutors and or teachers' assistants. Tutoring offers valuable experience in public speaking, problem solving skills and effective communication. These are very important in business with board meetings, group discussions, labor meetings and every time that your audience is not on your side. Students do not care that you are not a professor; you are there so they will test you. This will test how agile you are on your feet in front of an audience.

Some of you may like it for the entertainment. Consider tutoring business like the ultimate comedy hour: Chris Rock meets Warren Buffett.

You will be prompted to share your know-how when you start to work. Doing so in a simple manner like tutorials is great batting practice for the real thing. There is nothing more frustrating in a class when you have an employee who does not share information. There is also nothing more ineffective than confusing others or wasting their time. Tutoring teaches you how to retain audience attention, share key data and do so in an effective timeline.

"Your" students will constitute potential colleagues, employees, rivals and even bosses. Imagine if one of your students turns out to become a venture capitalist that may determine whether or not your entrepreneurial endeavor will receive financing. Someone you help today may end up helping you make your dreams come true. Your former students will remember your unselfishness, smarts and showmanship.

The other side of tutoring is that you forge strong ties with faculty and get a sense of the management (and mismanagement) in education.

This is also when you get a sneak preview of the politics in an institution. You see what happens when student needs are not met and their concerns are seen as burdens by the school. At times, what is best for students becomes secondary to faculty power struggles, turf battles, jealousy and envy.

No one said business was pretty – business school is no different.

PART FOUR: STATE OF MIND

No field of study is as transposable as psychology. Psychology examines the systematic knowledge of the powers and functions of the soul as well as how these are recognized at a conscious and subconscious level. The field allows students to read the behavior of others, assess the situation and take proper action.

Take the fundamentals of psychology, apply some sociology, add a dash of anthropology, a hint of geography and plenty of science and you have completed your SWOT analysis and found alternative plans of action.

One of the central theories in psychology is Sigmund Freud's Id, Ego and Superego division of the mind. It is a worthy parallel to make when examining what it takes to be successful in business.

Centuries before Freud, the great Greek philosopher Plato had broken up the human soul into three parts: a) the rational part or intellect; b) the will; and finally c) the appetite or desire.

He then extended this to the greater powers within society and argued that his ideal state would be composed of: a) the philosopher kings who govern society; b) the guardians who keep order and defend the society; and c) the ordinary citizens, farmers, merchants and craftsmen who provide for the society. The philosopher kings reflect the intellect (fitting since he was one), the guardians represent the will and finally the ordinary citizens represent the appetite. This was pronounced in *The Republic*.

One reason why this is interesting is that it shows that in discussions, debates and problem-solving exercises, it is always easier to look at something from both a macro-level and micro-level perspective. This can apply to the human mind and society or the economy and specific firm.

It is no coincidence then that Freud's analysis also divided the mind into three parts: the Id, Ego and Superego.

The Id is the mental representation of primal needs such as the drive to satisfy hunger and sexual needs. It does not however distinguish between the internal mind and the outside environment. While it stimulates carnal drives, it fails to separate the imagery from the actual stimuli; in other words, the actualization (or in business terms: the execution).

The Ego does distinguish between the internal mind and the external reality. Without it, images and thoughts remain just that: an unrealized dream or an unsatisfied need. It makes you take the necessary action to meet your desires, dreams and aspirations. This is where actualization or execution comes into play. Those who can execute over and over again may indeed develop an ego, in the loose sense of the term.

The Superego represents one's moral behavior. It is the mental reflection of a society's rule of ethics and code of conduct. It is the Superego that balances the Id's urges. These two are balanced by the more realistic Ego.

There are several nuances in Freud's theory that entail considerable discussion. But for the sake of this parallel, the main ones are that the Id is largely primitive and in Freud's analysis, subconscious. The Ego is conscious. The Superego is part of the unconscious that is formed through an internalization of moral standards set by parents, society and other such censors.

As you can imagine, applying this to business is thorny. Freud's theory was based on newborns. Business students entering school or those graduating can also be seen as newborns in a way. Moreover, in business settings, professors and employers replace parents while the business community replaces greater society.

In laymen's terms, the Id deals with oneself, our needs, our urges and our primal desires. One could call this our own private hedonism. Certainly, if we wished to satisfy all of them, these would clash with social mores and standards.

The Ego is the world in which we live. This is where realism, rationality and logic precede our desires (Id) and our idealism (Superego).

Finally, the Superego is our sense of idealism, a utopia where any excess is reprimanded, reckless conduct is shunned and the greater good is sought.

How does this apply to business?

If you chose to enter business, it means that you share one thing: a desire to succeed.

But to succeed at what?

Some of you may wish to reap personal gains such as money and power (Id). Others may have far more altruistic reasons to pursue business: to help others, share wealth and improve life – not "as we know it" but how we subconsciously dream it to be (Superego). Between these two extremes is reality (Ego).

Applying Freud to your claim to the throne can serve as a good balancing act when the road to take is unknown.

A very large nuance comes out in this analogy. While Freud argues that the Id is subconscious, most of your desires are not. You know you want money, that person that has caught your attention or that bag of chips that the person next to you is crunching on. This has more to do with the fact that you are a young adult. However, what is interesting is that as a young adult, there is still some element of subconscious that comes up. Either you desire something (or someone) and you do not know it or you know you want something (or someone) but do not know why.

In any case, if Freud's analogy is applied here, it is largely for the need component and not the subconscious element.

CHAPTER 8: RISE TO THE OCCASION & BREAK AWAY FROM THE PACT

Who lost in the finals at the World Cup, the female finals at Wimbledon or the Super Bowl?

Some of you may remember some of the answers to these questions. Other will not have a clue. Whatever the outcome, the bottom line is that winners are remembered and respected, while losers are forgotten and felt sorry for. Unless they stood out in an exceptional way, then they will be remembered or mocked.

For this and other reasons, we go to great lengths and shocking lows to win. We would also do anything to make sure that we attain everything we set out to accomplish. Winning means many things. It means getting the person we like, eating well, drinking the best wine, traveling to our dream destinations and working at the best company in the best position.

We want to win and would prefer to play fair, but depending on the situation, we would let fairness take a hit at the expense of victory.

The Id: Satisfying Our Personal Desires

Getting to the top and staying there entail far different personality traits. While the climb to the top takes more of an aggressive, focused and at times self serving personality, staying on top entails someone that is more of a diplomat and statesman.

Allow us to first consider some of the traits you will need to master to get to the top – fast. Consistent with Freud's theory, the following is not a definitive checklist to get to the top; it is simply a set of forces that will drive you to victory.

Be Passionate

Having a passion for something is essential to getting the job done and knowing what the job in question is. Otherwise, the project lingers on and the goal remains unrealized. More important, you find yourself asking where time flew by. Passion is the seed of victory.

Speculate

Take risks, be bold and do not be shy to get cut and bruised. Perhaps no one took a better risk than Tony Hawk. The skateboarder was leaps and bounds above his peers. He was the Tiger Woods, Wayne Gretzky or Michael Jordan of skateboarding. He knew that to bring his sport to the next level, he had to take his game to the next level. And he did at every chance he got. Tony Hawk flew so high for so long that it is impossible not to think of the Flight of Icarus. According to Greek Mythodology, Icarus was the son of Daedalus. Daedalus had made artificial wings for his son. As Icarus escaped from Crete, he flew so close to the sun that the wax with which his wings were fastened melted and he plummeted into the Aegean Sea. Legend notwithstanding, Tony Hawk became the superstar of his sport because he dared to fly so high. The risk and danger of falling into the sea, or crashing into the ramp should be taken but it should not be the driving force in your career. If it is this fear that drives you, then take the appropriate risk level that you are comfortable with to lead a balanced life.

Another man spoken of in the same breath as Icarus was Cisco Systems CEO, John Chambers. For the most part of the 1990s, Chambers, a former consultant, led San Jose-based Cisco Systems to the peak of the high-tech world, sitting for a brief moment atop the list of the world's most valuable firms, ranked by market capitalization. Market capitalization is equal to the number of shares outstanding times the share price. On that memorable day, what stock markets around the world were saying was that Cisco's future looked brighter than Microsoft's. Cisco weighed in at $555.4 billion, Microsoft closed with a value of $541.6 billion. Microsoft was king of software and computing, whereas Cisco was king of networks and the Web's infrastructure.

Did this last? Of course not. What ensued after April 2000 was a flight to quality. Not only was Cisco not the world's most valuable company, neither was Microsoft, the undisputed champion of the latter 1990s. Both slipped below that blue-chip icon of diversified American corporations, General Electric.

Over the following quarters Cisco Systems slipped precipitously in market capitalization and Microsoft leveled off. Thereafter, many shareholders, critics and analysts openly questioned Chambers' strategy. They had cheered when he raised the stock price by

40,000%. Did they forget that with increased return comes risk? They must have. A partner at venture investment company Sequoia Capital put it best, "it's not an easy call to go from 100 miles per hour to 20 miles per hour."

The lesson: never look back even though others will. If you choose a plan of action, understand the stakes. The higher the stakes, the higher you fly and the harder you fall. But without the flight, there is no escape.

Be Ambitious

Keyshawn Johnson was the 1996 first overall pick in the National Football League (NFL) by the New York Jets out of the University of Southern California Trojans. Keyshawn was expected to be the key player in the offense. While wide receivers do better when they have another solid receiver to keep the defense busy, Keyshawn did not share this perspective when teammate Wayne Chrebet was lining up alongside him.

Keyshawn felt the need to offer his perspective in a very public way after his rookie year. He went on to write a book called *Just Give Me the Damn Ball.*

This is not a knock against Keyshawn writing the book, although he showed reckless disregard for teammates, coaches and the media. He knew the latter would eat it up. After failing to prove himself and getting the ball in New York, Keyshawn ended up with the Tampa Bay Buccaneers.

The lesson is that as great as ambition is, you cannot wear it on your sleeve. In other words, even if you are thinking about a promotion, do not show more excitement for your next job than you do for your current one.

Be Impatient

The best thing to do is pace yourself. That being said, sitting there and pontificating over trends, projections, aspirations and potential outcomes will not get the job done. People that want things done yesterday get quite a bit done before you get out of bed. True, they may make some mistakes, but overall, they are ahead of the eight ball.

Think Of Yourself

Raymond Bourque ranks amongst the top players in the National Hockey League's (NHL) history. He broke into the league with the Boston Bruins, drafted 8th overall in 1979. He won the Norris trophy as the league's top defenseman five times. When he retired, he was the all-time leader for points and assists by a defenseman. He was the MVP of the all-star game in 1996 and was a 12-time first team All-Star.

But these were empty accomplishments until the 2000 season. After years of playing competitive hockey, the Boston Bruins were at the bottom of their division in the late 1990s. Despite his dream of winning the Stanley Cup, Bourque's loyalty to the Bruins and the city of Boston did not flinch. But in the second half of the 1999 season, Bourque was traded to the Colorado Avalanche. In 2000, he finally hoisted the Stanley Cup and capped off a truly remarkable career. No one is going to tell you to ditch your team to be part of a winner. You'd be a loser to ditch your teammates. But Bourque would not have been any less of a class act had he decided to put his own interests on top, especially given that Bruins management did not want to invest in building a winner. And while the following may sound ridiculous to some, it is not all that crazy after all. Some would argue that Bruins General Manager Harry Sinden traded Bourque away to let him play with a Cup contender. But it is not false to assume that Sinden traded Bourque away to salvage his salary. The same way that the Bruins were notorious for being selfish, Bourque should have done the same.

Be Quick Or Be Dead

One of the most respected and criticized individuals in business is Amazon's CEO and founder, Jeff Bezos. Bezos went all out to build Amazon at a steep price. Moreover, not everything was done as well as it could have been. Bezos was more interested in moving fast and fixing small mistakes later. Across the web, eBay CEO Meg Whitman also had something to say about speed: when you are down, act quickly so you do not find yourself out. Another example was Steve Case of AOL, who often said that managing the company was like trying to change a jet engine at high altitude and speed.

Blitz

As important as it is to focus on one task at a time, the most impressive attacks are outright blitzes. Blitzkrieg or lightning war is why Adolf Hitler took out France so easily in World War II. Study your enemy, understand the scope of their weaknesses, size up your strengths and blitz.

Be Driven – Better Yet, Drive!

Talking is part of the equation; you need to walk the walk and show action. Being a passenger means getting a view of the scenery, but it seldom means you get to go where you want.

Pride

Imagine that the answer to one of three questions at the start of this chapter was you. Yes you. Now, what is the question? The one about who finished second at the sporting events.

If you finish second, then congratulations. You must be tough as nails.

However, you should be asking yourself why you did not finish first. And if you do not care to ask yourself this question, then get yourself a coach, professor, mentor or partner that will.

Block The Noise Out

You may recall Luis Figo who left FC Barcelona to join rival Real Madrid. Upon his return to FC Barcelona in a Real Madrid uniform, Figo was pelted with everything from phones, fruits and bottles. Despite this, his play did not suffer.

Vince Lombardi must have been thinking ahead when he said: "Mental toughness is many things and rather difficult to explain. Its qualities are sacrifice and self-denial. Also, most importantly, it is combined with a perfectly disciplined will that refuses to give in. It's a state of mind - you could call it character in action."

Stick To Your Guns

Remy Shand hails from a "hotbed" of soul, Winnipeg, Manitoba. All right, maybe Winnipeg is as much of a hotbed of soul music as Corpus Christi, Texas, is a hotbed of hockey. Hockey defenseman

Brian Leetch was born in the town on the border of Texas and Mexico. Despite that fact, Leetch made his way to the National Hockey League (NHL) and won a Stanley Cup with the New York Rangers. As much as Leetch came from unorthodox roots to play hockey, Remy Shand may be even more of an enigma. Shand grew up listening to great influences such as soul artists Marvin Gaye and Otis Redding. By the time he was in his late teens, he had mastered many instruments and even penned some tunes. With his skills and songs, Shand was confident in getting a record deal. But his background and look did not help his quest. Even record executives who cared to give him a chance were not convinced enough to grant him a deal.

After years of sitting on his tunes and perfecting his skills, Shand had the opportunity of a lifetime come his way. He met and hooked up with Motown President Kedar Massenburg. Not only did he realize his dream; he was entrusted with the creative input he sought. When his album was released in 2002, the credits read "Written, Produced, Arranged, Performed and Mixed by Remy Shand". Massenburg was also instrumental in the careers of India.Arie and Erykah Badu.

The lesson is that Shand did not get such an opportunity by chance. He got it thanks to his skills, his writing and his confidence. When his first single played on the radio, no one cared what his skin color was or where he hailed from. All that mattered was that Remy kept the music going and the people moving.

Years before, Sylvester Stallone stuck to his guns and held out for the right deal for his now-classic boxing movie, *Rocky*. Not only did he write and act in the movie, he also retained the rights and reaped a substantial financial windfall.

Confidence Versus Arrogance

Hubris has killed many people. Thinking that you can do no wrong or thinking that no one can steal your thunder will lead to your downfall. Indeed only the paranoid survive, so expect to be unseated if you wish to keep the throne.

CHAPTER 9: UNDERSTANDING THE RULES OF ENGAGEMENT

Remember Steve Spurrier? He was the Florida Gators and Washington Redskins coach that never hesitated to run up the score. But for one moment forget about football and think of boxing. Imagine you have slowed down your opponent. You know that the next blow will be the last. You aim, lock, load and hit your target. The opponent starts to go down. Do you hit him again on the chin? What about the back of his head?

Hopefully neither choice is attractive. For one, you should not desire to harm anyone. Second, how would society perceive your actions? Do you want to become the subject of criticism and scorn?

While making it to the top at any cost (satisfying the Id) is attractive, social considerations will force you to pace yourself.

What are some these factors?

<u>The Superego: The Ideal Climb To The Top</u>

Be Unselfish, Generous And Share

You will work hard for your money but what is the point of stashing it under your mattress. Spread the wealth and enjoy the fruits of your labor while you still can.

Two forms of generosity exist. The first one is akin to altruism. This is when you give and expect nothing, except perhaps a sense of personal satisfaction.

The second form of generosity is self-serving. Some may help others to reap benefits such as goodwill, good press and good fortune.

The bottom line though is that either is better than not giving.

Share Your Knowledge

Everything that you know, you have learned from somewhere or someone. You may also hold these sources in high esteem and revisit them to further advance your knowledge. When your time comes, do not be shy to share what you know. Not only will you gain credibility, you will be surprised at how much you learn when you teach.

Joe Satriani ranks among the best guitarists in recording history. Part of his legacy is having taught the great Steve Vai as well as Kirk Hammett of Metallica. Vai returned the favor by introducing his mentor to executives at Relativity Records, which led to a recording contract for Satriani.

Be Humble And Modest

Never let success get to your head. For one, others do not care about your accomplishments as much as you think. Second, no matter how big you think you are, you can (and should) aspire to do more. Third, people support a humble leader. Their approval will serve as wind in your sails. History is not kind to arrogant individuals.

Soon after his professional basketball career began, Philadelphia 76ers guard Allen Iverson signed a lifetime endorsement deal with Reebok. While this could have been viewed as vain, Iverson said that he was trying to be a better human being and a better basketball player. His modest reaction to such a special deal silenced his critics.

None other than Al Pacino demonstrated humility and modesty to Jamie Foxx during the taping of the film *Any Given Sunday*. The loud and boisterous comedian stated in a *Playboy* magazine interview that he was impressed by Pacino's professionalism. While other lesser-known (and less successful) actors display a cocky attitude as a result of fame and success, Pacino went about his job with calm dignity. Foxx confided: "That's what I learned - to be modest no matter what the accomplishment. (...) To see him be just a normal cat made me go: Oh it's cool to be normal. You don't have to do all the antics. That's the true talent." Indeed.

Give To Take

The sooner you start to help others, the sooner they will pass you the ball, so to speak. Instead of refusing to involve others and ending up on the bench, control the game even when you do not control the puck. Wayne Gretzky did just that because he anticipated the action on the ice even when the puck was not in his possession. Even if you are in the middle of a heated debate, let the other side speak, it is then that you can catch them off guard and crush their argument.

Respect Everyone

No matter what your age, creed, ethnicity, gender, position, rank, title or IQ, respect everyone and anyone if you want the same in return. This means not taking anyone for granted. One of the original and toughest crime bosses, Joe Bonnanas, avoided a Hollywood-style mafia death and passed away in his sleep at the tender age of 97. When asked how come he lived so long, people had one word: respect. Joe "Bananas" respected everyone and gave no one reason to whack him.

Be Responsible

Whether you do so for political, personal or professional reasons, look beyond the tip of your nose. Everything you do has repercussions that could come back to haunt you if you do not look ahead.

Understand Your Competition

No matter how confident you are, never underestimate your competition. AOL Time Warner is the largest media company in the world. Everyone including Bill Gates underestimated AOL Chairman Steve Case. Legend has it that back in 1993, before most knew what Yahoo!, eBay, MSN and AOL were, Gates met Case, looked him straight in the eye and said: "I can buy 20% of you or I can buy all of you. Or I can go into this business myself and bury you."

Seven years later, AOL bought Time Warner to form a behemoth of a firm, AOL Time Warner. This helped balance AOL's online presence with Microsoft's operating system dominance. Each has a legitimate claim as the leading architect of the New Economy.

Measure The Return, Take Acceptable Risk

It is good to want to win at everything, but sometimes, winning is not necessary. Imagine that you were a sprinter with dreams of attending the Summer Olympic Games and run the 4 by 100 meters sprint. If five athletes show up to the qualifications and only four will make the cut, you need not risk an injury to finish first.

Too Smart For Your Own Good

If ever there was an unexpected explosion, the fate of Long Term Capital Management would come to mind. What is LTCM? Well, what do suppose would happen if a trader at Salomon Brothers decided to start his own fund? If he was none other than star John Meriwhether, wealthy individuals and resourceful institutional investors would ask to whom they should make out the check.

Add to this fund a couple of Nobel Prize winners in Robert H. Merton and Myron S. Scholes (of the Black-Scholes options pricing model). LTCM took unconventional trading strategies that appeared to carry little, if any risk. Given that the lower the risk, the lower the return, the fund would only be profitable if LTCM borrowed money to leverage and amplify returns. If you thought that Archimedes was a fan of leverage, wait until you get to know Meriwhether.

According to different sources, the debt-to-capital ratio of the fund stood anywhere from 25 to 1 or 125 to 1. Either scenario would mean that if the strategy did not go according to plan, the fund would blow up.

And blow up it did after Russia defaulted on debt obligations. While LTCM partners knew what the consequences would be, they never imagined how bad it would get. As a result, they never drew an accurate risk profile for investors.

Give Praise

Your mother always said that unless you could say something nice, you should not say anything at all. Remember the Olympic trials where five tried out but only four made the cut? Well, you were the odd one out. Do you complain and make excuses?

If you want to win in business, avoid saying anything bad at all and take this one step further: even if you cannot say anything nice about someone, try harder and find something decent to say.

If you consider the previous example, what would you do all year as the Olympics approach and you know that you will be watching from the comfort of your living room? Do you hibernate while the chosen four prepare or do you remain positive, congratulate and encourage them throughout them?

You congratulate, you encourage, but most important, you leave the spotlight to them. You will not always get what you want and you should always learn something when you fail.

Learn From Failures

What happens when the Id gets out of control and counters everything that the Superego stands for? Errors in judgment lead to mistakes. In life as in business, mistakes can harm you. Most of the times, mistakes slow you down from realizing your full potential. Try to learn from other people's mistakes so you do not repeat them.

But there is a lesson to be learned from everything. Swedish tennis legend Mats Wilander went on record as saying that young Russian pro Marat Safin would be better served to lose than to win the 2002 Australian Open. His position was that Safin would learn little by winning but a tremendous amount if he were upset in the tournament.

If this is true, then you will learn as well by watching the four finalists go to the Olympics. They won because they recognized the stakes and played their cards accordingly. It was not a direct result of nepotism, nor was it a result of a conflict of interest in judging. They out hustled you when they had to and avoided injury. They visualized the outcome and measured what the immediate return was. As a result, they took on the appropriate level of risk and sacrifice.

If the team goes on to win gold, then at least your anthem will play at the end of the event. And if you had noble goals, whether you stand under that flag on the podium or watch from home should not make a real difference, should it? An oversimplification perhaps, but you should get the point.

Now, what if they do not win the gold? Would a silver or bronze do? Be a champion and a professional and do not under any circumstance question their effort. You owe them that much at least.

Show Class

Classy is an underrated trait. Never in short supply, either in your interaction with people or how you conduct yourself.

It would be easy to give an example of class by someone lauded by the media. To many, Sun Microsystems CEO Scott McNealy is aggressive, arrogant and a bully. To a minority, he is assertive and confident. Opinions vary depending on whom you ask.

When the technology sector was hot, Sun's market capitalization soared to over $200 billion. McNealy's swagger increased as it rose with the market. When the market cooled and Sun was hit hard, he did not change. Even in conference calls with blood hungry analysts and reporters, McNealy remained true to form, staying optimistic and always poking fun at his archenemy Bill Gates. Was this a lack of class? Was McNealy bitter because Sun's market value cratered to $30 billion in 2002 while Bill Gates kept Microsoft on even footing?

You could argue that McNealy was staying true to himself. The competitive side of him could not allow him to soften his stance. To his credit, he has shown class. When he beat General Electric Chairman Jack Welch at golf, he named the cup the Welch Cup because he acknowledged that golf notwithstanding, Welch was the top executive in the world. Welch beat McNealy the next year. Welch on the other hand bragged to the entire world about beating Greg Norman, even if it was at a charity game.

Demonstrate Etiquette

Social Settings

Be Polite

An obvious point to start with: do not be rude or profane. Do not stoop to the lowest common denominator.

Speak Softly

There are times when a loud booming voice is appropriate but in most social settings, people do not like to be screamed at.

Temper Temper

Depending on the situation, you may be tempted to lose your cool. The chance may increase with alcohol in your system. Remain calm and collected. Perception is everything so if you control your temper, you will gain support.

Let Others Finish

Bite your tongue when tempted to jump in and make your point. If you clamor to gain the last word, people will assume you do not listen. To hear "I told you so" is bad enough. Imagine how "I told you so but you cut me off" feels?

Never Get A Laugh At Someone Else's Expense

If you need to take a jab at other people to get a laugh, you have a problem, not a sense of humor. Get laughs on your own merit and never try to win an argument by poking fun at someone else.

Letter Etiquette

Message

What is the message of the letter? Be concise and devise an effective outline to guide you. This is crucial for cover letters.

Short And Sweet

No need to ramble on. Procter & Gamble made the 1-page memo a part of corporate lore. As a general rule, your business correspondence should adhere to the same standard.

Mistakes

Make sure that the spelling, grammar and syntax are correct.

Hook

Give your audience a reason to get in touch with you. Close with a powerful sentence or ask a question if you must. Just make sure that they care about the answer.

Phone Etiquette

Do Unto Others What...

Follow the golden rule: do unto others what you wish to be done to you. If you hate to be bothered at weird hours, you are not alone. Look at the time before you pick up the phone and dial. If you hate the sales pitch cold call, well, at least be warm when you reach out and touch someone.

Speak To The Right Person

Chances are that you will have to speak to more than one person before getting through to the right one. Do not bore them with your trivial details.

Short And Sweet

Whatever you need to say, do not ramble. People lose focus fast.

Introduce Yourself

Once on the phone with the right person, make sure to introduce yourself. If someone referred you, mention it.

Refer To Previous Meeting(s)

In the event you have met before, refer to the meeting.

Summarize

At the end of the conversation, recap to ensure that both of you are on the same page.

Message Etiquette

Short And Sweet

If the person is not there, you may be asked to leave a message. You are better off not leaving one most of the time. The best time to leave a message is when a deadline is imminent or if you need to impart information. If calling for employment, chances are that your message will be ignored and you will have to call again.

Hook

Like the hook at the end of your letter, you should leave off with something interesting to make them want to call you back. Give reason for them to be curious.

Email Etiquette

Unless you are firing away dozens of emails with a friend, there are some very important considerations when you send messages via the Web. Email may have replaced the phone and snail mail but not without a considerable downside. While a letter can be shredded and a phone call disputed in real time, your reputation and intelligence serve as permanent attachments when you press send.

Do Not Use Slang

Avoid slang in business email unless the recipient is a personal acquaintance.

Relax There With The Punctuation Cowboy!!!

Nothing is as irritating as receiving an email with a dozen exclamation points and question marks where they have no place. Make your words deliver the punch and leave the exclamation points alone!"/$%?!

Be Careful

You never know who will read it. People can forward your email to others on purpose or by accident.

Include Previous Correspondence

If you are following up on a point, leave the previous exchange or any other relevant correspondence in your message. This will expedite a decision and signal your ability to synthesize information.

Leave Your Signature

If you have one, it is always good to use it. Beware of what you put there though. For example, if you prefer an email reply as opposed to a phone call, then do not leave your number.

Address It To Those That Care

People have way too much email to go through as it is. If A needs to know something and B does not, then only email it to A. Otherwise B will pay less attention when he is the lead recipient.

Use The Bcc: When It Makes Sense

The great thing about email is that you can copy people or send a message out to multiple recipients without divulging private email addresses to everyone.

Beware Of The Bcc:

A Bcc: is like a confidential wiretap. It can help you just as it can backfire and lead to a mistrial. So exercise caution.

To: Versus Cc:

The distinction is relevant in a large firm but universal in importance nonetheless. Think about who the primary recipient is in addition to who is being copied. This has a lot to do with the chain of command and from whom you expect an answer or action.

Subject Matter

With regard to the subject line, make sure that it will help both you and the recipient sort it at a later date. Avoid bland subjects that do not stand out in the Inbox.

Fear

No matter our age, we all fear something or someone. Fear can lie in the subconscious or the conscious level of the mind. Wherever fear may be stored, it impacts our daily life and future development.

As you can imagine, fear is an obstacle to success. While it is not hard to identify, it is quite hard to overcome. How can you overcome it? Identify your fear and face it. If it involves a person, find out what it is about their personality that makes you cringe. Study them from a distance and up close, one-on-one and in groups.

Do not forget that a lack of understanding and familiarity are major factors in fear. Once you get to know someone, the fearsome façade is shattered. If anything, you should now see their weakness and beat them at their own game.

In business, computer systems are upgraded often. It is normal to be wary as long as you do not hide from the new technology.

Break up the scenario in small steps. In other words, is it the system? Are the textbooks hard to follow? Could it be that the training officers are inadequate?

Whatever you do, remain positive and do not blame anyone throughout the process, not even yourself. Imagine you are a tutor and there is one chapter that keeps you up at night. Attack it head on and do many problems to improve your aptitude. If you are still not sure, go ask the professor.

While some may be cynical and suggest that you do not grasp the material, you will master it when class starts. The students are your clients and as long as they are satisfied, then you have succeeded.

Kobe Bryant turned heads even before he made it to the National Basketball Association (NBA). In fact, he skipped college and made the transition from high school. From day one, Kobe pushed the envelope and embraced the challenge of playing with older, more experienced players. He said that Babe Ruth hit prodigious home runs but also struck out. Even teammate Shaquille O'Neal said that Kobe took shots that no one else dared take. Kobe stated that he had "no fear and no qualms about shooting for the fences."

Intimidation

When we first looked at Keyshawn Johnson, he was busy keeping the ball away from fellow receiver Wayne Chrebet. Instead of an ally, Johnson saw him as a threat. Johnson would never admit to being intimidated by anyone, but he must have been. Otherwise he would have done his thing when the ball was thrown and let his talent tell the story.

The knock against Keyshawn is that he failed to realize that Chrebet could have created room for him on the field.

Criticism

Accepting, understanding and utilizing criticism is the mark of a champion. Criticism can be positive, harsh, direct or subtle – it comes in many forms but it signals that you could be doing something better.

Even if you are confident that your methodology is beyond reproach, when someone gives you their input, do not disregard it but take it with a grain of salt just the same.

Study The Source

Who is doing the criticizing is critical. If the greatest sommelier in the world criticizes your home made wine, take notes. If Homer Simpson tells you that your beer is flat, toss him an Alka-Seltzer.

Why The Criticism?

Some criticize to put you down while others criticize to improve you. In the professor example, good and even bad teachers often criticize in good faith. Others criticize to boost their own confidence. This is key: you must learn to accept criticism but understand why it is there.

Negative Is Destructive

If you feel that the motive is unfair and stems from jealousy, envy and bitterness, shrug it off but do not erase it from memory. Everything has some element of truth, no matter how minute or beneath the surface it may be.

Positive Is Constructive

An important element in life and in business is understanding your audience. And if someone criticizes you, your message has failed. The key here is to ask yourself whether their reproach offers something tangible for improvement or whether it is needless. After all, a coach can point out an error in mechanics or timing but should offer some hint of a resolution. Otherwise, he is adding to the problem.

Personal

Another important consideration is whether the criticism is a personal attack or whether it is impartial. Not all personal criticism is bad. It can be constructive. If your professor tells you that your inability to work in group settings will hamper your career, he is trying help you address a shortcoming. This can be useful down the road.

Professional

In the same example, imagine your professor tells you that you failed because you did not research the project enough and missed some points. This should be taken as a prompt to improve your ability.

Punishment?

Criticism does not come in words only. Sometimes there are forms of punishment attached. As long as there is no physical component, punishment is fine if it is constructive and helps eradicate the cause and not just the consequence.

Shutting Up The Critics

Perhaps no story better illustrates how someone handled criticism than Mr. Nike, Michael Jordan. Critics bashed MJ when he started his career and said that MJ would never be a good defensive player. Jordan could have been arrogant and said: "If I can play such good offense, why waste my time on defense." But he never said such a thing, at least not on record. What he did do, for the record, was go on to win the 1988 Defensive Player of the Year Award and go down as one, if not the best players in the history of the National Basketball Association (NBA).

Leave The Bitter For The Gin

Be the greater person and do not hold a grudge. Appreciate what others have done for you, no matter how minute. A relevant example is what happened to Ozzy Osbourne years after he was kicked out of Black Sabbath. On his double-live album, *Live and Loud*, he included live tracks from performances with his old bandmates. The same bandmates that kicked him out. His drunken antics and drug abuse led to his departure but Ozzy was suffocated in Sabbath. Tony Iommi, Geezer Butler and Bill Ward were the main songwriters while Ozzy was in the background. When Ozzy went solo, he recruited Randy Rhoads and released one chart-topping album after another. His tours were legendary and served as a launching pad for the careers of Def Leppard and Motley Crue, to name two. But years after the water had passed under the bridge, Ozzy acted like a grown man and invited his original bandmates to join him on stage. They played a handful of songs together, pleasing the fans and showing that they had come full circle.

Maintain Your Credibility

What students seek in the short term is recognition and respect. In time a position of power and prestige can be attained.

In the late 1990s, the heir to Henry Ford, William Clay Ford Jr., succeeded Jacques Nasser as CEO of the Ford motor company. While the two were working side by side, Nasser was involved with the firm's day-to-day operations, product development and sales. Ford took care of government issues, public relations and figurehead responsibilities. Despite his famous last name, Ford did not want to be seen as an "Irrelevant Dilettante." Ironically, it was Nasser who was axed after a Ford Explorer scandal involving tire manufacturer Bridgestone caused embarrassment, financial losses and lawsuits. After all, the problem was operational and thus fell on the head of Nasser.

Remain Diplomatic

A challenge in life is to balance toughness with diplomacy. We often want to do the right thing, but not at the expense of others.

In the 2002 Salt Lake City Winter Games, Canadian figure skaters Jamie Salé and David Pelletier were robbed by Olympic bureaucracy and closed-door politics. In case you were living in a cave, the French judge was to vote for the Russian pair in exchange for a vote in another contest. As media inquiries and fan outrage began to pile up, incoming International Olympic Committee President Jacques Rogge had the tough decision of setting a precedent and awarding two gold medals in one event to compensate for the scandal. Rogge was courageous enough to do what was right and awarded the Canadians a [second] gold medal. However, he stopped short of reversing the result because it would have been unfair to the Russian skaters. Yes, maybe they did not deserve the gold, but it was because of judge misconduct and not their skating.

In the corporate world, Roberto Goizueta was regarded as a statesman when he was anointed to establish Coca-Cola as a global brand. His successor, Doug Ivester, was far from diplomatic. When consumers got sick in Belgium, it hurt Coca-Cola and sank Ivester. He refused to accept the notion that a company product was to blame for making some consumers sick.

Apologize When You Must

Do apologize, but not too often. It could lose effect fast and when you are sincere, people will think you are crying wolf.

Now that you have identified your inner desires and faced the constraints of society, you must act. But how do you balance personal drive with the greater good? How do you execute?

The Ego: Take Your Vision And Build Your Dream

Learn

Learn as much as you can, ask questions and absorb everything. Knowledge is the tip of the iceberg. Each person has a comparative advantage: something that they specialize in. This is the area you should focus on. How fast can you pick up new tricks? The key is to develop complementary tools that you can apply from one situation to another.

Separate Need To Know From Nice To Know

There are certain tools that you need to know like the back of your hand. There are other tools that you can access at any time to make an informed decision. Learn how to differentiate the two in order to avoid information overload.

Be Versatile And Remain Independent Of Others

Remember the Game you have been engaged in? Hope you do. Make sure that your tools can help you win. Can you upgrade them?

Open Your Blinders

Be aware of open doors beyond your field of vision. Develop a multicultural knowledge base and outlook. Whatever you do, do not be ethnocentric.

Opening up to other people is important because in the complex world of international deals and cross border agreements, you will often have to rely on foreign jurisdictions. Shutting off the line of communication will mean that when you do need foreign assistance, your call will fall on deaf ears. When George W. Bush came into office in 2001, he stated that the US was back on a path toward isolationism. He stated that he was not interested, like his predecessor Bill Clinton was, in nation building and foreign policy.

He voiced his desire to get out of the Kyoto resolution despite international criticism.

His worst nightmare was realized on 9/11. On that day, his isolationist policy no longer made sense and he had to make a 180-degree shift. When he needed international support, he expected foreign governments to forget about his previous position.

Remain Determined

It is easy to lose focus and motivation. There is also nothing wrong with doubting yourself from time to time. When you put your head down on your pillow though, keep in mind that when you wake up, you can change everything based on your performance on that day. Your best performance is always yet to come.

Stay Positive

The number one rule in life is to be positive. No one should be overly optimistic. After all, a realistic look at life is key to surviving and thriving in business. However, the same way that most people prefer sun to rain, most people you meet and interact with will prefer positive individuals that brighten up a room.

Since group settings are common in business, you have no choice but to put the right foot forward and walk into the room with a positive outlook.

Change is inevitable in corporate life. Sometimes it is a technical innovation or upgrade that requires extensive training. Many experienced employees accustomed to the old technology may seek to throw in the towel and give up. This is when it is twice as important to remain positive and lead by example.

You will be seen as an agent of change and a candidate for a promotion. A reality of the business world though is that seniority is king. So no matter how quick you may be, your ride to the top could be slow. You can compensate for the lack of experience by mastering technology.

Successful people can take a negative and turn it into a positive. If someone has become a thorn in your side, take their negative energy and do something bold so that they have no choice but to change their attitude.

Negativity Breeds Negativity

If there was ever a case where negative spin came to haunt someone, perhaps it was President George W. Bush. When Bush entered the White House, he spent considerable time putting the economy down. Perhaps the economy was well on the path to recession. Perhaps he wanted to criticize the Clinton administration. Whatever it was, he felt that the economy was poor. The economy was indeed showing signs of fatigue but his role as Chief Executive requires him to remain upbeat. The same way that the Federal Reserve Chairman can influence the market, so can the President.

With his negative spin on the economy, Bush was sideswiped by the adverse economic effect of the 9/11 attacks. No one could have ever predicted the tragedy but when Bush went out afterwards and tried to convince consumers, analysts and the media that all was well with the economy, political pundits were skeptical based on his previous assertion. The lesson is simple: always be positive because it can get worse, much worse.

Stay Confident

Being positive is easier said than done for a leader. Soldiers may remain positive in war because they have a great military commander motivating them.

While confidence is key in business and sports, it is crucial in the military. Lives are at stake. Having any hint of trepidation may prove to be your undoing. Throughout the Afghanistan War, Secretary of Defense Donald Rumsfeld came across as the most confident player in the conflict. Even when facing tremendous uncertainty, he remained stalwart. Was his confidence born after 9/11?

His pedigree would suggest not. Educated at Princeton, Rumsfeld moved up the political rank throughout the Richard Nixon and Gerald Ford era. He then excelled in business as the Chairman and CEO of General Instrument Corp. and as the Chairman of the Board of Directors of Gilead Sciences, before being appointed as the head of Defense.

If you want others to have confidence in you, then you must have confidence in yourself first. Vince Lombardi said it best:

"Confidence is contagious and so is lack of confidence, and a customer will recognize both."

New England Patriots backup quarterback Tom Brady is a good model to follow. When the 2001 season began, the former Michigan University Wolverine and 6th round draft pick caught the attention of many offensive players. Despite being the backup to the highest paid athlete in the league, Drew Bledsoe, Brady walked around like he owned the team. This is not to imply that he was cocky, but it signaled that he was confident in his ability and comfortable in his own skin.

His confidence reverberated within the locker room and onto the field to the extent that some players speculated how he would respond if he were at the helm of the offense. When starter Drew Bledsoe went down to injury, Brady led the team to the playoffs. In a game against the Oakland Raiders, his confidence was clear. Playing in snowy New England, Brady engineered a brilliant comeback against the favored Raiders in the second half to lead his team to the AFC Championship game.

Grace Under Fire

The quality of grace under fire will play a large role if you intend to make it to the top. Yahoo! hired Terry Semel because he was one of the few people who could keep his composure when the rug was being pulled from beneath his feet.

But perhaps no one better exemplified grace under fire than President George W. Bush after the 9/11 attacks. Cynics can claim that the event could have brought out the best in any leader, but President Bush managed to reassure the country while prescribing an effective response.

Trust

If you have faith in yourself, others will put their faith in you. This is the lesson you need to learn in business. No one will hand you the ball on Super Bowl Sunday if you doubt yourself.

The key is not to have blind confidence in anything you do but rather to determine where your comparative advantage lies in order to exploit it.

Whatever you choose to do, the key is to study your potential new city first to make sure that you know what you will be getting yourself into.

Letters Of Recommendation

Many of you will need letters of recommendation for either that new job or graduate education. Here are some points to consider.

Important In The Overall Evaluation

Even if it is not required, 60% of recruiters put considerable emphasis on letters of recommendation, so if you are in a position to secure one, go for it.

This Is Your Candidate, I'm Telling You...

A letter of recommendation is important because it is a third party assessment of you. This is very important so make sure that this person has clout, is respected and knows you.

Don't Be Shy

It may be intimidating to ask someone for a recommendation, but get used to it. You will have to ask for favors throughout your life. The trick is to make it sound like you are doing *them* a favor.

What Is Important?

Letters of recommendation should emphasize your role in class, school and the greater community.

Familiarity

It is best to ask someone that knows you in another context, not just school. If there is one thing that you were interested in and dedicated yourself to, try to get a letter of recommendation from people in power that you interacted with.

Be Clear

When you do ask someone for a letter, do not be shy to explain what it is for and who will read it.

Break It Down

Make it so simple that they cannot say no. Explain what the recipient of the letter is looking for but do not tell them what to write (even though a few may ask).

Why Them?

Do not be afraid to let the person know why you are asking them. This will help them in formulating the letter. Why that teacher? Why this advisor?

Thank Them

While it may be seen as part of their job description, the truth is that a professor is going the extra mile for you, so thank them.

Do The Legwork

Get the necessary forms ready and consider the red tape. Chances are that this will fall on you so get your act together and do not wait until the last second.

Let's Get Personal

As a general rule, personal letters work better than generic ones.

<u>Take It Easy!</u>

Whatever happens, or does not happen in the days, weeks and months after you begin your job search, you should realize that it is better to take it easy and not worry too much. After all, it is not like you control the situation.

<u>Choosing The Right Career Path For You</u>

A good example of big thinking is Ted Leonsis. A lieutenant of Steve Case at AOL Time Warner, Leonsis is part owner of the National Hockey League's (NHL) Washington Capitals and the National Basketball Association's (NBA) Washington Wizards. Although very successful, he keeps a list of the 101 things he has yet to accomplish. While it is not always recommended to have such a list carved in stone, reading some of the items will widen your mental scope and keep you hungry for more.

Being Comfortable In Your Own Skin.

Choosing the right path is not about money, status or the size of your cubicle. It boils down to being comfortable in your own skin, in the words of the great Jack Welch.

Are You An Entrepreneur?

Are you a do it yourself kind of person? If you are, then you can imagine that nothing any employer could offer, no matter how promising or lucrative will come close to pursuing your dream.

Independent of what your dream is, the reality is that you will not have the opportunity to develop your dream by yourself. You will need backers. If you can tap into love money, guilt money or pity money, good for you. Grab it, say thanks and tell them that this is the best investment they will ever make. Jeff Bezos has been able to retain considerable ownership of Amazon because of such financing.

Unless you have a Dell-like business model where your cost of capital is nonexistent, then you will have to find a more original way to raise financing. Despite the traumatic market meltdown that followed the Internet rush, there is still ample money to raise for your fledgling operation. But where to start?

Dilution

The first fact you must deal with is dilution or giving up a piece of the pie. The general rule is that as you seek investors and look to hire, you will have to part with a piece of your dream, your business and your proverbial pie. This applies to financing as it does to marketing. To some extent, this was what Palm was doing by licensing its operating system to rival Handspring. By sharing the technology, Palm accelerated the development of handheld devices, offered additional selection, decreased prices and increased the number of users. Although Palm lost market share, the size of the pie grew. The pie that we are now salivating over is your company. When is it ok to give up a piece?

The first obvious answer is when an investor is interested in sharing the risk – and upside – of your business. The "how much" question depends on many factors:

- your chances of securing alternative sources of financing
- timing (how fast do you need the money)

- the market's appetite for risk
- your product or service
- your business plan
- the cycle time between investment and profitability (or at least once you are cash flow positive)
- your background, and mostly
- whether or not you will be tapping into smart money or not

Smart Money

Smart money is valuable and worth diluting for if the investors can bring management expertise, technological knowledge and open doors down the road. But smart money is also highly overrated, so be vigilent.

Respect

The second, less obvious factor is how much you respect them. You may be desperate enough for financing that you give up 25% of your firm for $100,000 in seed capital. You just valued your firm at $400,000. In one year, as this investor sits idle and you grow your business tenfold, your company may be worth $4 million dollars and his stake will be pegged at $1 million. Can you handle that? The answer should be yes because it was thanks to his investment that your stake is now $3 million. But this is easier said than done, so keep this in mind.

Whenever someone wants a piece of the pie, you should have no problem with this so long as they grow the size of the pie. If they do not, then do not even think of giving them the fork. Of course, sometimes you have little choice.

Communication Breakdown

Whether you intend on building the next Dell or General Electric, chances are that your idea will be lumped into the same category as Pets.com and Enron. In other words, taken with a *pound* of salt.

You must stand out and deliver your best elevator pitch. The investor community, like any other, is a tight one. If one of them does not like your idea, the rest will follow.

If one of them does like your idea, consider yourself invited to the Game. But what you fail to see is that you are a pawn in a game of chess while they are sitting at the table next to you playing a game of high stakes poker.

Stick to your guns if you have any intention of joining the big boys. At that point, hold on to your pair of deuces and bide your time in the hope that the cards are reshuffled until you find yourself with a full house. And even then, one group of investors may be carrying four aces while another is holding a royal flush. Which one of the two groups will emerge victorious is unknown to you. What is certain is that they both can beat you hands down. So the goal is to match up with the group of investors that will yield 5 jokers. This is what Google managed to do by getting in the proverbial bed with both Michael Moritz of Sequoia and John Doerr of Kleiner Perkins Caufield & Byers.

No one said it would be easy, especially when the cards are held to their chest. Playing in this high stakes game of poker is like shaving in the dark with a machete.

What Do Investors Look For?

Understand one thing: human capital is the single most important factor that investors look at. Yes the product must be right and the timing even better, but it takes a team to succeed.

The first ever American President to have an MBA was George W. Bush, who obtained his from Yale. Bush understood the value of building a team around him so he surrounded himself with the right people. His supporting cast all brought political experience to the table along with business experience. Secretary of Defense Donald Rumsfeld hailed from big business and Vice President Dick Cheney was the CEO of Halliburton, the world's largest oil services company. Bush knew that it would take the right team to win "investor" (voter) support.

While great firms such as Hewlett-Packard and Yahoo! have been built on friendship, keep in mind that friendship and business are like oil & vinegar: they make a great dressing when the salad is just right, but if the balance is off, you will be left with a bitter taste in your mouth.

AIDA

AIDA is a simple acronym that stands for Attract, Inform, Desire and Action. The challenge is to put the two A's side by side. If your goal is to get the investor to *act*, then you must first *attract* their attention. Only then will they be interested enough to have the desire to invest in your dream, in you and in your company.

How Do You Attract?

Attracting investors will require you to do far more than to think out-of-the-box. This is simply not enough. You need to shatter the box and create a new shape that piques their interest. At the same time, you have to keep it simple so that your users will have no choice but to go with you.

Raise Their Interest Level

Once you have attracted them, the challenge is to test their level of interest. You need to go in with a solid contingency plan to come out with one course of action. Even if you have nothing more than a prayer, extend your wing and make them think that you are doing them a favor by sharing your dream.

Brew The Desire

Two down, two to go. Investors need the desire to take the risk to invest in your project. Become familiar with their risk profile and study their investment criteria until you know it like the palm of your hand. Anyone can see the back of your hands, but only those that you open up to should know the composition of your palm.

Action, Not Words!

As you get close to sealing the deal, how can you ensure that they act? You stay the course, focus on the message and keep it simple.

<u>KISS</u>

Warren Buffett is the quintessential investor because he puts his money into what he understands. At the other end of the spectrum, Enron went bankrupt partially as a result of complicated financial maneuvering and obscure deals.

Jerry Seinfeld will go down in showbusiness history for popularizing a show about nothing. But the reason why *Seinfeld* was brilliant was because it was so simple. Everyone could relate to it.

ABC

Just as you put the Dom Perignon on ice, you will probably realize that the financing process is taking too much time. How do you adhere to the mantra coined *Glengarry Glen Ross* by writer David Manet: Always Be Closing?

You make it simple. You leave no question unanswered. You provide the exact amount of information, not an inch more and not an inch less. You follow up. You always leave a door open (for yourself) so that they can let you back in the poker game.

It's All About The Company

Is Bigger Better?

One fork in the road you will encounter is whether you wish to take the safe, stable road that large firms take or choose the volatile ride that smaller firms offer.

By now, you should know which path best suits your personality and your aversion to risk.

Company Goals Versus Personal Ones...

The company you aspire to work for should have a mission and vision that is an extension of your personality. If it does not, you will never make the most of it.

Level Of Control

Whatever you choose to do, it is imperative to control your own destiny.

Seeing Beyond, Seizing The Opportunity, Winning And Staying On Top

Imagine you are playing hockey and time is counting down. You are down by one goal and the third period is ticking away. You are eager to grab the puck and make something happen, whether you set someone up for a goal or score yourself.

What bothers you more: knowing that you are about to lose or the fact that you were on the ice for 20 minutes of the game and controlled the puck for a meager 4 of those 20 minutes? This scenario is typical in the game of hockey, with a fast pace and line changes.

With a couple of minutes left, you see the puck trickle to the boards. Your coach has told you never to leave your position at center. But you know that you can skate in, scoop the puck and break in front of the net. You see the play unfold the way Wayne Gretzky would.

What do you do? Do you listen to what your coach said or do you break from the pack, hit the corner, take a hit and go in for the tie?

It depends on the Game. If you are up by a goal, you should never sacrifice victory for a personal statistic. But you are losing now. For that reason, why wait for someone else to take control of the puck if you know it is there for the taking? Go for it.

The analogy here is that you need to do what is right for the greater good. But you must also take charge and be in control of your destiny.

Recognize An Opportunity

Before Steve Case was at AOL, he was trying to sell hair conditioner and pizza, first at Procter & Gamble and then at Pizza Hut. Regardless, Case was never the type to overachieve. The Hawaii native earned a degree not in business, but Political Science from Williams College.

The year in question was 1985 when Steve Jobs had just launched the Macintosh. The World Wide Web looked more like a dark back alley than an information superhighway. If that were not enough, the Pentagon had a tight noose around the neck of the Internet. The only thing that investors, consumers and citizens alike were sure of was that cars would fly by Y2K.

Despite the less than glamorous situation, Case realized that the potential of a global network where people could interact, exchange ideas and buy goods from all over the world was too large to let go of. After a couple of detours, his vision took shape in the form of Quantum Computer Services Inc., as AOL was first known. The small operation, far from New York City and Silicon Valley, was headquartered in Virginia. But Case was not interested in producing tobacco – he wanted to get Americans online.

Deeper pockets and more experienced managers backed larger rivals like Compuserve and Prodigy. But AOL was unfazed. Microsoft had not even yet released the Windows 95, but once it did, it would have a window into the lives of over 80 million customers. IBM and Sony were also running feasibility tests and licking their chops at the potential of the WWW.

The irony was that Case knew that the elite Web surfers – the ones who had built it into what it was – hated the awkward simplicity of AOL. After all, Silicon Valley had never pandered to the masses. Three firms that did shoot for the masses were located far from California: AOL in Dulles, Virginia; Microsoft in Redmond, Washington and Dell in Austin, Texas.

AOL never catered to the techno crowd, but rather the everyday consumer who was interested in the Internet but could not be troubled to go through the confusion of getting online.

Remember what Craig Barrett of Intel said about the US representing 4% of the world and the importance of having the other 96% in mind? The same numbers can be transposed to the elite versus the masses in the consumer market.

Musicians are often criticized for moving away from their original fan base to please a mainstream audience. Paul Oakenfold has done more to increase the reach, penetration and credibility of electronica music that anyone else. To the knowledgeable fan, John Digweed may be the better mixer, Paul van Dyk may be a better producer, but no one moved audiences like Oakenfold did from 1996 onwards. As a result, many new fans were introduced to the genre. The more fans who tuned in, the greater the windfall and exposure for his cohorts.

When the same question about "selling out" was put forth to Roc-a-Wear co-founder and Jay Z confidant Damon Dash, the outspoken and successful entrepreneur stated: "I could care less about that."

In the end, you cannot set out to please such a small group and expect to come out on top. You need to think big. Very big.

Steve Case represents the best example of thinking big. He sought to exploit his big idea but he always saw the greater picture. He was confident that he was on to something major. He may have had a couple of strikes against him, but he was willing to run with his newfound baby for as long as he could. He viewed it as having a significant impact in everyday life.

This confidence was highlighted when Microsoft co-founder Paul Allen sold his stake in AOL after senior management refused his offer to buy them out. Case stuck to his guns even after Gates made his now legendary claim of putting him and AOL out of business.

In your pursuit of personal, professional, spiritual and material "profit," you will cross paths with people that may or may not share your vision and goals. It is up to you to take the proper course of action.

In the most ironic of twists, back in 1995, then editor of *Information & Interactive Services Report* Roc Kuckro, called Case the "Ted Turner of online services." The analogy made sense. After all, what made Turner successful was bundling content and distributing it over his cable empire.

The irony however is that Turner sold his company to Time Warner, later acquired by AOL. When it was all said and done, Turner was on Case's payroll.

Case was not interested in building content. He wanted to offer a platform that would allow a critical mass to have access to the Web, with a clean and simple interface. Case bundled the network, interface and content better than anyone, including himself, could have ever imagined. He offered "best of breed" content before anyone thought of the term. He made comments that went against the institutional imperative by claiming for example that "content is no longer king, building a community around that content is the key." He also managed to drive the commerce component of the Web fast enough to allow him to buy Time Warner.

While Case may not be considered a great visionary, his business sense inspired many mergers in media and technology. After all, convergence may be a buzzword, but Case was doing it way back, stating: "we weren't smart enough to call it convergence" in 1985. While many felt that customizing content (or "disintermediation") would be too hard for most, Case made it happen.

He has taken his one big idea and shown that you need not be the first to market – you need to be the first to scale. Case managed to scale his business far and wide without being the smartest, but by being the one who made it simple.

The most successful business ideas endeavor to make something so simple that your consumers and investors scratch their head and ask how come they did not think of it. This was the AOL model, the same model that allowed the young firm to buy the venerable Time Warner.

Case and company went on to combine speed, offense and a killer instinct to come out on top in a competitive and lucrative field. He shared Intel co-founder Andy Grove's notion that "only the paranoid flourish" and he wasted little time. He rode up the stock market and used his newfound stability and traction to weather the storm when the market cratered. It was that paranoia that drove him to acquire Time Warner.

AOL never cared about building the best technology: they cared about finding out what consumers wanted. Bob Pittman served as a very important Head of Operations in AOL's rise. Perhaps the best hire Case ever made, Pittman put it best when he said: "I remind people all the time that Coca-Cola does not win the taste test. Microsoft is not the best operating system. Brands win."

At the time, the AOL/Time Warner merger was the biggest of all time. Was bigger better?

It looks like the answer is: it depends. After years of promoting the "bigger is better, we need size" mantra, many of the mergers that were hailed as necessary were subsequently questioned. Pittman resigned from AOL Time Warner in July 2002. By then, Time Warner executives were holding many of the top positions at the new firm. But the fact remains: AOL – the first Internet company to crack the *Fortune 500* list – hedged itself by acquiring Time Warner. Using its stock as currency was the icing on the cake. Steve Case naming himself Chairman was the cherry on the icing. While analysts raved about convergence, Case was more interested in hedging himself.

Centuries earlier, Alexander The Great's dream of uniting a global empire that spanned from Europe to Asia to Africa led to his demise. Alexander conquered Asia by defeating the mighty Persian Empire, but on his way to conquer Africa, he fell ill and died. What made the tragedy worse was that his reluctance to select a successor made it impossible for future generations to build on his accomplishments. By spreading himself too thin, he never fully attained his enormous potential.

Never underestimate the importance of execution *and* focus. No one should hold you back from thinking big. But to achieve greatness, you need to focus on one thing and execute - only then will you keep the inning going and have a chance to step up to the plate again. That is how you build an empire.

CONCLUSION

When you rest your head on your pillow tonight, be grateful for what you have. When you wake up tomorrow, you must believe that there is the possibility of a better job, a stronger love and a greater life. The key term however is *possibility*.

In life, idealism and optimism must be balanced with realism and execution for nothing in life falls on your lap. Determine what you are good at, do your best, keep on trying and hedge yourself well.

Then, and only then, will all of your dreams come true. And *that* is the definition of success.

THE END.